ATOM
AND ION SOURCES

by

L. VÁLYI C. Sc. (PHYS.)

Central Research Institute for Physics of the Hungarian Academy of Sciences

A WILEY—INTERSCIENCE PUBLICATION

JOHN WILEY & SONS

LONDON—NEW YORK—SYDNEY—TORONTO

Published as a co-edition of John Wiley & Sons — London
and Akadémiai Kiadó, Budapest

Translated by Magda Monori Kovács

Copyright © by Akadémiai Kiadó, Budapest 1977

Library of Congress Cataloging in Publication Data
Vályi, László, Atom and ion sources.

Bibliography: p.

1. Ion sources.   2. Atomic beams.   I. Title.
  QC702.3.V34   530.4   76-44880

ISBN 0 471 99463 4

Printed in Hungary

PREFACE

The utilization of atom and ion beams has recently become much more extensive in science and technology. The variety and wide range of the possible applications necessitates, to an increasing extent, a comprehensive monograph on the production of atom and ion beams. This book has been written with the intention of meeting this demand. The aim has been to provide researchers and technicians concerned with the utilization and development of atom and ion sources with not only a textbook on the physical processes taking place in these sources, but also with a manual on the available types of beams permitting a choice of the most appropriate source type for a given task. Workers are thereby enabled to set the best operational parameters for the chosen apparatus.

Consequently, the book deals with the elementary physical processes occurring in atom and ion sources taking into account the physical and ion-optical properties of the atom or ion beams extractable from these sources. The different methods of production of particle beams, in addition to the types of sources employed by these methods in order to obtain atom or ion beams from different elements in the gaseous, the liquid, or the solid state, are discussed in detail.

An entire chapter is devoted to the theoretical and the practical aspects of the production of ion beams with special properties. In this chapter the methods for the production of negatively charged, multiply charged, nuclear spin-polarized, and pulsed ion beams are considered together with the discussion of the construction of particle sources required by these methods.

Within the scope of this book it has obviously been impossible to consider each detail of every process which may take place in particle sources or every method which can be used in each type of source employed. Even so, every endeavour has been made to include all the important processes, methods and apparatuses needed for the understanding and proper utilization of this powerful tool at an advanced level.

L. Vályi

CONTENTS

CHAPTER IV

Special ion sources

CHAPTER V

Distribution of charges and masses in ion beams

CHAPTER I

ELEMENTARY PHYSICAL PROCESSES
IN ATOM AND ION SOURCES

1.1 EXCITATION OF ATOMS AND MOLECULES

Transformations from one to another discrete atomic or molecular state by which the atom or molecule is brought to a higher energy level are called excitation. By the excitation of atoms the electron state is brought to a higher energy level. By the excitation of molecules not only the electron states in the molecule but also the vibrational and rotational states of the molecule can be changed.

The excitation process takes place only if the conditions are satisfied which are imposed on the transition by the selection rules and the conservation laws [1–3].

Atoms and molecules can be excited mostly by electron, ion or neutral particle impact and by the absorption of light quanta.

1.1.1 EXCITATION OF ATOMS AND MOLECULES
BY ELECTRON IMPACT

The impact of electrons on atoms or molecules leads to an excitation process only if the energy needed for the transition is imparted to the atomic or molecular system by the kinetic energy of the electron in the inelastic collision. After impact the kinetic energy of the system of colliding particles is lower than before because of the inelastic collision. The difference between the kinetic energies before and after impact is expended on the increase in the internal energy of the particle struck by the electron, that is, on the excitation of this particle to a higher energy level.

Let $v_{e(m)}$ and $v_{e(n)}$ be the electron velocities before and after the inelastic collision, respectively, while the atom is brought from the state m at energy E_m to the state n at energy E_n. The internal energy increment E_{mn} can then be expressed as

$$\frac{m}{2}(v_{e(m)}^2 - v_{e(n)}^2) = E_m - E_n = E_{mn} \qquad (1.1)$$

This energy increment is called the excitation energy of the given transition. In other words, this amount of energy is needed for an allowed transition to occur. It follows that if the value of the imparted collision energy E_e is lower than the excitation energy, the excitation process does not take place, that is, the excitation cross section σ_g is equal to zero. Thus, the excitation function $\sigma_g(E_e)$ has a pronounced sharp threshold at $E_e = E_{mn}$. The values of the excitation energy E_{mn} are specified for a few atoms of interest in Appendix, Table 1.

If the impact energy E_e is higher than the excitation energy E_{mn}, the excitation cross section is higher than zero.

In terms of quantum theory the excitation cross section can be expressed for an energy close to the threshold value, as [1–3]

$$\sigma_{0n} = C(E_e - E_{0n})^{1/2} \qquad\qquad (1.2)$$

if

$$E_e - E_{0n} \ll E_{0n}$$

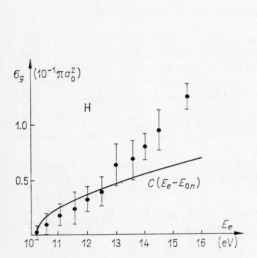

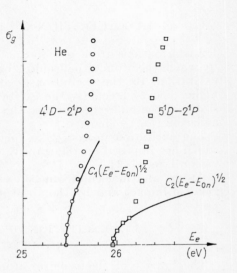

Fig. 1.1 (1s–2p) excitation cross section values of hydrogen atoms for various impact electron energies, as calculated (solid line) and measured at energies close to the threshold of excitation

Fig. 1.2 (4¹D–2¹P) and (5¹D–2¹P) excitation cross section values of helium atoms for various impact electron energies, as calculated (solid line) and measured at energies close to the threshold of excitation

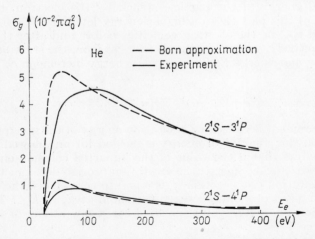

Fig. 1.3 Excitation cross section values of helium atoms for various impact electron energies, as calculated and measured for singlet–singlet transition

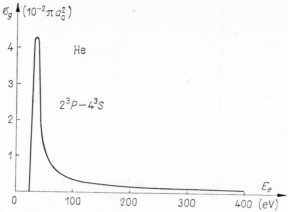

Fig. 1.4 Excitation cross section values of helium atoms for various impact electron
energies as calculated for triplet–triplet transition

where C is a constant. However, in practice the expression (1.2) holds in
a very narrow range only and few experimental verifications have actually
been reported. This is probably due either to the spread of the impact energy
of the electrons or to the inaccuracy of the measurements. The experimental
values — in agreement with predictions — obtained for hydrogen and helium
are shown in Figs 1.1 and 1.2, respectively [4, 5].

Figure 1.1 shows the $1s–2p$ excitation cross section of hydrogen for electron
impact. In this case the measured values agree with those obtained from
expression (1.2) in a somewhat broader range of impact energies than in the
case of helium (Fig. 1.2).

Beyond the threshold energy the values of the excitation function show
a linear increase with increasing impact energy up to a maximum after
which the function starts to decrease. The shape of the excitation function
can substantially change with the type of transition involved (Figs 1.3, 1.4,
1.5). In Fig. 1.3, a broad maximum can be seen for the singlet–singlet transitions $(3^1P \rightarrow 2^1S, 4^1P \rightarrow 2^1S)$ while in Fig. 1.4 there is a sharp maximum
for the triplet–triplet transition $(2^3P \rightarrow 4^3S)$ in the helium atoms. Similarly
different shapes can be seen in Fig. 1.5 for the optically allowed $(1s \rightarrow 2p)$
and the forbidden $(1s \rightarrow 2s)$ transitions in hydrogen [1–7].

The excitation cross section can be calculated by classical and quantum
mechanical approximation methods or by the use of semi-empirical formulae.

The classical approach to the inelastic collisions of electrons is the two-
particle collision model which describes the impact of the bombarding
electron only on the orbital electrons of the atom, either ignoring [10] or
taking into account [14, 15] the motion of these electrons.

The classical model is the most simple and most crude approximation
to the inelastic collision between electrons and as such it is not expected
to give quantitatively accurate data on the position and value of the maximum excitation cross section. It yields, however, a reasonable qualitative
picture on the impact energy dependence of the excitation cross section.

A more accurate description of excitation processes is obtainable in

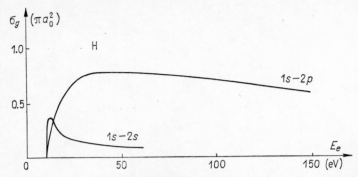

Fig. 1.5 Excitation cross section values of hydrogen atoms for various impact electron energies as calculated for (1s–2p) and (1s–2s) transitions

quantum mechanical representation. It should be stated, however, that due to the substantial difficulties of the calculations in terms of quantum mechanics, it is preferable to apply various simplifying approximations.

The approximations which have been applied by different workers are known as the Born approximation [7, 8, 21], the distorted wave method [17, 22], the assumption of strong coupling between states [16, 18, 19, 24, 35], and some modified versions of these methods.

In terms of the Born approximation, in which the electron–atom interaction is taken to be a perturbation corresponding to the first approximation in perturbation theory, the excitation cross section for the transition mn is given as

$$\sigma_{mn} = \frac{8\pi e^4}{\hbar^2 v_e^2} \int_{K_{\min}}^{K_{\max}} \mid \sum_s (e^{-i\boldsymbol{Kr_s}})_{mn} \mid^2 \frac{dK}{K^3} \tag{1.3}$$

where $\boldsymbol{r_s}$ is the coordinate of the sth electron, $\boldsymbol{K} = \mid k_n \boldsymbol{n}_1 - k_m \boldsymbol{n}_0 \mid$, $K^2 = k_m^2 + k_n^2 + 2k_m k_n \cos \Theta$ is the impulse imparted by the striking electron with Θ being the scattering angle with $k_m k_n \cos \Theta = K dK$ and denoting the wave numbers k_m and k_n of the electron before and after the inelastic impact, respectively [1, 3]. It follows from eq. (1.1) that the wave numbers satisfy the condition

$$k_m^2 = k_n^2 + \frac{2m(E_n - E_m)}{\hbar^2}$$

The integration has to be performed over the range of possible impulse changes from $K_{\max} = k_m + k_n$ to $K_{\min} = k_m - k_n$.

Formula (1.3) predicts very low values for σ_{mn} if those of the imparted impulse K are high, and the cross section becomes negligible if

$$K^2 > \frac{2m \mid E_m \mid}{\hbar^2} \tag{1.4}$$

Now, if the values of K are low and formula (1.4) does not hold, the term $(e^{-i\boldsymbol{Kr_s}})$ in eq. (1.3) can be expanded into a power series with respect to K.

This gives for the excitation cross section of the optically allowed transition mn in hydrogen, assuming dipole transition and the case of $m = 0$, the formula [2, 25]

$$\sigma_{0n} \approx \frac{4\pi e^4}{\hbar^2 v_e^2} \mid X_{0n} \mid^2 \ln \frac{2m_e v_e^2}{\mid E_n - E_0 \mid} \tag{1.5}$$

For a quadrupole transition, when the dipole transition is forbidden, the excitation cross section is given by

$$\sigma_{0n} \sim \frac{8\pi e^4}{\hbar^4 v_e^2} \mid (X^2)_{0n} \mid^2 \mid E_0 \mid \tag{1.6}$$

where x_{0n}, $(X^2)_{0n}$ are matrix elements which can be evaluated from the expression

$$(X^i)_{0n} = \int r^i \psi_0 \psi_n dr$$

The matrix element of the dipole moment for the $1s \rightarrow np$ transition in hydrogen is obtained as [1]

$$X_{0n}^2 = \frac{2^8 n^7 (n-1)^{2n-5}}{(n+1)^{2n+5}}$$

At higher impact energies the Born approximation yields cross section values in good agreement with experimental data (Fig. 1.3) and even at impact energies corresponding to the maximum of the excitation cross section, the predictions do not deviate by an order of magnitude from the experiment. However, the agreement is substantially worse at values near the excitation threshold.

Figure 1.3 shows the predictions from the Born approximation [28, 29] together with the measured values [9, 30, 31] of the $3P$ state excitation cross section in helium atoms, as normalized to 400 eV impact energy. It is apparent from the figure that for the optically allowed excitation the Born approximation yields reliable values at impact energies above 200–300 eV.

At the low range of impact energies the agreement with experimental data can be improved by the use of a semi-empirical method. In this case the approximation (1.5) is written in the form [3]

$$\sigma_{0n} \approx \frac{4\pi}{\hbar^2 E_e} \mid X_{0n} \mid^2 \varphi \left(\frac{E_e}{E_{0n}} \right) \tag{1.7}$$

The universal function $\varphi \left(\dfrac{E_e}{E_{0n}} \right)$ has been calculated, by introducing into eq. (1.7) the values measured on hydrogen and alkali metal atoms [7, 26, 27], as

$$\varphi \left(\frac{E_e}{E_{0n}} \right) = \frac{\ln \left[\dfrac{E_e}{E_{0n}} - 0.9 \left(\dfrac{E_e}{E_{0n}} - 1 \right) \right]}{\dfrac{E_e}{E_{0n}} - 0.7} \tag{1.8}$$

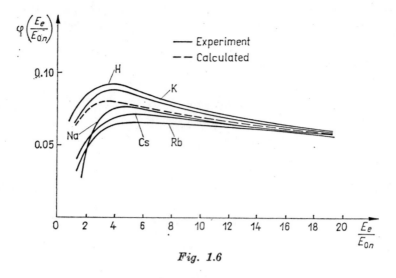

Fig. 1.6

The experimental values and those calculated from (1.8) for the function $\varphi\left(\dfrac{E_e}{E_{0n}}\right)$ are compared in Fig. 1.6.

A number of other attempts have been made to improve the agreement between the theoretical and experimental values in the range of low impact energies. These include the distorted wave method [17, 22] and other methods in which the exchange interaction and the strong coupling between the $1s-2s$, $1s-2p$, and the $1s-2s-2p$ states are taken into account [16, 18, 19, 24, 32, 33, 35, 36, 37]. The results of the different calculations compared with

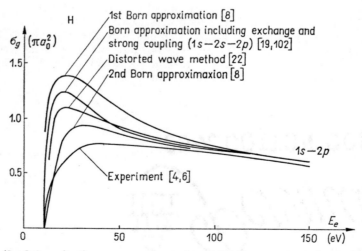

Fig. 1.7 ($1s-2p$) excitation cross section values of hydrogen atoms for various impact electron energies, as measured and calculated by different approximation methods

the experimental data [4, 63] are shown in Fig. 1.7 for the excitation cross section of the $1s \rightarrow 2p$ transition in hydrogen atoms by electron impact.

As is apparent from expressions (1.5), (1.6) and (1.7), at low values of the imparted impulse K, the cross section for the optically allowed transitions is primarily determined by the matrix element of the dipole or quadrupole interaction of the striking electron and the atom involved. If the atom is brought by the electron impact into a state from where the transition to the ground state is optically forbidden the transition cross section is determined mainly by the exchange interaction of the striking electron with the atomic electrons and the strong coupling between the $1s$–$2s$–$2p$ states.

Figure 1.8 shows the experimentally obtained excitation cross section for the $(1s \rightarrow 2s)$ transition in hydrogen atoms [7] as compared with the reported predictions by the use of different methods [8, 16, 18, 19, 24, 35]. The best agreement with the experimental data has been achieved by the calculations which take into account the strong coupling between the three $(1s$–$2s$–$2p)$ states [16, 19, 35].

The cross sections for the optically forbidden transitions are smaller than those for the optically allowed transitions (Fig. 1.5) except in the range of impact energies near to the excitation threshold. In this range there is a considerable contribution from electron exchange interactions to the excitation cross section. At low impact energies the striking electron is captured by the atom and the atomic valence electron is subsequently emitted by the atom. Since the electron exchange probability is restricted to a very narrow energy interval, the excitation cross sections of the forbidden transitions give plots with steeper rise time and sharper maxima (Figs 1.5, 1.8) compared with those for the allowed transitions.

Similarly to the $1s \rightarrow 2s$ transition in hydrogen, the $1^1S \rightarrow 2^3S$ transition in helium is characterized by an excitation cross section with a sharp maximum both in the experimental [40, 41] and the theoretical [39] results.

The lifetimes of the optically forbidden transitions may exceed those of the allowed transitions by several orders of magnitude, since the atom

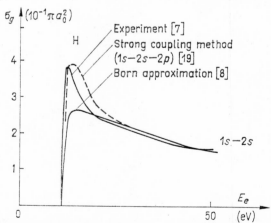

Fig. 1.8 ($1s$–$2s$) excitation cross section values of hydrogen atoms for various impact electron energies, as measured and calculated by different approximation methods

cannot get out of the forbidden state by electric dipole — or quadrupole — radiation. This state is called metastable.

The atom can get out of the metastable state by magnetic dipole radiation, the simultaneous emission of two photons, or because of an impact. The atoms can also be brought out of the excited metastable states by induced transition, that is, by the absorption of a quantum which raises the atomic energy to a level from which there is an optically allowed transition to the ground state (Paragraph 4.4.3). The atoms of many elements can be excited to metastable levels which correspond generally to the lowest excitation level (see Appendix, Table 2). Thus, e.g. the $2^2S_{1/2}$ state in hydrogen atoms and the $2^3S_{1/2}$ state in helium atoms are metastable with lifetimes τ(H, $2^2S_{1/2}) > 2 \cdot 10^{-3}$ sec [45, 47] and τ(He, $2^3S_{1/2}) > 10^{-3}$ sec [48], respectively. These lifetimes are very long compared with the average lifetime ($\tau \sim 10^{-9}$–10^{-7} sec) of the allowed transitions. This difference of several orders of magnitude makes it possible to utilize the metastable states for nuclear physics and plasma physics studies, the production of nuclear spin polarized particle beams and for many investigations and uses.

If the electron strikes a molecule, the latter interacts with the electron as a unit system and consequently not only the electron states, but also the vibrational and rotational states of the molecule may undergo a change. In this case the energy imparted by the striking electron may contribute partly to the excitation of electrons, partly to the change of the atomic motions within the molecule.

The energy spectrum of the electron states in a molecule is thus a system of vibrational (v) and rotational (j) levels, as illustrated in Fig. 1.9. The figure clearly shows the band structure characteristic of a diatomic molecule.

In a diatomic molecule the bound electron states can be excited by electron impact with no change in the distance r between the nuclei [58, 59], that is, the molecule is brought by the impact energy — according to the Frank–Condon principle — from the point of the ground state potential curve at r to the point of the potential curve for the excited state at the same value of r (Fig. 1.10). Owing to vibration, the atomic spacing within the molecule oscillates in the interval Δr around the equilibrium spacing r_0 of the stable molecular state (Fig. 1.11). It follows that the transition from the ground state to a higher electron state takes place in the interval Δr. The transition probability has the highest values at the limit of this interval, i.e. at the reversal point of the vibration, where the nuclei are most likely to be.

The excitation of the electron states in a molecule occurs by electron impact essentially in the same way as that of the atoms.

The cross sections for the excitation of the electron states are again the largest in the case of allowed transitions. The excitation function has a maximum at energies near the threshold value for excitation. This maximum varies from 10^{-17} to 10^{-16} cm² (Fig. 1.12).

Figure 1.12 shows the values of the cross section versus impact energy for the electron transitions in a nitrogen molecule excited by electron impact. The function was determined from the electron velocities measured in nitrogen gas [49].

If the rotational and vibrational states of the molecule are excited simultaneously with the electron states, the energy dependence of the excitation

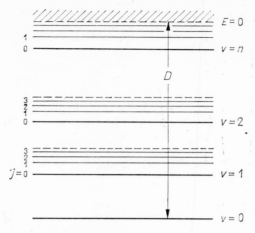

Fig. 1.9 Characteristic band structure of energy levels consisting of a system of vibrational (v) and rotational (j) levels in diatomic molecules

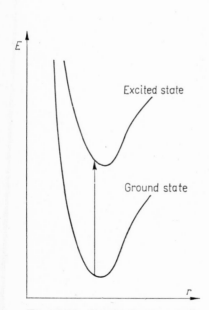

Fig. 1.10 Frank–Condon-type transition from ground to excited state in diatomic molecules

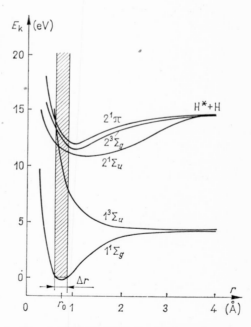

Fig. 1.11 Potential curve for electron states in hydrogen molecules plotted as a function of the distance r between the atomic nuclei

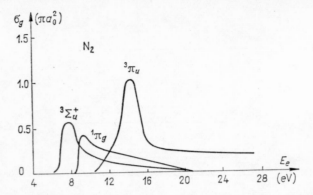

Fig. 1.12 Excitation cross section values of electron transitions in nitrogen molecules against impact electron energy

function will be of a different shape compared with the pure electron excitation function [50].

The excitation of the molecular rotational level by electron impact — a process which is largely responsible for the stopping of slow electrons in molecular gas — causes the internal rotational motions of the molecule to change. If the moment of nuclear rotation changes by an even number,

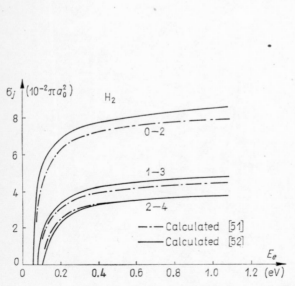

Fig. 1.13 Cross section values for transitions between rotational states in hydrogen molecules as calculated by two different methods

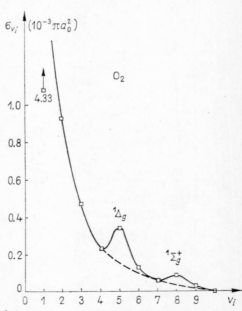

Fig. 1.14 Excitation cross section values of vibrational levels in oxygen molecules at electron impact energy $E_0 = 0.16$ eV for various vibrational quantum numbers v

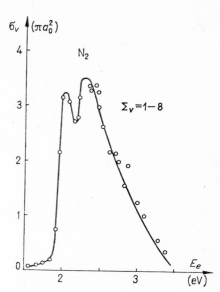

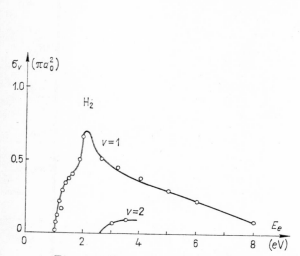

Fig. 1.15 Excitation cross section values of vibrational levels with $v = 1$ and $v = 2$ in hydrogen molecules for various impact electron energies

Fig. 1.16 Total cross section values for excitations of the vibrational levels of $v = 1$–8 in nitrogen molecules for various impact electron energies

the cross section for the transition from one to another rotational state in a diatomic molecule can be calculated in terms of the Born approximation [51] as

$$\sigma_{j,\,j+2}(E_e) = \frac{8\pi Q}{15}\,\frac{(j+1)(j+2)}{(2j+1)(2j+3)}\left[1 - \frac{2B(2j+3)}{E_e}\right]^{1/2} \qquad (1.9)$$

$$\sigma_{j,\,j-2}(E_e) = \frac{8\pi Q}{15}\,\frac{j(j-1)}{(2j-1)(2j+1)}\left[1 - \frac{B(4j-2)}{E_e}\right]^{1/2} \qquad (1.10)$$

where Q is the electric quadrupole moment and B is the rotational constant given by $B = \dfrac{\hbar}{4\pi}\,\mu r^2$ with μ standing for the reduced molecular mass $\dfrac{m_1 m_2}{m_1 + m_2}$ and r for the distance between the two atoms in the molecule.

The values of the excitation cross section predicted for hydrogen from formulae (1.9) and (1.10) by taking $Q = 0.464$ are comparable to those obtained from more precise calculations [52] and this proves that the approximations yield values of reasonable accuracy (Fig. 1.13) in practice. The excitation cross section of the rotational level evaluated from electron mobility measurements in hydrogen and nitrogen molecules [49, 53, 54] shows a good agreement with that predicted from (1.9) for the transition of the state with moment of rotation 4 to that with 6 at $Q = 1.04$ in nitrogen while that predicted for hydrogen agrees with the experimental value only by taking another value for Q [105, 106].

Similarly to the rotational level, the cross section for the electronic excitation of the vibrational level in a molecule can be predicted in terms of the Born approximation [55, 56]. In agreement with the experimental data [57] it can be seen from Fig. 1.9 that there are many fewer vibrational than rotational levels in a molecule.

Figure 1.14 shows the result of the experimental determination of the cross section for the excitation of the vibrational level in oxygen molecules [57]. The excitation cross section measured as 10^{-21} cm^2 at 0.63 eV electron energy for the first vibrational level in hydrogen molecules [60] agrees in order of magnitude with the prediction by Born approximation [55, 56].

If vibrational levels are excited by electron impact on the molecule, an increase in the cross section at a given value of electron energy is always observed. This increase is due to the contribution from negative molecular ions formed in the electron–molecule collision and rapidly decaying to an electron and an excited molecule.

This contribution can increase the cross section of the vibrational level to maximum values from $3 \cdot 10^{-17}$ to $5 \cdot 10^{-16}$ cm^2. Figures 1.15 and 1.16 show the excitation cross sections of the vibrational levels in hydrogen and nitrogen molecules, respectively, if there is a contribution from negative ion formation [63].

1.1.2 EXCITATION BY ATOM OR ION IMPACT

In the inelastic collisions of atoms with atoms, or atoms with ions, a far greater number of processes are likely to take place than in inelastic atom–electron collisions. In the former cases the state of both colliding particles can change. If both of the atoms are hydrogen atoms in the ground state, their inelastic atom–atom collision can lead to the transitions

$$\text{H}(1s) + \text{H}(1s) \rightarrow \begin{array}{c} \text{H}(1s) + \text{H}(nl) \\ \text{H}(nl) + \text{H}(1s) \\ \text{H}(nl) + \text{H}(n'l') \end{array} \tag{1.11}$$

while in the case of atom–ion collision to

$$\text{H}(1s) + \text{H}^+ \rightarrow \text{H}(nl) + \text{H}^+ \tag{1.12}$$

The cross sections for the processes described by (1.11) have been evaluated by the use of the Born approximation [64–66] as

$$\sigma_{(1s-2s,\ 1s-1s)} \sim \frac{0.72}{E_a}\,\pi a_0^2 \tag{1.13}$$

$$\sigma_{(1s-2p,\ 1s-1s)} \sim \frac{1.8}{E_a}\,\pi a_0^2$$

if the state of the striking atom is left unchanged, and as

$$\sigma_{(1s-2s,\ 1s-\Sigma nl)} \sim \frac{4.3}{E_a} \pi a_0^2$$

$$\sigma_{(1s-2p,\ 1s-\Sigma nl)} \sim \frac{2.1}{E_a} \pi a_0^2$$

(1.14)

if the energy level changes in both atoms. Here E_a is the energy of the striking atom in keV units.

The cross section for the transition defined by (1.12) can be expressed [2] by analogy with formula (1.5) in terms of the Born approximation as

$$\sigma_{(nl,\ n'l')} = \frac{4\pi Z_i^2 e^4}{\hbar^2 v_i^2} \mid X_{(nl,\ n'l')} \mid^2 \ln \frac{2m_e v_e^2}{E_{n'l'} - E_{nl}}$$

(1.15)

where Z_i is the charge and v_i the velocity of the ion. Calculations have also been made [64] on the values of the transition matrix. It is apparent from eq. (1.15) that the excitation cross section exceeds by a factor of Z_i^2 that of an electron impact at the same electron velocity, provided the latter is higher than the velocity of the electron bonded in the atom. At these impact energies the excitation cross section values of the optically allowed transitions vary with E_a^{-1} for atom–atom, and with $E_i^{-1} \ln E$ for atom–ion collisions (see formulae (1.13), (1.14) and (1.15)). Both the predicted and the measured values [2, 64–66, 114–117] show larger excitation cross sections for atom on ion than for atom on atom impacts (Figs 1.17 and 1.18). At low atomic or ionic impact energies, when the Born approximation fails to

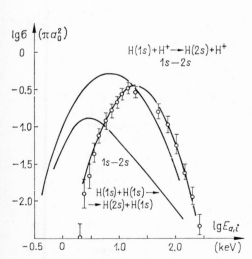

Fig. 1.17 (1s–2s) excitation cross section values of hydrogen atoms for collisions with hydrogen atoms in ground state or with hydrogen ions

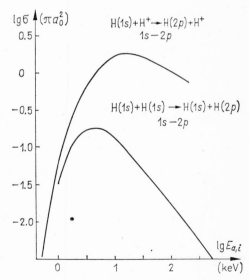

Fig. 1.18 (1s–2p) excitation cross section values of hydrogen atoms for collisions with hydrogen atoms in ground state or with hydrogen ions

work, the approximation to the perturbation of a stationary state [2, 67–69] can be employed. If the striking atom is slow, the two atoms are treated (at small distances from each other) as a quasi-molecule, consequently the wave functions of the two atoms transform to that of the quasi-molecule [311].

Since a molecule has more possible states than an atom, the quasi-molecule can be brought to any of the atomic states, thus the excitation by slow atom impact can be treated as a transition from one to another state of the quasi-molecule.

This method can be used also for the approximation to the excitation energy transfer process of slow atoms [68, 69].

At low impact energies, i.e. under the condition that

$$E_{mn} \ll E_i \ll 100 \, ME_{mn}$$

(here E_{mn} is the excitation energy of the given state, E_i is the relative kinetic energy and M is the mass of the striking particle in ^{16}O units), the excitation cross section of the processes described by (1.11) can be calculated from the formula [64]

$$\sigma_{(1s-2s)} = 1.3 \cdot 10^{-6} \frac{Z_i^2 E_i^4}{M^4 E_{(1s-2s)}^7} \pi a_0^2$$

$$\sigma_{(1s-2p)} = 7.4 \cdot 10^{-9} \frac{Z_i^2 E_i^4}{M^5 E_{(1s-2p)}^7} \pi a_0^2$$

(1.16)

1.1.3 EXCITATION BY LIGHT QUANTA

Atoms or molecules can also be excited by the absorption of light quanta. This type of excitation occurs only if the energy $h\nu$ of the 'absorbed' light quantum is equal to the difference between the energy level E_n of the excited state and that of the state at energy E_m, that is, if the Bohr frequency requirement

$$h\nu = E_n - E_m \tag{1.17}$$

s me t and if there is an optically allowed transition between the two states. This means that to evaluate the transition probability, it is sufficient to determine in first approximation the matrix element of the electric dipole momentum since the probabilities of the also possible magnetic dipole and electric quadrupole transitions are lower by factors of 10^{-5} and 10^{-8}, respectively, compared with the probability of electronic dipole transition [1, 3]. The absorption of a light quantum corresponding to resonance can occur in a very narrow range of energies only.

An atomic or molecular system can also be excited by the scattering of light quanta on the atom or molecule. Part of the energy $h\nu$ of the scattered quantum may be expended on the excitation of the atom or molecule. In this case the energy of the scattered light quantum decreases by the value of the excitation energy $E_{mn} = E_n - E_m$ and reduces to

$$h\nu' = h\nu - (E_n - E_m)$$

1.2 IONIZATION OF ATOMS AND MOLECULES

Atomic or molecular ionization can be regarded, similarly to the excitation processes, as a change of the atomic molecular states.

Ionization is a process in which the atomic or molecular electrons are brought from the discrete energy state to a state contributing to the continuous spectrum, that is, one or more electrons escape from the atom or molecule. As a result of ionization, the originally neutral atom or ion with charge n transforms to an ion with charge m or $n + m$ with the simultaneous emission of m free electrons.

Atoms or molecules can be ionized by electron, atom or ion impact, by irradiation, or by exposure to an electric field.

1.2.1 IONIZATION OF ATOMS OR MOLECULES BY ELECTRON IMPACT

Inelastic collisions of atoms or ions with electrons can lead to ionization if the energy needed for the ejection of one or more electrons is imparted in the collision by the striking electron to the atom or molecule. The amount of energy needed for electron ejection depends on the given energy state of the colliding atoms or molecules.

Let E_{nl}^i be the minimum energy required for the ionization of an atom or a molecule in either the ground or any excited state (n is the main and l the orbital quantum number). This energy corresponds to the binding energy of the electron at the energy level in question. The energy E^i (see Appendix, Tables 3 and 4) required for ionization from the ground state is called the ionization potential.

Ionization occurs if the energy of the electron which makes an impact on the atom or molecule exceeds the value of E_{nl}^i.

It should be noted that it is possible for ionization to take place even at impact electron energies below the value of the ionization potential. This happens when the atoms or molecules are brought to an excited state by electron impact, then struck again by electrons during the lifetime of the excited state and eventually becoming ionized.

Similarly to the excitation cross section, the ionization cross section σ_{nl}^i is zero at impact electron energies $E_e \leq E_{nl}^i$. The cross section for single ionization σ_i is a function of the impact electron energy. This function is called the ionization function of the molecule. If $E_e = E_{nl}^i$ the ionization function shows a sharp threshold.

The ionization cross section increases with increasing impact energies and its value close to the ionization threshold can be predicted for single ionization of the atoms from the formula [70]

$$\sigma_i = C(E_e - E_0^i) \tag{1.18}$$

where C is a constant. In this expression the interaction between the striking and the ejected electrons has been ignored. A more precise formula has been reported [71] in order to take account of this interaction by modifying the exponent in (1.18) from 1 to 1.1269 for the ionization of atoms, and to 1.056

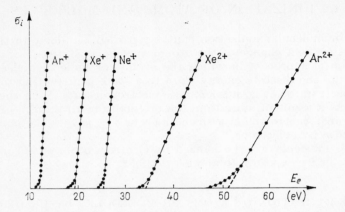

Fig. 1.19 Variations of ionization cross section values of argon, xenon and neon atoms with impact electron energies at values near the threshold energy

for that of ions. This small difference cannot be established from the experimental results available to date, thus for practical purposes it is reasonable to assume that the ionization cross section for a multiple ionization by, for example, a factor of n, can be predicted at impact energies near the ionization threshold as

$$\sigma_n^i = C(E_e - E_0^i)^n \tag{1.19}$$

The energy dependence of σ_n^i at impact energies near the threshold value as described by (1.19) has been confirmed in a number of experiments [4, 72–80]. For hydrogen atoms the prediction (1.19) proved to be correct in the range of 5 eV above the threshold energy [4].

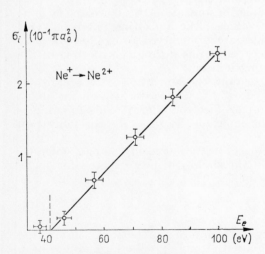

Fig. 1.20 Variation of ionization cross section values of Ne$^+$ ions with impact electron energies near the threshold energy

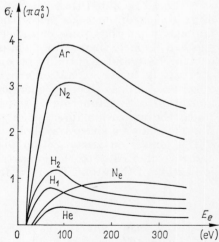

Fig. 1.21 Variation of the ionization cross section values of H$_1$, H$_2$, He, Ne, N$_2$ and Ar with impact electron energies

For the double ionization of sodium atoms the quadratic, for their triple ionization the cubic dependence was shown in an energy range of ~ 40 eV above the threshold energies for double and triple ionizations, respectively [76].

The validity of formula (1.19) was also confirmed by measurements on the ionization of noble gas atoms and ions [72–75, 77–80]. The ionization cross sections against impact electron energies, as measured at values close to the ionization threshold energies are shown for neon, argon and xenon in Figs 1.19 and 1.20.

The ionization function increases with increasing impact energies up to a maximum after which it decreases with increasing energies as shown for some molecules and atoms in Fig. 1.21.

For the evaluation of ionization cross sections both classical and quantum mechanical approximations or semi-empirical formulae can be used.

In the classical approach the inelastic collision leading to ionization can be treated as a collision of a pair of particles if the interaction between the striking and atomic orbital electrons only is considered. If in this case the motion of the atomic electrons is ignored, that is, the atomic electron is considered to be at rest at the time of the impact, then the energy imparted to the bound electron by the striking free electron is given by [94]

$$ E = \frac{4m_1 m_2}{(m_1 + m_2)^2} \frac{E_{e_2}}{1 + \dfrac{\varrho^2 v_{e_2}^4}{e_1 e_2} \left(\dfrac{m_1 m_2}{m_1 + m_2} \right)^2} \tag{1.20} $$

where m_1, m_2; e_1, e_2; v_1, v_2; E_{e_1}, E_{e_2} stand for the masses, charges, velocities and energies of the free and bound electrons, respectively. For electron–electron collisions we have $e_1 = e_2 = e$ and $m_1 = m_2 = m$ and therefore formula (1.20) reduces to the Thomson expression [10] of the form

$$ E = \frac{E_{e_2}}{1 + \dfrac{\varrho^2}{e^4} E_{e_2}^2} \tag{1.21} $$

Hence the collision parameter ϱ can be obtained as

$$ \varrho^2 = \frac{e^4}{(E^i)^2} \left[\frac{E}{E_{e_2}} - \left(\frac{E}{E_{e_2}} \right)^2 \right] = \frac{e^4}{(E^i)^2} f\left(\frac{E}{E_{e_2}} \right) \tag{1.22} $$

Assuming that each transfer of energy $E > E^i$ leads to ionization, i.e. to a transfer of energy higher than E^i, then the cross section for collisions leading to ionization can be expressed for a number N of electrons in the outer shell of the atom to be ionized [10] as

$$ \sigma_i = N\pi\varrho^2 = N\pi \frac{e^4}{(E^i)^2} f\left(\frac{E}{E_{e_2}} \right) \tag{1.23} $$

The Thomson formula (1.23) generally gives a qualitatively correct picture of the energy behaviour of the ionization cross section but the position and value of the maximum do not agree well with the experimental data. It is apparent from Fig. 1.22 that the maximum value of the cross section was observed at higher energies than the Thomson prediction. It follows that the values calculated from (1.23) are overestimated at lower and under-estimated at higher electron energies, by comparison with the experimental values.

In a further refinement of the classical model the motion of the atomic electrons is also taken into account [11, 13–15, 96]. In this formulation the experimental values are more comparable.

The best agreement with experiment is obtained from the classical two-particle collision model formulated by Gryzinski [11]. The ionization cross section for the transfer of impact energy $E > E^i$ is expressed in Gryzinski's formulation as

$$\sigma_i = \frac{\sigma_0}{(E^i)^2} \, f_1 \left(\frac{E_{e_2}}{E^i} \, , \, \frac{E_{e_1}}{E^i} \right) \qquad [\text{cm}^2] \qquad (1.24)$$

where

$$f_1 = \left(\frac{E_{e_2}^2}{E_{e_2}^2 + E_{e_1}^2} \right)^{3/2} \begin{cases} \dfrac{2}{3} \dfrac{E_{e_1}}{E_{e_2}} + \dfrac{E^i}{E_{e_2}} \left(1 - \dfrac{E_{e_1}}{E_{e_2}} \right) - \left(\dfrac{E^i}{E_{e_2}} \right)^2 \text{ if } E^i \le E_{e_2} - E_{e_1} \\[2ex] \dfrac{2}{3} \left[\dfrac{E_{e_1}}{E_{e_2}} + \dfrac{E^i}{E_{e_2}} \left(1 - \dfrac{E_{e_1}}{E_{e_2}} \right) - \right. \\[2ex] \left. - \left(\dfrac{E^i}{E_{e_2}} \right)^2 \right] \left[\left(1 + \dfrac{E^i}{E_{e_1}} \right) \left(1 - \dfrac{E^i}{E_{e_2}} \right) \right] \text{ if } E^i \ge E_{e_2} - E_{e_1} \end{cases}$$

and

$$\sigma_0 = \frac{m_2}{m_1} (Z_1 Z_2)^2 \, 6.56 \cdot 10^{-14} \qquad (\text{eV})^2 \, \text{cm}^2$$

where E^i is given in electron volts and Z_i is the charge of the colliding particle in electron charge units.

In the case of an electron–electron collision, when $m_1 = m_2 = m$ and $Z_1 = Z_2 = Z$ and assuming that $E_{e_1} \approx E^i$, which strictly holds only in the ground state of hydrogen atoms, formula (1.24) reduces to

$$\sigma_i = 6.56 \cdot 10^{-14} \frac{Z^4}{(E^i)^2} f_2 \left(\frac{E_{e_2}}{E^i} \right) \qquad (1.25)$$

where

$$f_2 = \left(\frac{E_{e_2}^2}{E_{e_2}^2 + E_{e_1}^2} \right)^{3/2} \begin{cases} \dfrac{E^i}{E_{e_2}} \left(\dfrac{5}{3} - \dfrac{2E^i}{E_{e_2}} \right) & \text{if } 2E^i \le E_{e_2} \\[2ex] \dfrac{2}{3} \dfrac{E^i}{E_{e_2}} \left(2 - \dfrac{2E^i}{E_{e_2}} \right)^{3/2} & \text{if } 2E^i \ge E_{e_2} \end{cases}$$

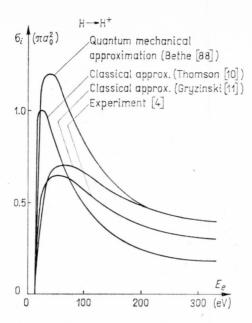

Fig. 1.22 Comparison of ionization cross section values obtained from experiment as calculated from different theories

It should be noted, however, that the limitation of the two-particle collision model is apparent even in these refined forms at higher impact energies where neither (1.24) nor (1.25) gives a satisfactory agreement with measured values (see Fig. 1.22). The reason for this is that the interaction during the time of collision between the striking electron and the residual atom is not accounted for in this model.

A wholly reliable result would be obtainable by accurate quantum mechanical calculations, but, as has already been mentioned, these are too difficult to perform.

At higher impact electron energies it is preferable to apply the Born approximation which permits the ionization cross sections to be predicted in good agreement with the experimental data [88–91]. However, at lower electron energies the predictions from the Born approximation are higher than the measured cross section values.

The deviation at the low energy end can be reduced to some extent by taking account of the exchange interaction [2, 31, 33, 70, 92] or by the distorted wave approximation [97, 98].

The Born approximation to the ionization cross section was formulated by Bethe for higher impact electron energies in the form [88]

$$\sigma_{nl}^i = \frac{2\pi e^4}{m_e v_e^2} \frac{C_{nl}}{|E_{nl}^i|} Z_{nl} \ln \frac{2n_e v_e^2}{O_{nl}} \qquad (1.26)$$

where

$$C_{nl} = \frac{Z_{\text{eff}}^2}{n^2 a^2} \int | X_{nl,k} |^2 \, dk$$

Z_{nl} is here the number of electrons on the shell nl, the value of the parameter O_{nl} varies from 0.05 to 0.3, but can generally be approximated by 0.1. Z_{eff} is the effective charge of the nucleus the field of which acts on the shell nl of the electron, $X_{nl,k}$ is the matrix element proportional to the oscillator strength of the electric dipole transition.

The value of C_{nl} has been calculated for hydrogen atoms with respect to different nl shells [88] as

$1s$	$2s$	$2p$	$3s$	$3p$	$3d$	$4s$	$4p$	$4d$	$4f$
0.28	0.21	0.13	0.17	0.14	0.07	0.15	0.13	0.09	0.04

If hydrogen atoms are ionized from the ground state, (1.26) can be written in the form

$$\sigma_{1s}^i = 0.285 \frac{2\pi e^4}{| E_{1s}^i | \, m_e v_e^2} \ln \frac{2 m_e v_e^2}{0.48 \, | E_{1s}^i |} \qquad (1.27)$$

The ionization function predicted from (1.27) agrees well with the experimental values in the energy range $E_e > 200$ eV, at lower energies the formula does not hold (see Fig. 1.22).

In order to improve the quantum mechanical approximation at lower impact energies a term arising from exchange interaction was introduced into the Born approximation [33]. The thus predicted values showed a better agreement with the experimental data on hydrogen atoms (Fig. 1.23).

As has been seen, the classical model describes the ionization function reasonably well at low impact energies and fails at high energies while the quantum mechanical approximation ceases to hold at low energies. Thus, attempts have been made to extend the validity of the classical model to higher energies by introducing different coefficients [12, 13, 81, 84, 86, 99], or to extend that of the quantum mechanical formalism to the low energy end [15, 82, 83, 93].

Among these attempts the semi-empirical formulae of the classical approach, which include a logarithmic term, have proved to be the most satisfactory. These formulae are a combination of the properties favourable in the classical formulae at lower energies with those of the Born approximation advantageous at higher energies.

For ionization by electron impact one of the two most satisfactory formulae was proposed by Gryzinski [12] as

$$\sigma_i = \frac{\sigma_0}{(E^i)^2} f_3 \left(\frac{E_{e_2}}{E^i} \right) \qquad (1.28)$$

where in the case of electron–electron collision $\sigma_0 = 6.56 \cdot 10^{-14} Z_e^2$ (eV)^2 cm^2

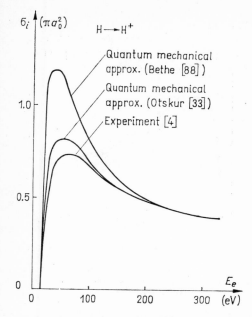

Fig. 1.23 Comparison of experimental ionization cross section values as calculated from the Born approximation formula (1.27) and from the expression accounting for exchange interaction

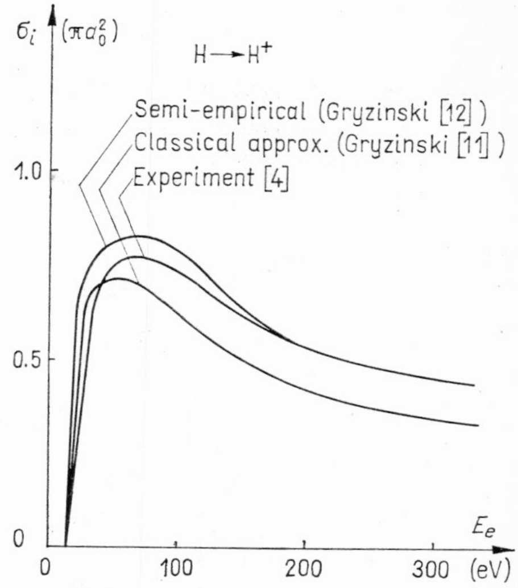

Fig. 1.24 Comparison of ionization cross section values of hydrogen atoms obtained from experiment as calculated from the classical approximation and from the semi-empirical formula

if E^i is given in eV units and

$$f_3 = \frac{E^i}{E_{e_2}} \left(\frac{E_{e_2} - E^i}{E_{e_2} + E^i} \right)^{3/2} \left[1 + \frac{2}{3} \left(1 - \frac{E^i}{2E_{e_2}} \right) \ln \left(2.7 + \sqrt{\frac{E^i}{E_{e_2}} - 1} \right) \right]$$

The other expression, proposed by Drawin [86], reads

$$\sigma_i = 2.66 \frac{\pi e^4 Z_n}{(E^i)^2} f_a \frac{E^i (E_{e_2} - E^i)}{E_{e_2}} \ln 1.25 f_b \frac{E_{e_2}}{E^i} \qquad (1.29)$$

where $f_a = 0.7, ..., 1.3$; $f_b = 0.8, ..., 3.0$ which can generally be approximated by $f_a = 1.0$ and $f_b = 1.0$.

The cross section values calculated from expression (1.28) show a better agreement with the experimental data than those predicted from (1.24) or (1.25) (see Fig. 1.24).

The logarithmic correction in (1.29) reduces the value of the cross section to compare with that obtained from the Thomson formula at low energies, the maximum is smoothed by this term and the formula transforms at higher energies to that derived from the quantum mechanical approximation.

In Figs 1.25–1.28 the calculations from (1.28) and (1.29) are compared with the experimental values.

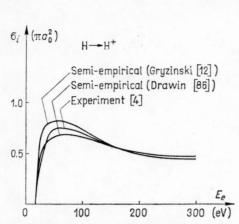

Fig. 1.25 Comparison of ionization cross section values as calculated from Gryzinski's and Drawin's semi-empirical formulae for hydrogen atoms with the experimental values

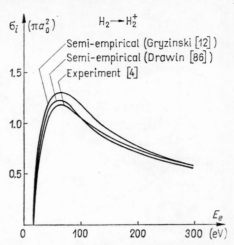

Fig. 1.26 Comparison of ionization cross section values as calculated from Gryzinski's and Drawin's semi-empirical formulae for hydrogen molecules with the experimental values

It is apparent from the figures that both expressions yield values in reasonably good agreement with the experimental data. Thus, both formulae can be used for the calculation of the ionization cross section when the error of the results involved in these relatively simple expressions is permissible in the given application.

If atoms are ionized from their excited states, the threshold energies for ionization are lower than required for ionization from their ground state

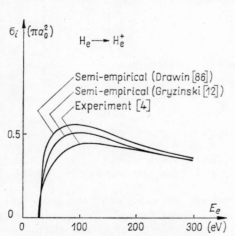

Fig. 1.27 Comparison of the ionization cross section values as calculated from Gryzinski's and Drawin's semi-empirical formulae for helium atoms with the experimental values

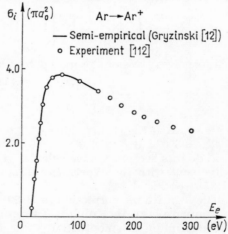

Fig. 1.28 Comparison of ionization cross section values as calculated from Gryzinski's semi-empirical formula for argon atoms with the experimental values

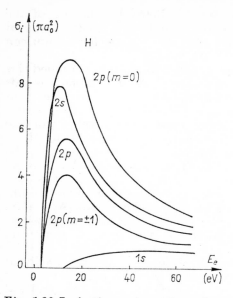

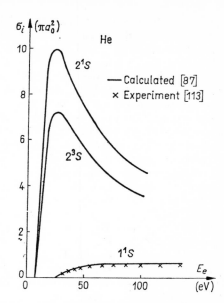

Fig. 1.29 Ionization cross section values as calculated from the Born approximation [107] for hydrogen atoms in the $1s$ ground state and in the $2s$, $2p$ excited states

Fig. 1.30 Ionization cross section values as calculated from Gryzinski's semi-empirical formula for helium atoms in the 1^1S ground state and in the 2^1S and 2^3S excited states. Values calculated for the 1^1S state are compared with the experimental data

because the binding energies of the electrons are lower in the excited than in the ground state.

The ionization cross section values of excited atoms can be calculated from the formulae obtained by the classical and quantum mechanical approximations and the semi-empirical method for ionization from the ground state.

Both the calculated and the measured values [24, 107–110] indicate that the ionization cross sections of excited atoms are larger than those of atoms in the ground state.

This fact can be interpreted also in terms of the Bohr model. If the ionization cross section is considered to be delimited by the orbit of the electron assigned to the main quantum number n of the atom, then the ionization cross section of the atom in an excited state will be n^4/Z^2 times that of the ground state assigned to $n = 1$ since the radius of the Bohr orbit is given by

$$a_n = \frac{\hbar}{m_e e^2} \frac{n^2}{Z}$$

The ionization cross sections calculated from the quantum mechanical Born approximation for the $n = 1$ and $n = 2$ levels of hydrogen atoms [107] are shown in Fig. 1.29.

Ionization cross sections have also been predicted by the use of the Gryzinski formula (1.28) for the ground states and some excited levels of helium, neon, argon and mercury atoms [24]. The results for helium can be seen in Fig. 1.30.

These calculations and the conclusions of other workers who studied this problem [108–110] show steeper slopes and a shift of maxima towards lower impact energies in the ionization functions at excited levels as compared with those observed at ground states. Consequently, there is a considerable difference between the ionization cross section values for the excited and ground states. Multiple ionization can be brought about by electron impact on atoms either by a single impact on an atom in the neutral state, or by inducing gradually a transition from the (n)-th ionization state to the $(n + 1)$-th state.

The threshold energy for multiple ionization by single impact, symbolized by E_n^i, increases as the degree of ionization, n, increases (Appendix, Table 4), whereas the cross section for multiple ionization substantially decreases as the degree of ionization increases. This can be seen from Fig. 1.31 and Appendix, Table 5 [103, 111, 112, 118].

The maximum cross section value for multiple ionization can be predicted to an accuracy within an order of magnitude if the value of the ionization cross section in the state $n = 1$ is known. One can then use the empirical formula [119]

$$\sigma_n^i = \sigma_1^i \exp\left(-\frac{6.78}{A^{1/4}}\sqrt{n-1}\right) \tag{1.30}$$

where A is the atomic number.

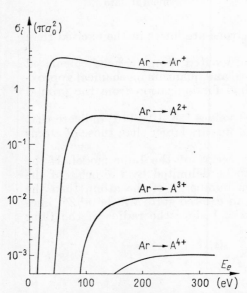

Fig. 1.31 Single and multiple ionization cross section values calculated for argon atoms

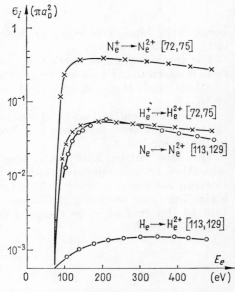

Fig. 1.32 Ionization cross section values calculated for helium and neon atoms and molecules

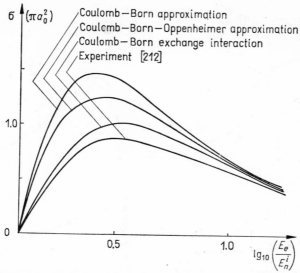

Fig. 1.33 Comparison of ionization cross section values of ions as calculated from different approximations taking into account the Coulomb field with values obtained from experiment

The maximum cross section values for multiple ionization calculated from (1.30) are compared with the measured values [103, 111, 112, 118] in Table 5. The alternative process, i.e. gradual ionization, can occur only if the lifetime of the n times ionized state is longer than the time interval between the inelastic ionizing collisions. The cross section for electron impact-induced transition from the n times ionized to the $n + 1$ times ionized state is generally larger than that from the neutral to the $n + 1$ times ionized state by a single impact [87, 91, 102, 111–113, 118, 120, 121]. Figure 1.32 shows the cross section values for the $He \rightarrow He^{2+}$, $He^+ \rightarrow He^{2+}$, $Ne \rightarrow Ne^{2+}$ and $Ne^+ \rightarrow Ne^{2+}$ transitions.

The ionization of positive ions differs from that of neutral atoms because the Coulomb field of the n times ionized atom affects the interaction due to the attractive force acting on the striking electron.

The ionization cross sections of ions can be evaluated to the best agreement with experimental data by the quantum mechanical method in terms of the Coulomb–Born approximation which also accounts for the exchange interaction [120, 212] (Fig. 1.33).

Reasonable agreement with experimental data has also been obtained for the ionization cross sections of ions predicted from the semi-empirical formula (1.28) [87], as is apparent from the comparison of the measured and calculated values for $He^+ \rightarrow He^{2+}$ and $Ne^+ \rightarrow Ne^{2+}$ processes in Fig. 1.34.

If atoms with higher mass number are struck by electrons, ionization can take place by electron ejection from the outer or inner electron shell of the atom. At low impact energies only the outer shell can be stripped of its electron whereas at higher energies either process can take place. The

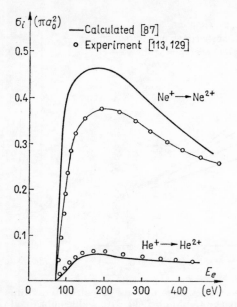

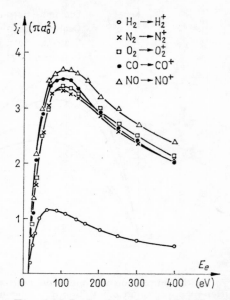

Fig. 1.34 Ionization cross section values as calculated from Gryzinski's semi-empirical formula for He$^+$ and Ne$^+$ ions as compared with experimental values

Fig. 1.35 Ionization cross section values for H$_2$, N$_2$, O$_2$, CO and NO molecules obtained from experimental data [103, 126–131]

electron ejection probability is much lower for the inner than for the outer shell. If the ionization occurs by electron ejection from the inner shell, the ejected electron is replaced by one from the outer shell and the excess energy of the latter is emitted in the form of radiation.

The calculation of the ionization of atoms with several electrons is rather complicated. It is difficult to compare the theoretical values with experimental data since the ionization from the inner shell has to be considered in addition to the simple process occurring from the outer shell, moreover, for atomic systems with several electrons the frequently occurring autoionization cannot be left out of consideration.

Calculations can therefore be performed above all for heavy atoms with a single valence electron which can be treated similarly to the ionization of hydrogen atoms, since the non-valence electrons form a closed shell and the valence electron can be regarded as a particle moving in the central field of the nucleus [91, 122, 125].

The ionization of molecules by electron impact occurs, similarly to excitation, according to the Frank–Condon principle. The threshold energy for ionization varies with the molecular state and corresponds to the difference between the energies of the potential energy curves for the transition from one electron state to another.

In the case of molecules the ionization process can be approached by considering the potential energy curves for the electron state of the molecule and the Frank–Condon principle.

The ionization cross sections of molecules have been studied by a number of authors [103, 126–131]. The results of measurements on hydrogen, nitrogen, oxygen, carbon monoxide and nitric oxide molecules are shown in Fig. 1.35.

1.2.2 IONIZATION OF ATOMS OR MOLECULES
BY ATOM OR ION IMPACT

The ionization of atoms or molecules due to inelastic collision with atomic particles depends, similarly to excitation, on the states of the colliding particles and on the impact energy and it can occur in several ways.

The inelastic collision between atomic particles can lead to ionization if the impact energy is higher than

$$E_u = \frac{E_0^i(m_1 + m_2)}{m_2}$$

where m_1 and m_2 are the masses of the colliding particles [133, 134]. It follows from this expression that for $m_1 = m_2$, we have $E_u = 2E_0^i$, that is in the processes $H + H$ or $H^+ + H_1$ the impact energy must exceed by at least twice the value of the ionization potential to induce ionization.

At lower impact energies ionization can take place only if the colliding atoms or molecules are in an excited state and the excitation energy of one particle is transferred to the other during the inelastic collision. This process may happen in various ways, such as

$$A^* + B^* \rightarrow \begin{matrix} A^+ + B\ + e \\ A\ + B^+ + e \end{matrix} \qquad (1.31)$$

$$A\ + B \rightarrow A\ + B^+ + e \qquad (1.32)$$

In the reaction of type (1.31) the excitation energies of both atoms are imparted to one of the particles and the combined energies are sufficient for ionization.

The process of type (1.32) can occur only if the ionization potential E_B^i of particle B is lower than the excitation energy E_A^e of particle A, that is,

$$E_B^i < E_A^e$$

These processes are of particular importance if the excited atoms are in a metastable state. If the excited state of the A atom is metastable in the reaction (1.32), the process is usually known by the name 'Penning effect'. Appendix, Table 6 lists the processes of type (1.32) observed by different authors [136–141] together with the reported values of the cross sections.

Besides the reactions described by (1.31) and (1.32), ionization can take place at much lower energies than defined above if one of the colliding atoms is in an autoionized and the other in the ground state. In this case the autoionized state decays, during or after the collision, to an ion and an electron. The energy spectrum of the electrons released in the ionization process contains a set of resonances which correspond to the decays of the autoionized atomic states. This permits the autoionized states to be studied by observation of the released electron spectrum. Thus, a number of auto-ionized states were identified in helium and argon atoms from the resonances in the electron spectrum [201–204].

Theoretically, this process can be interpreted in terms of the 'quasi-molecule' formed by the colliding particles [205–206], in the act of collision, when the basic term of the quasi-molecule is crossed by the boundary of the continuous spectrum. This can be followed by the decay of the quasi-molecule and the emission of a free electron. If the basic term of the quasi-molecules is crossed by the terms of the autoionization state of one of the colliding atoms, the collision process can lead to ionization.

Similarly to the energy dependence of the cross section for excitation, the energy curve for the ionization cross section on atom or ion impact exhibits a maximum. The ionization functions measured for some atoms and molecules can be seen in Fig. 1.36.

In the case of ion–atom collision the ionization cross section can be evaluated in terms of the quantum mechanical approximation by use of the Bethe formula

$$\sigma_{nl}^i = \frac{2\pi Z_i^2 e^4}{m_e v_i^2} \frac{C_{nl}}{|E_{nl}^i|} Z_{nl} \ln \frac{2m_e v_e^2}{O_{nl}} \tag{1.33}$$

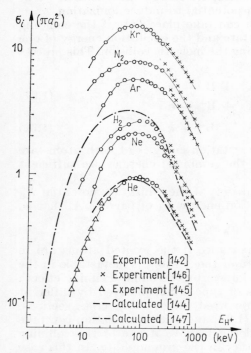

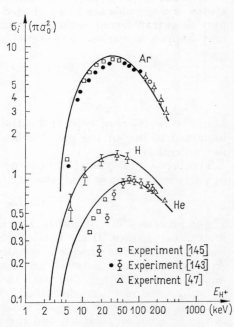

Fig. 1.36 Comparison of calculated and experimental ionization cross section values for He, Ne, Ar, Kr atoms and H₂, N₂ molecules on collision with H₁⁺ ions

Fig. 1.37 Comparison of ionization cross section values as calculated from Gryzinski's formula (1.34) with the measured values for hydrogen, helium and argon atoms

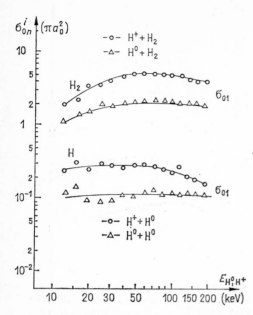

Fig. 1.38 Cross section values for single ionization measured on hydrogen atoms and hydrogen molecules bombarded with hydrogen atoms or hydrogen ions

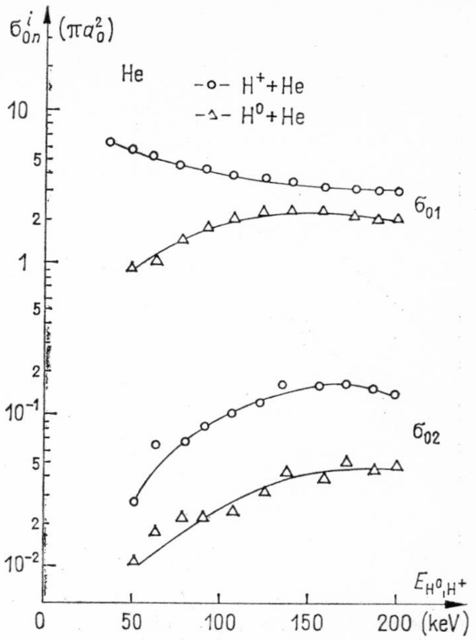

Fig. 1.39 Measured values of single and double ionization cross sections of helium atoms for collisions with hydrogen atoms or hydrogen ions

where Z_i is the charge of the colliding particles, while the other symbols are the same as those used in (1.26).

A good agreement with experimental data can also be obtained from the expression proposed by Gryzinski [12] as

$$\sigma_i = \frac{\sigma_0}{(E^i)^2} f_4 \left(\frac{v_r}{v} \right) \tag{1.34}$$

where

$$f_4 = \frac{v_r}{v} \left\{ \left(\frac{v_r}{v} \right)^2 \left(\frac{v_r^2}{v_r^2 + v^2} \right)^{3/2} \left[\frac{v_r^2}{v_r^2 + v^2} + \frac{2}{3} \left(1 + \frac{1}{\beta} \right) \ln \left(2.7 + \frac{v_r}{v} \right) \right] \cdot \right.$$
$$\left. \cdot \left(1 - \frac{1}{\beta} \right) \left[1 - \left(\frac{1}{\beta} \right)^{\left(1 + \frac{v_r}{v} \right)^2} \right] \right\}$$

with

$$\beta = 4 \left(\frac{v_r}{v} \right)^2 \left(1 + \frac{v}{v_r} \right)$$

and v_r is the velocity of the bombarding particle.

The values predicted from (1.34) are compared with the measured values in Fig. 1.37.

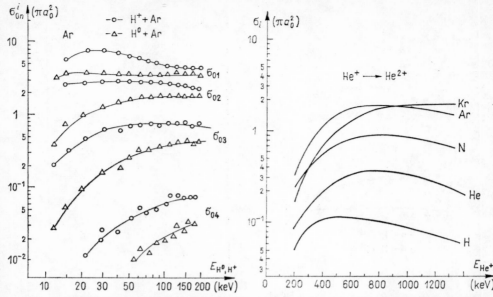

Fig. 1.40 Experimentally determined cross section values for single, double, triple or quadrupole ionization of argon atoms in collisions with hydrogen atoms or with hydrogen ions

Fig. 1.41 Experimentally determined ionization cross section values of He^+ ions for collisions with hydrogen, helium, nitrogen, argon and krypton atoms

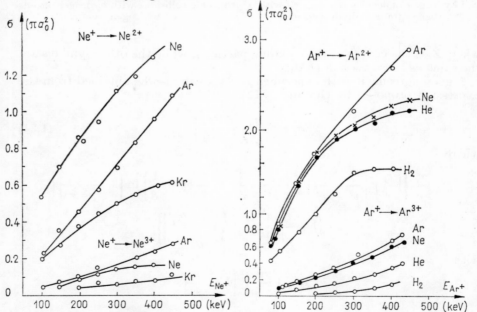

Fig. 1.42 Experimentally determined values of single or double ionization cross section of Ne^+ ions for collisions with argon, neon and krypton atoms

Fig. 1.43 Experimentally determined values of single or double ionization cross section of Ar^+ ions for collisions with hydrogen molecules and with helium, neon, or argon atoms

Numerous experiments and calculations have been made on the ionization cross sections of various gas atoms and molecules by H, H^+, He, He^+ or He^{2+} particles [142–149].

The measurements show that the ionization cross section values are higher for ion than for neutral atom impact and the maximum value exceeds in both cases that of the ionization cross section in the case of electron impact.

Figures 1.38–1.40 show the cross section values for single and multiple ionizations measured on hydrogen, deuterium, helium and argon bombarded with H^+ or H [142, 143]. The difference between the values obtainable by ion and neutral atom impact are easily apparent from the figure.

Figures 1.41–1.43 show the cross section values for the production of multiply charged ions, measured on hydrogen, helium, nitrogen, neon, argon and krypton gas bombarded with He^+, Ne^+ or Ar^+ ions [384, 385].

1.2.3 IONIZATION OF ATOMS AND IONS BY LIGHT QUANTA

Atoms and molecules can also be ionized upon the absorption of photons. This process is called photoeffect or photoionization. Ionization by photons differs from the excitation by light quanta in that the final state of the electrons taking part in the process lies in the continuous and not in the discrete energy spectrum.

Photoionization can occur if the energy $h\nu$ of the photon captured by the atom or molecule exceeds the ionization potential of the capturing particle, that is, if

$$h\nu > E^i_{nl}$$

Photons of any frequency which meet this requirement can be absorbed and thereby induce ionization. This means that the absorption spectrum is continuous in the case of photoionization.

The kinetic energy E^k_e of the electron released in photoionization is given by the Einstein equation

$$E^k_e = h\nu - E^i_{nl} \tag{1.35}$$

The photon energy $h\nu$ can be related to the length, through the equation

$$eV = h\nu = h\,\frac{c}{\lambda}$$

hence

$$\lambda = \frac{hc}{eV} = \frac{12\,400}{V}\,\text{Å} \tag{1.36}$$

where V is to be understood in volt units.

The wavelengths corresponding to the ionization potential E^i_0 are listed for some atoms in Appendix, Table 3. It can be seen from the table that the ionization of gas atoms requires light quanta with fairly large energies and wavelengths lying in the ultraviolet or soft X-ray range of the wavelength spectrum.

The photoionization cross section, i.e. electron ejection by photons, can be evaluated for hydrogen atoms or for the K shell of ions analogous to hydrogen (with nuclear charge $Z \ll 137$) from the equation formulated from the known eigenfunctions of the continuous spectrum [150] as

$$\sigma_f^i = \sigma_T^i \frac{2^7 \pi \alpha^{-3}}{Z^2} \left(\frac{E_0^i}{h\nu} \right)^4 \frac{e^{-4\xi \text{arc cot} \xi}}{1 - e^{-2\pi\xi}} \qquad (1.37)$$

where

$$\sigma_T^i = \frac{2^2 \pi r_0^2}{3} = 6.65 \cdot 10^{-25} \, \text{cm}^2 \qquad (1.38)$$

is the scattering cross section for Thomson radiation with $r_0 = \dfrac{e^2}{mc^2} =$ $= 2.818 \cdot 10^{-13}$ cm, i.e. the classical electron radius, while $\alpha = \dfrac{1}{137} = \dfrac{e^2}{\hbar c}$ is the fine structure constant and

$$\xi = \sqrt{\frac{E_0^i}{E_e^k}} = \frac{Ze^2}{\hbar\nu}$$

In formula (1.37) it is assumed that the energy of the incident light quantum is high compared with the ionization energy of the electron on the K shell and that the energy of the outflying electron is low compared with its rest energy mc^2, that is,

$$E_0^i \ll h\nu \ll mc^2$$

which means that the relativistic correction is small.

Close to the ionization threshold, when $h\nu \to E_0^i$, that is, $\xi \to \infty$, formula (1.37) can be reduced to

$$\sigma_{f_0}^i = \sigma_T^i \frac{2^7 \pi \, 137^3}{(2.71)^4} \qquad (1.39)$$

Now, if the condition $E_0^i \ll h\nu \ll mc^2$ is satisfied and the Born approximation can be applied $\left(\xi = \dfrac{Ze^2}{\hbar\nu} \ll 1 \right)$, expression (1.37) takes the form

$$\sigma_{f_1}^i = \sigma_T^i \frac{2^6 \cdot 137^3}{Z^2} \left(\frac{E_0}{h\nu} \right)^{\frac{7}{2}} \qquad (1.40)$$

The values calculated from the exact formula (1.37) and from the approximation (1.40) are compared in Fig. 1.44 with those determined from experiments [154, 155]. It can be seen that the values predicted by use of the exact formula (1.37) agree well with the experimental data.

It has to be noted, however, that at lower photon energies the photoionization cross section expressed exclusively for the K shell may include a contribution from photoionization on the outer shells which cannot be ignored.

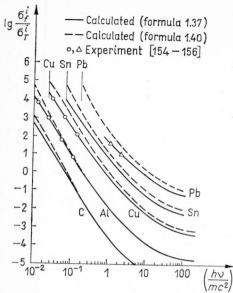

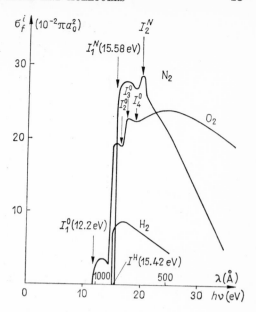

Fig. 1.44 Comparison of logarithmic values of the ratios of photoionization cross section σ_f calculated from the exact formula (1.37) or from the approximation formula (1.40) to the Thomson radiation cross section σ_T with measured experimental values

Fig. 1.45 Photoionization cross section values calculated for hydrogen, nitrogen and oxygen molecules by use of formula (1.42)

In the relativistic case, when the energy $h\nu$ of the incident quantum is much higher than the ionization potential, i.e.

$$E_0 \ll h\nu \geq mc^2$$

the photoionization cross section can be evaluated from the formula [151, 152]

$$\sigma^i_{f_2} = \sigma^i_T \frac{3Z^5\alpha^4}{2} \frac{(\gamma^2 - 1)^{3/2}}{(\gamma - 1)^5} \cdot$$

$$\left[\frac{4}{3} + \frac{\gamma(\gamma - 2)}{\gamma + 1} \left(1 - \frac{1}{2\gamma\sqrt{\gamma^2 - 1}} \ln \frac{\gamma - \sqrt{\gamma^2 - 1}}{\gamma + \sqrt{\gamma^2 - 1}} \right) \right] \qquad (1.41)$$

where

$$\gamma = \frac{1}{\sqrt{1 - \dfrac{v^2}{c^2}}}$$

If the system is analogous to hydrogen, the photoionization cross section is given for light atoms also by the function of the form [157, 158]

$$\sigma^i_f(v, n) = \frac{g 2^5 \pi e^6 Z^4 R}{3^{3/2} h^3 v^3 n^5} \qquad (1.42)$$

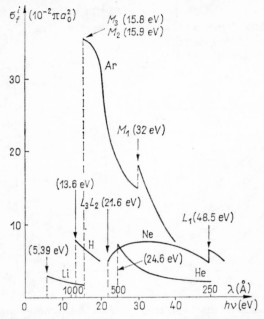

Fig. 1.46 Photoionization cross section values calculated for hydrogen, lithium, helium, neon and argon atoms by use of formula (1.42)

where $R = \dfrac{2\pi^2 e^4 m}{\hbar^2 c}$ is the Rydberg constant, n is the main quantum number of the initial state, and g is the multiplying factor evaluated in [158] which can generally be taken as $g = 1$ (the deviation from this value being not more than 10–20 per cent).

Photoionization cross section values calculated for given atoms and molecules from formula (1.42) are shown in Figs (1.45) and (1.46). The measured values obtained in numerous experimental studies of photoionization [160–173] show a good agreement with those calculated from (1.42).

1.2.4 AUTOIONIZATION

Autoionization takes place if two of the atomic electrons are simultaneously excited by impact or radiation with the sum of the two excitation energies being higher than that required for the ejection of one of these electrons.

The energy level corresponding to the autoionization state belongs to the continuous spectrum. The autoionization state may cease if one of the excited electrons is brought by the interaction between electrons to a lower level while the other is released and a positive ion is formed.

There are several ways in which autoionization states can be produced. One of them is autoionization initiated by two electrons excited to nearly the same degree, in another case the residual atom, in addition to one of the electrons, is excited to such an extent that the combined excitation energies exceed the ionization energy of an electron.

The difference between the above mentioned two types of autoionization originates from the electron shell structure.

Another characteristic property of the autoionization states is their lifetime. The long-lived states which depend on the interaction of the electron spins can be distinguished from the short-lived autoionizations which are maintained by the electric interaction between electrons.

Autoionized states have a long lifetime if their decay is forbidden by the law of total electron spin conservation. States of this type can decay as a result of magnetic interaction between the spins or because of photon radiation. The lifetime of these autoionized states is given by $\left(\dfrac{\hbar c}{e^4}\right)^4$ and it is thus $\left(\dfrac{\hbar c}{e^4}\right)^3$ times longer than that of the states decaying because of electrostatic interaction between electrons.

Autoionized states of such long lifetime are, for instance, the $Li(1s2s2p)^4P$ state of lithium atoms and the $He^-(1s2s2p)^4P$ state of negative helium ions where all three electron spins have the same direction.

The stripping of one of the three electrons leads, in the negative helium ions, to the formation of an atom in the ^3S triplet state the electron of which has a lower binding energy than that of the electron in the negative helium ion [174]. In lithium atoms the removal of an electron causes the formation of a lithium ion in the ^3S state which is at a higher level than the $Li(1s2s2p)^4P$ state [175].

Experiments were made to determine the excitation energies of the long-lived autoionized states of lithium atoms [176]. The thus measured values agree reasonably well with those predicted by the method accounting for the screening constant [175], as is apparent from the data listed in Appendix, Table 7. The calculation method [175] made it possible to explain the results of the experimental investigations of transitions in the autoionized states of lithium [177] and helium [178] atoms.

The autoionization cross section in the case of electron impact was measured for the $Li(1s2s2p)^4P$ state at 58 eV electron energy and it was found to be 10^{-19} cm^2 [176]. Theoretically, the lifetimes of the long-lived states for $He^-(4P_{3/2})$ and $Li(4P_{3/2})$ were estimated as $1.7 \cdot 10^{-3}$ sec and $1.7 \cdot 10^{-5}$ or $1.6 \cdot 10^{-5}$ sec, respectively. The experimental value $(5 \pm 1) \cdot 10^{-6}$ sec agrees with the predictions in order of magnitude.

Autoionization states with short lifetimes, which are substantially shorter than the time characteristic of the radiation, cause resonances to appear in the cross sections for elastic or inelastic scattering on atoms and for photon absorption. This phenomenon permits the parameters of such autoionization states to be experimentally determined from the observation of the cross sections of these processes. In the case of electron scattering by atoms these parameters can be calculated from the data on the scattering phase [184–187] or by choosing the appropriate wave function system [181–183]. Actually, theoretical calculations have been made in connection with the resonance in the electron scattering on hydrogen atoms due to autoionization [184–187]. The scattering phase was determined and this made it possible to evaluate the energies of the autoionized state of negative hydrogen ions. The results of the calculations [187, 198] are listed in Appendix, Table 8. In the experimental investigation of electron scattering by hydrogen atoms the broadest resonance was observed at 9.7 ± 0.15 eV [188] and at 9.56 eV energy [190] while the resonance width was measured as 0.048 eV in good agreement with the predictions [187] to be seen in Appendix, Table 8.

Experimental and theoretical investigations have also been performed on the energies and level widths of autoionized states in helium atoms [191–195]. The values calculated by the variation method [194, 195] and those evaluated from elastic and inelastic electron scattering experiments on helium ions [191, 192] are compared in Appendix, Table 9.

Resonances attributable to autoionization states were also observed in the cross sections for electron scattering on noble gases [199, 200].

The resonance phenomenon caused by autoionization also appears in inelastic electron scattering experiments on molecules. The formation of a negative molecule ion in an autoionized state as a result of an electron–molecule collision leads either to excitation of the molecular vibrational level or to dissociation of the molecule.

Resonances, similar to those observed in the cross sections for elastic or inelastic collisions between electrons and atoms in the presence of autoionized states, can also be detected on photoionization functions [207–209]. If the time characteristic of the photon absorption is much longer than the transition time (10^{-13} sec) from the autoionization state to the state lying in the continuous spectrum, the photoionization cross section is affected by the transition process [210].

1.2.5 SURFACE IONIZATION

Atoms or molecules incident on hot metal surfaces leave the surface partly in the form of negative particles, partly as positive or negative ions. This process is called surface ionization. This phenomenon was first observed in the case of caesium atoms on hot tungsten surfaces [213, 214].

Surface ionization can be induced either by surrounding the hot metal surface with the gas or vapour of the experimental element at low pressure when thermal motion causes the atoms or molecules of the element to impinge on the metal surface, or by bombarding the hot surface with atomic or molecular beams of the element to be ionized. If the atoms or molecules impinged on the surface remain adsorbed on the hot metal for a sufficient time, they get in thermodynamical equilibrium with the surface.

The process of surface ionization can be characterized by the degree of ionization, expressed as

$$\alpha = \frac{n_i}{n_a} \tag{1.43a}$$

or the surface ionization coefficient given by

$$\beta = \frac{n_i}{n_0} \tag{1.44a}$$

where n_i and n_a are the numbers of ions and atoms leaving the surface, respectively, and n_0 is the number of particles incident on the surface. In the stationary state, when

$$n_0 = n_i + n_a$$

the degree of ionization is related to the ionization coefficient as

$$\alpha = \frac{\beta}{1 - \beta} \qquad (1.43\text{b})$$

and

$$\beta = \frac{\alpha}{1 + \alpha} = \frac{1}{1 + \dfrac{1}{\alpha}} \qquad (1.44\text{b})$$

If the ions are positive, the value of α varies with the surface temperature T, the ionization potential E^i of the impinged atom or molecule, the work function φ of the surface and the electric field V which causes the ions emerging from the surface to leave the surface.

In the absence of an applied electric field, the degree of ionization can be evaluated from the equation [216]

$$\alpha_0^+ = A \exp\left(\frac{Z_i(\varphi - E^i)}{kT}\right) \qquad (1.45)$$

where $A = \dfrac{g_i}{g_a}$ is the statistical weight ratio of ionic to atomic states of the particles leaving the surface (in the case of alkali metals $g_i = 1$ and $g_a = 2$), Z_i is the charge of ions and k is the Boltzmann constant.

This formula has been derived with the assumption that the hot metal surface can be regarded as homogeneous with respect to the work function and that the incident and leaving particles are in thermodynamical equilibrium with the surface.

In the case of possible backscattering of atoms or ions in the surface ionization process, the degree of ionization is given by the equation which contains the contributions from the reflection factor r_a and r_i of the atoms and ions, respectively, in the form [220, 221]

$$\alpha_0^+ = A \frac{1 - r_i}{1 - r_a} \exp\left(\frac{Z_i(\varphi - E^i)}{kT}\right) \qquad (1.46)$$

Under the action of an applied field, the kinetic energies λ_i and λ_a required for the removal of the ions and atoms, respectively, may be changed and the equation for α takes the form [217]

$$\alpha^+ = A \exp\left[\frac{Z_i(\varphi - E^i) + (\lambda_a' - \lambda_a) - (\lambda_+' - \lambda_+)}{kT}\right] \qquad (1.47)$$

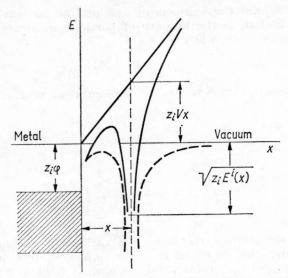

Fig. 1.47 Shift of valence electron level of atoms exposed to an external homogeneous electric field V

If the process takes place in a homogeneous field which accelerates the ions, we have for atoms $\lambda'_a = \lambda_a$ while $\lambda'_i < \lambda_i$, that is, the applied electric field reduces by $Z_i\sqrt{Z_i V}$ the force binding the ions to the metal surface (Fig. 1.47).

Since $\lambda'_i - \lambda_i = -Z_i\sqrt{Z_i V}$, expression (1.47) can be written in the form

$$\alpha^+ = A \exp\left\{\frac{Z_i[(\varphi - E^i) + \sqrt{Z_i V}]}{kT}\right\} \tag{1.48}$$

The increase in the degree of ionization, as expected from eq. (1.47), has been confirmed by the experimental data [218].

In the case of an inhomogeneous applied field, we find

$$\lambda'_a - \lambda_a = \frac{1}{2} a_a V^2$$

$$\lambda'_i - \lambda_i = \frac{1}{2} a_i V^2 - Z\sqrt{Z_i V}$$

$$E^i(x) = E^i - Vx$$

and eq. (1.47) takes the form [215, 217]

$$\alpha' = A \exp\left\{\frac{Z_i[(\varphi - E^i) + Vx + \sqrt{Z_i V} + \frac{V^2}{Z_i}(a_a - a_i)]}{kT}\right\} \tag{1.49}$$

where a_a and a_i are the polarizations of atoms and ions, respectively and x is the distance of the adsorbed atoms from the metal surface.

The above equations for the degree of ionization can be applied to any atom in thermodynamical equilibrium with the homogeneous metal surface having uniform work function. As a consequence of thermodynamical equilibrium the composition of the particles leaving the surface is determined by the state of the adsorbed atomic layer and not by the atomic states prior to adsorption.

It should be noted that while the number n_0 of atoms or atomic ions incident on the surface is independent of the temperature depending only on the intensity of the atom or atomic ion beam, the number of incident molecules varies with the surface temperature because of the temperature dependence of the molecular dissociation on the surface (see Paragraph 1.6).

If the incident particles do not remain on the surface for the time needed for the onset of thermodynamical equilibrium, that is, the surface is struck by fast atoms or ions which are reflected so that they keep most of their energy which enables each of the particles to leave the surface, then α_0 can be evaluated at any surface temperature from the equation [219]

$$\alpha_0^+ = A \exp \left\{ \frac{Z_i \left[(\varphi - E^i) - \frac{1}{Z_i} (\lambda_a(x_0) - \lambda_i(x_0)) \right]}{kT} \right\} \qquad (1.50)$$

where x_0 is the critical distance of the metal surface from the adsorbed atoms at which the electron transition from the adsorbed atoms to the metal surface is equal to zero, while $\lambda_a(x_0)$, and $\lambda_i(x_0)$ are the energies required for the removal of atoms and ions at the critical distance, respectively.

The equations considered up to now can be applied to surfaces with homogeneous work function, thus do not hold for ionization processes on polycrystalline surfaces. As a matter of fact, a surface with uniform work function is only the boundary surface of an ideal monocrystal in the absence of any adsorbed atom of another element.

Polycrystalline surfaces are composed somewhat like a mosaic and consist of different crystal planes with unlike work functions φ_k. It is extremely difficult to evaluate the work functions for the different crystal planes. For example in experiments on tungsten monocrystals, only the range of the possible values of φ_k could be determined as

4.6 to 6.0 eV for the $\langle 110 \rangle$ plane, and

4.2 to 5.3 eV for the $\langle 116 \rangle$ plane.

The metal surface involved in the surface ionization can be regarded as a set of homogeneous surface elements with individual work functions in which case the degree of ionization α_0 can be evaluated from the sum formula [222]

$$\alpha_0^+ = \left\{ \left[\sum_k \frac{F_k A}{A + \exp \frac{Z_i}{kT} (E^i - \varphi_k)} \right]^{-1} - 1 \right\}^{-1} \qquad (1.51)$$

where F_k is the homogeneous surface element with work function φ_k.

The treatment of surface ionization emitting negative ions differs from the above formulation only in that the E^i ionization potential has to be replaced by the electron affinity energy S in the equation for the degree of ionization. The affinity energy is equal to the energy needed for the removal of the electron coupled to the atom (see Paragraph 1.3).

Accordingly, in the case of negative surface ionization, α_0^- is given by

$$\alpha_0^- = \frac{n}{n_a} = A \exp \frac{Z_i(S - \varphi)}{kT}$$

and if $V \neq 0$

$$\alpha_0^- = A \exp \left[\frac{Z_i(s - \varphi) + (\lambda_a' - \lambda_a) - (\lambda_-' - \lambda_-)}{kT} \right]$$

The intensity of the ion current obtainable from the surface can be evaluated by making use of eq. (1.44a) in the form

$$I = Z_i n_0 F \beta = \frac{Z_i n_0 F}{1 + \alpha^{-1}}$$

If a monocrystalline surface is used, the positive ion current from the surface can be evaluated by making use of equation (1.45) for α in the absence of applied electric field, writing

$$I^+ = \frac{Z_i n_0 F A}{A + \exp \left[\dfrac{Z_i(E^i - \varphi)}{kT} \right]} \tag{1.52}$$

and for negative ion current if $V = 0$

$$I^- = \frac{Z_i n_0 F A}{A + \exp \left(\dfrac{Z_i(\varphi - s)}{kT} \right)}$$

The positive ion current obtainable from surface ionization for an inhomogeneous electric field in the case of polycrystalline surfaces can be calculated [223, 224] as

$$I^+ = \sum_k \frac{Z_i n_0 F A}{A + \exp \left\{ \dfrac{Z_i \left[(E^i - \varphi_k) - V_x - \sqrt{Z_i V} - \dfrac{V^2}{2Z_i}(a_a - a_i) \right]}{kT} \right\}} \tag{1.53}$$

In the case of polycrystalline surfaces the negative ion current obtainable from surface ionization for an inhomogeneous electric field is given by

$$I^- = \sum_k \frac{Z_i n_0 F_k A}{A + \exp \left\{ \dfrac{Z_i \left[(\varphi_k - s) - V_x - \sqrt{Z_i V} - \dfrac{V^2}{2Z_i}(a_a - a_i) \right]}{kT} \right\}} \tag{1.54}$$

It is apparent from the ion current formulae that the current increases with the surface temperature and reaches the maximum $I = Z_i n_0 F$ with $T = \infty$ if the bracketed value in the exponent is positive. If this bracketed value is negative, the maximum ion current is given by the formula for $T = 0$.

Experimental investigations carried out by exposure of the surface to gas, vapour, atomic or molecular beams, covered most thoroughly the alkali metal atoms of caesium, rubidium, potassium, sodium, tungsten, molybdenum, tantalum, rhenium, platinum and iridium polycrystals [213, 216, 220, 221, 225–232, 611, 748–753], and sodium atoms on the surfaces of the ⟨110⟩ and ⟨116⟩ planes of a tungsten monocrystal by measuring the variations of the generated ion current with temperature and applied electric field.

The results of these experiments showed that the predicted and observed curves for $I = f(T)$ are in qualitative agreement at surface temperatures above a given value T_m. A more accurate, quantitative comparison of the theoretical and experimental values is hardly possible because of the dissimilar experimental conditions; nevertheless, attempts at such a comparison have been already reported [220, 221, 229].

The predicted and experimental values of $I = f(T)$ at surface temperatures $T_m > T > T_c$ do not agree because of the substantially greater variations of the measured ion currents compared with those calculated from the formulae. Figure 1.48 shows the measured temperature curves for ion currents obtained at two different pressures. It is apparent from the figure that for caesium, rubidium, potassium alkali atoms and a tungsten surface the ion current steeply changes between 0 and the maximum at surface temperatures between 800° and 1000 °K — which does not follow from the theory.

This deviation can be attributed to the fact that the atoms impinging on the surface at surface temperatures $T_c > T$ do not attain sufficient kinetic energy to enable them to leave the surface. Consequently, larger fractions of the atoms which should be ionized are absorbed and also the value of the work function φ_k undergoes a change. This change then results in a sudden reduction of the ion current.

The kinetic energy needed by the ions for leaving the surface must be higher than the evaporation energy λ_i. The value of λ_i can be calculated

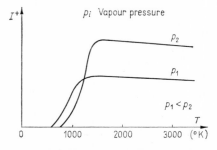

Fig. 1.48 Dependence of ion current generated from caesium, rubidium and potassium atoms on tungsten ionizer on surface temperature of the ionizer at different values of the metal vapour pressure p_i

at threshold temperature T_c in the absence of electric field from the formula

$$\lambda_i = akT_c,$$

and if an electric field is applied, from the expression [748–753]

$$\lambda_i - Z_i\sqrt{Z_i V} = akT_c$$

It is apparent from the expression which holds in the presence of an applied electric field that the threshold temperature T_c decreases as the applied field increases. This effect has actually been observed in the experimental investigation of the surface ionization of potassium atoms on a tungsten surface [753].

1.3 NEGATIVE IONS

If an electron is coupled to a neutral atom, a negatively charged particle will be formed. Such particles having a negative charge are called negative ions. There are many atoms which combine with an electron to form a stable negative ion, and to remove the electrons bound to neutral atoms a given energy has to be imparted to the system. This energy is called the binding energy E of the electron, or the electron affinity energy S of the atom. The binding energy of the electron is one of the most important characteristics of negative ions, related to the energy level occupied by the electron in the negative ion.

The electron affinity of atoms depends on their position in the periodic system. As can be seen also from Appendix, Table 10, the electron binding energy has the highest value in the negative halogen ions of noble gas electron shell configuration and the lowest value in the negative alkali metal ions, if the negative $He^-(1s, 2s, 2p)^4P$ ion is disregarded. The electron affinity energies of the other atoms have values ranging between those of the two above mentioned groups.

The electron affinity energies of atoms can be calculated by the variation method. This method was used for the evaluation of the electron binding in H^- [233–236], $He^-(1s, 2s, 2p)^4P$ [174] and Li^- [237] ions. The electron affinity energies in negative ions containing several electrons can be evaluated by using either the Hartree–Fock approximation to the variation method [238–240] or the statistical method [241–242].

The semi-empirical isoelectronic extrapolation method [243–248] is also useful for the determination of electron affinity energies in atoms. The method utilizes the principle that atomic particles with the same number of electrons and the same electronic shell structure have identical properties. Such isoelectronic systems are, for example, H^-, He, Li^+, Be^{++} which always have two electrons in addition to the atomic nucleus.

This method makes it possible to use the measured values of the ionization potential of isoelectronic atoms ($E^i_{(He\rightarrow He^+)}$, $E^i_{(Li^+\rightarrow Li^{++})}$, $E^i_{(Be^{++}\rightarrow Be^{+++})}$) to express the ionization potential function for the isoelectronic system in terms of the nuclear charge ($Z_{He} = 2$, $Z_{Li^+} = 3$, $Z_{Be^{++}} = 4$) since the three values of the ionization potential correspond to three points of the function. The extrapolation of the thus obtained relation to $Z = 1$, permits the ionization potential of the H^- ion, i.e. the electron affinity energy of the hydrogen atom, to be estimated. In a similar way, other isoelectronic systems have also been utilized for the estimation of the electron affinity energies in a number of atoms [243–248]. It can be seen from Appendix, Table 10 that the estimates made by the isoelectronic extrapolation method do not considerably deviate from those calculated by other methods so that the former, which is the most simple of the available methods, proves to be satisfactory in many cases.

Experimentally, the electron affinity energy can be most accurately determined from the photon absorption spectrum of the negative ion [249–252, 293–295, 301] or by the surface ionization method [253–260, 285, 288, 295, 297]. In the former case the minimum photon energy is measured at which the photo-decay of the negative ion is still possible and this energy is taken to be the binding energy of the electron in the negative ion. In the latter type of experiments the hot metal surface is bombarded with a beam composed of a known ratio of two different elements. The measured ratio of the two types of negative ion currents permits the ratio of the electron affinity energies in the atoms of the two bombarding elements to be determined [259, 260].

Negative ions can be produced by radiative electron capture, three-particle collision, electron–molecule collision, surface ionization and charge transfer processes.

In the radiative electron capture the electron is captured by the neutral atom and the excess energy is emitted by radiation. This reaction occurs in the sequence

$$A + e \rightarrow A^- + h\nu \tag{1.55}$$

The cross section σ_{rc} for this process can be given by utilizing the principle of partial equilibria and the σ_{pd} cross section for photo-decay [251–263] in the form

$$\sigma_{rc} = \frac{g^-}{g_a} \frac{(h\nu)^2}{(Kc)^2} \sigma_{pd} \tag{1.56}$$

where g_a and g_- are the statistical weights of the atom and the negative ion, respectively, $h\nu$ is the photon energy, K is the impulse of the electron and c is the velocity of light.

Experimentally, the radiative capture cross section was evaluated for hydrogen in a shock wave tube [264], for oxygen in a gas discharge tube [265].

In three-particle collision the possible reactions for collisions between atoms and electrons can be described as

$$A + e + B \rightarrow A^- + B$$

$$A + e + e \rightarrow A^- + e \tag{1.57}$$

By electron impact on molecules negative ions can be produced in the reactions

$$e + AB \rightarrow A^- + B$$
$$e + AB \rightarrow A^- + B^+ + e \qquad (1.58)$$

Reactions (1.58) lend themselves to the determination of the electron affinity in atoms. To do this, the minimum electron energy $E_{e(min)}$ needed for inducing the reaction in question must be determined. The binding energy of the electron E^A in the negative ion can be expressed in terms of the thus obtained value as

$$E^A = D - E^A_{e(min)}$$
$$E^A = D + E^i_B - E^B_{e(min)}$$

where D is the dissociation energy of the AB molecule and E^i_B is the ionization potential of the B atom.

The methods of negative ion production by surface ionization and by charge transfer are discussed in Paragraphs 1.2.5 and 1.4, respectively.

Negative ions can decay due to photon capture by photoemission or as a result of collision with atoms.

The cross section for photo decay can be predicted to a reasonable agreement with experimental data in terms of the single-electron quantum mechanical approximation [251, 266, 270, 304–306].

For negative ions with an s electron the cross section of photo-decay is given by the single-electron approximation [266] as

$$\sigma_{pd} = \frac{4\pi}{3c} B^2 \frac{\varepsilon K^3}{(h\nu)^3} \qquad (1.59)$$

where B is the parameter of the approximated wave function for a valence electron of negative ions, K is the impulse of the outflying electron, $h\nu$ is the photon energy, defined as

$$h\nu = \frac{\varepsilon^2}{2} + \frac{K^2}{2m_e}$$

and

$$\varepsilon = \sqrt{2E_-}$$

The parameter B has been calculated from theoretical formulae for various negative ions. These calculations give $B_{H^-} = 1.63$ [266]; $B_{Li^-} = 2.2$ [268]; $B_{Na^-} = 1.7$; $B_{K^-} = 1.9$ [267]; $B_{Rb^-} = 2.4$; $B_{Cs^-} = 2.2$ [269].

The photo-decay cross section values calculated from formula (1.59) for H^- ions are shown along with the measured values [270] in Fig. 1.49 and prove that for $h\nu < 2.5$ eV the agreement is quite good.

The photo-decay cross sections of negative ions with a p electron predicted in terms of the single electron approximation for O^-, C^-, Ce^- and F^- [305, 306] were found to be in good agreement with the experimental data. Figure 1.50 shows the calculated values [306] compared with those determined from experiment [251].

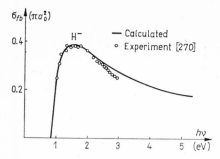

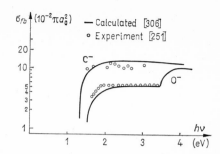

Fig. 1.49 Comparison of photodecay cross section values as calculated from formula (1.59) for H^- ions with experimental data

Fig. 1.50 Comparison of calculated with measured values of photodecay cross section for C^- and O^- ions

In the case of negative ion–atom collisions the negative ion can decay if the colliding particles remain for a sufficiently long time at a short distance from each other since the term of the electron of negative ions crosses the limit of the continuous spectrum if the nuclei of the colliding particles get very close to each other in the slow collision process. It was shown by theoretical calculations [271–275] that the decay probability of negative ions is equal to unity if the minimum distance r_{min} of their nucleus is less than a distance r_n from that of the struck atom; r_n stands for the distance between nuclei at the crossing point between the electron term and the limit of the continuous spectrum. It follows that for $r_{min} > r_n$ the decay probability is equal to zero.

The cross section for the decay of negative ions in slow collision with atom can be evaluated from the formula [271, 272]

$$\sigma_{ab} = \pi r_k^2 \left(1 - \frac{U(r)}{E_c} \right) \tag{1.60}$$

where $U(r)$ is the interaction potential of the negative ion with the atom

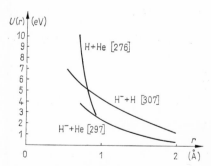

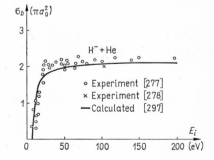

Fig. 1.51 Interaction potential $U_{(r)}$ as a function of distance r between the colliding particles

Fig. 1.52 Dependence of decay cross section of negative ions produced in the collision of H^- ions with helium atoms on the impact energy E_i

[272, 276, 297, 307] (Fig. 1.51), and E_c is the energy of the colliding particles in the centre of mass system.

If a slow negative ion with energy E_i and mass m_i makes an impact on atoms with mass m_a which can be regarded as at rest in the laboratory system, E_c can be related to the energy in the laboratory system as

$$E_i = \frac{m_a + m_i}{m_a} E_c$$

It follows from this relation that for negative ions with a large mass the threshold energy of the reaction can be much higher in the laboratory than in the centre of mass system. Thus, for example, $E_{c(J^-)} = 8.6$ eV, $E_{i(I^-)} = 280$ eV and $E_{c(Br^-)} = 7.2$ eV, $E_{i(Br^-)} = 150$ eV [271].

Figure 1.52 shows the calculations from formula (1.60) [297] together with the measured values [277, 278].

1.4 CHARGE TRANSFER

In the process termed charge transfer, the colliding atoms or ions transfer electrons to each other and thus their charges will change according to the number of electrons lost or gained in the transfer process. The general formula describing these types of process reads

$$A^i + B \rightarrow A^k + B^{(i-k)} + \Delta E \tag{1.61}$$

where 'i' and 'k' are the charges of the atomic particles, given in electron charge units, before and after the charge exchange, respectively, while ΔE is the difference between the ionization energies of the atoms A and B.

The processes described by (1.61) can be grouped according to the number of transferred electrons, e.g.

$$A^i + B \rightarrow A^{i-1} + B^+ + \Delta E_1 \tag{1.62}$$

$$A^i + B \rightarrow A^{i-2} + B^{2+} + \Delta E_2 \tag{1.63}$$

$$A^i + B \rightarrow A \quad + B^i \quad + \Delta E_3 \tag{1.64}$$

If the charge exchange occurs at $\Delta E = 0$, the process is called resonance charge transfer and described by

$$A^i + A \rightarrow A + A^i \tag{1.65}$$

In the following a simplified notation of the form

$$A^i \rightarrow A^k$$

will be used for any charge transfer process and the cross section for charge transfer will be symbolized by σ_{ik}.

The cross sections σ_{ik} can be theoretically calculated in terms of the quaneum mechanical approximation [67–69, 312–324] or by the use of semitmpirical methods [324, 366–368].

The most simple and the most easily treated processes are those of the resonance charge transfer type. Charge exchange by resonance can occur between the atoms and ions of the same element in the process described by (1.65). The atomic particles involved in the collision can be regarded during the act of collision as a quasi-molecule, thus the charge transfers can be treated as transitions between different quasi-molecular states and their cross sections can be evaluated by using the perturbed stationary state or the collision parameter method. The resonance charge transfer cross section values are relatively large, and substantially exceed the values of the charge transfer cross section between atoms and ions of different elements. Resonance charge transfer cross sections can be as large as and even exceed the gas kinetical cross sections of atomic particles, thus, charge transfer can take place even if the distance between the colliding particles is greater than the size of the particles.

If the interaction energy of the colliding particles is low by comparison with the electron energy, charge transfer across the above mentioned great distance will occur only if $\Delta E = 0$, or if it is extremely low. Consequently, the charge transfer cross section is determined essentially by the resonance charge transfer process occurring for the higher collision parameter and the other transitions can be ignored. This permits the cross section calculations to be simplified as the infinite set of equations reduces to two equations defining the two states involved in the charge transfer.

The solution to this set of two equations gives for the resonance charge transfer probability [314, 315, 325]

$$P = \sin^2 \xi \tag{1.66}$$

The cross section for resonance charge transfer is given by

$$\sigma_r = 2\pi \int\limits_0^\infty P \varrho d\varrho \tag{1.67}$$

where ϱ is the collision parameter and P is the charge transfer probability for the inelastic collision of two atomic particles in the interval of collision distances $(\varrho, \varrho + d\varrho)$.

Making use of formula (1.66), the resonance charge transfer cross section can be expressed as

$$\sigma_r = 2\pi \int\limits_0^\infty \sin^2 \xi \varrho d\varrho$$

where $\xi(\varrho)$ is a function of the collision parameter. If the value of ϱ varies from 0 to R_0, (R_0 is the distance between the two colliding nuclei at which the energy levels are crossed in the quasi-molecular state), then the averaging over the phase shifts [68, 69, 314] gives

$$P = \overline{\sin^2 \xi} = 1/2$$

and eventually we can write

$$\sigma_r = \pi \int\limits_0^\infty \varrho d\varrho$$

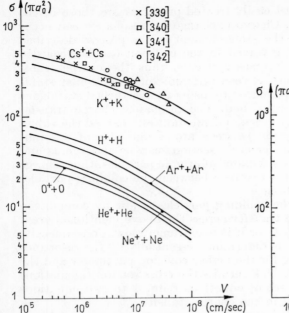

Fig. 1.53 Resonance charge transfer cross section values calculated from expressions (1.68) and (1.69) for hydrogen, oxygen, helium, neon, argon, potassium and caesium ions and atoms. For the process $Cs^+ + Cs \rightarrow Cs + Cs^+$ the calculated values (solid line) are compared with experimental data of other authors

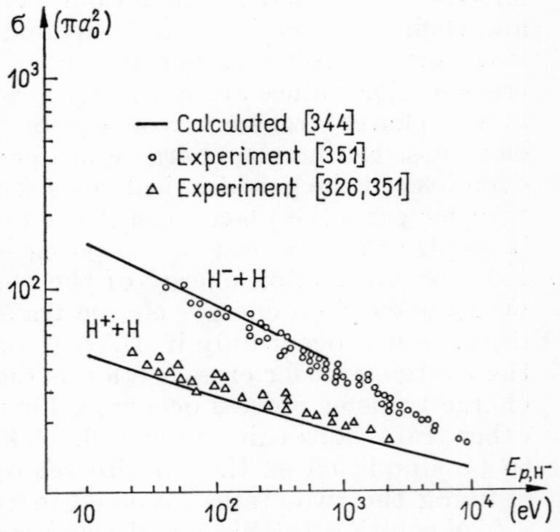

Fig. 1.54 Comparison of resonance charge transfer cross section values calculated for $H^+ + H \rightarrow H + H^+$ and $H^- + H \rightarrow H + H^-$ processes with those obtained from experimental data

Some investigators [315, 318] proposed a resonance charge transfer cross section of the form

$$\sigma_r = \frac{1}{2} \pi R_0^2 \tag{1.68}$$

This formula was derived with the assumption that charge transfer can occur only if the collision distance $\varrho \leq R_0$, while the value of ξ is given as $\xi(\varrho) = \dfrac{1}{\pi}$ if $\varrho = R_0$.

R_0 can be evaluated from the equation

$$\alpha R_0 - \left(2\beta - \frac{1}{2}\right) \ln \alpha R_0 - \frac{\beta - 0.125}{\alpha R_0} = \ln A - \ln \left[\frac{(2\pi^3)^{1/2} \, \alpha \hbar}{(2\beta)! \, m_e v} \right] \tag{1.69}$$

where

$$\alpha = \left(\frac{2m_e E^i}{\hbar^2} \right)^{1/2}$$

$$\beta = \left(\frac{m_e e^4}{2\hbar^2 E^i} \right)^{1/2}$$

if the electron is in the Coulomb field while in other cases $\beta = 0$. A is a normalizing factor which can be taken as unity for ions with unit charge, v is the relative velocity.

The resonance charge transfer cross section values calculated by use of formulae (1.68) and (1.69) are shown for some atoms in Fig. 1.53. The predicted and measured [339–342] values are compared for the process $Cs^+ + Cs \rightarrow Cs + Cs^+$. The agreement is seen to be fairly good. Similar good agreements with the predictions were obtained for the experimental values of the processes $He^+ + He \rightarrow He + He^+$ [329—336], $Ne^+ + Ne \rightarrow$ $\rightarrow Ne + Ne^+$ [329–332, 337], $Ar^+ + Ar \rightarrow Ar + Ar^+$ [329–337] and $K^+ +$ $+ K \rightarrow K + K^+$ [338, 339]. For the resonance charge transfer $H^+ + H \rightarrow$ $\rightarrow H + H^+$ the agreement is less good but still useful for practical applications. A better agreement was achieved with quantum mechanical calculations [343, 344], Fig. 1.54.

It can be seen from Fig. 1.53 that the resonance charge transfer cross section is strongly dependent on the ionization potential. This dependence on the ionization potential can also be observed for double and triple resonance charge transfer processes, see e.g. the experiments $He^{2+} + He \rightarrow$ $\rightarrow He + He^{2+}$ [345], $Ne^{2+} + Ne \rightarrow Ne + Ne^{2+}$ [337, 345, 347], $Ar^{2+} +$ $+ Ar \rightarrow Ar + Ar^{2+}$ [337, 345], $Kr^{2+} + Kr \rightarrow Kr + Kr^{2+}$ [337, 345, 347], $Xe^{2+} + Xe \rightarrow Xe + Xe^{2+}$ [337, 345], $Ne^{3+} + Ne \rightarrow Ne + Ne^{3+}$ [345, 346], $Kr^{3+} + Kr \rightarrow Kr + Kr^{3+}$ [345, 346].

In addition to the above listed atoms, the resonance charge transfer cross sections were investigated experimentally on the processes $H^- + H \rightarrow$ $\rightarrow H + H^-$ [344, 351], $K^- + K \rightarrow K + K^-$, $Na^- + Na \rightarrow Na + Na^-$, $Rb^- + Rb \rightarrow Rb + Rb^-$, $Cs^- + Cs \rightarrow Cs + Cs^-$ [280], $Li^+ + Li \rightarrow Li +$ $+ Li^+$ [352], $Rb^+ + Rb \rightarrow Rb + Rb^+$ [330, 331, 337, 339, 347], $Xe^+ +$ $+ Xe \rightarrow Xe + Xe^+$ [330, 337, 339], $Hg^+ + Hg \rightarrow Hg + Hg^+$ [349], $Cd^+ + Cd \rightarrow Cd + Cd^+$, $Zn^+ + Zn \rightarrow Zn + Zn^+$ [350].

If the charge transfer takes place between different atomic particles, when, usually, $\Delta E \neq 0$, the prediction of the cross section value becomes difficult. This is partly due to the fact that the nuclear interaction potential is still unknown.

In the few cases when the interaction potentials in the initial and final states of the colliding particles are known, the method defined by (1.68) can be used for the prediction of the charge transfer cross section [69, 314].

In other cases if the relative velocity v_a of the colliding atomic particles is lower than v_{max}, i.e. the velocity at which the charge transfer cross section has the maximum value, the cross section can be calculated under the adiabatic condition [321]

$$v \ll \frac{a\,|\,\Delta E\,|}{h} \tag{1.70}$$

In formula (1.70) a is the distance between the colliding particles at which charge transfer can occur and it can be evaluated from the expression $v_{max} = \dfrac{a\,|\,\Delta E\,|}{h}$, if the values of ΔE and v_{max} are known (e.g. in the case of one electron charge transfer $a \sim 7 \cdot 10^{-8}$ cm [278, 321]).

Under condition (1.70) the charge transfer cross section can be calculated at low collision energies from the expression

$$\sigma = \mathrm{A}e^{-\mathrm{B}\frac{a|\varDelta E|}{hv}} \tag{1.71}$$

where A and B are constants of the given charge transfer process. It is apparent from the formula that the cross section increases with increasing v up to a maximum given by

$$\frac{a|\varDelta E|}{hv_{\max}} = 1 \tag{1.72}$$

and that it monotonically decreases as v continues to increase.

It has been shown [309, 335, 353] that under adiabatic conditions (1.70), expression (1.71) can be applied to a single electron charge transfer and that the maximum of the cross section values for the process can be calculated from (1.72) if $\varDelta E$ is known. Figure 1.55 shows the values of the charge transfer cross section σ_{10} for processes $\mathrm{He^+ + Ne \rightarrow He + Ne^+}$, $\mathrm{H^+ + Ne \rightarrow H + Ne^+}$ and $\mathrm{H_2^+ + Ne \rightarrow H_2 + Ne^+}$.

In most cases the cross section values for charge transfer between different types of atomic particles are determined from experimental data. The experimental methods for the study of charge transfer cross sections are the measurement of the decrease in beam intensity [354–359] or a mass spectrometric method [357–363]. In both cases the beam transmitted through the target is considered to be a three-component system. In the most simple case, that of a proton beam, these components are $\mathrm{H^+}$, $\mathrm{H^0}$ and $\mathrm{H^-}$. In a three-component system 6 types of charge transfer process are possible with the cross sections denoted as σ_{10}, σ_{1-1}, σ_{01}, σ_{0-1}, σ_{-10}, σ_{-11}.

In the case of the method when the decrease in beam intensity is measured, the measurement always gives only the total attenuation cross section since

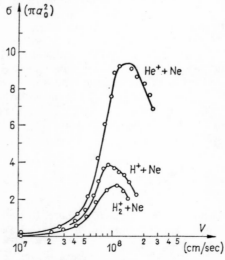

Fig. 1.55 Values of charge transfer cross section σ_{10} for processes $\mathrm{H^+ + Ne \rightarrow H + Ne^+}$, $\mathrm{He^+ + Ne \rightarrow He + Ne^+}$ and $\mathrm{H_2^+ + Ne \rightarrow H_2 + Ne^+}$

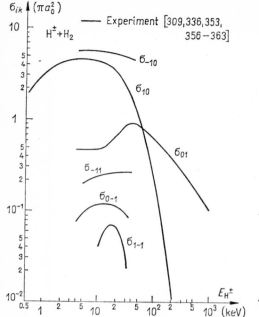

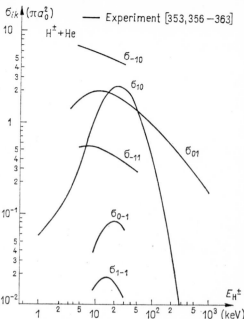

Fig. 1.56 Values of charge transfer cross section σ_{ik} for H$^\pm$ + H$_2$ collisions

Fig. 1.57 Values of charge transfer cross section σ_{ik} for H$^\pm$ + He collisions as a function of hydrogen ion energy

simultaneously two charge transfer processes are responsible for the decrease in the intensity of a beam crossing a thin target in the same charge state.

The mass spectrometric method permits individual cross sections σ_{ik} to be evaluated if the beam crosses a thin target so that the particles can undergo only one collision as with this method the different components can be measured separately.

With a thick target it is possible to reach a charge equilibrium in the beam [355] when the value of the component $N_{i\infty}$ in the charge state i does not vary any more with increasing target thickness. The values of $N_{i\infty}$ can be directly measured or it can be calculated if the values of the charge transfer cross sections σ_{ik} are known. For the measurement of $N_{i\infty}$ the charge equilibrium for e.g. helium and hydrogen up to 450 keV can be attained already with a target thickness of $(10^{-5}-10^{-4})$ mg/cm^2. In the case of hydrogen with the three components H$^-$, H^0 and H$^+$, $N_{i\infty}$ is related to σ_{ik} for $N_{-1\infty} + N_{0\infty} + N_{1\infty} = 1$ [356] as

$$N_{-1\infty} = \frac{\sigma_{0-1}\sigma_{10}}{\sigma_{-10}\sigma_{01} + \sigma_{-10}\sigma_{10} + \sigma_{0-1}\sigma_{10}}$$

$$N_{0\infty} = \frac{\sigma_{-10}\sigma_{10}}{\sigma_{-10}\sigma_{01} + \sigma_{-10}\sigma_{10} + \sigma_{0-1}\sigma_{10}}$$

$$N_{1\infty} = \frac{\sigma_{-10}\sigma_{01}}{\sigma_{-10}\sigma_{01} + \sigma_{-10}\sigma_{10} + \sigma_{0-1}\sigma_{10}}$$

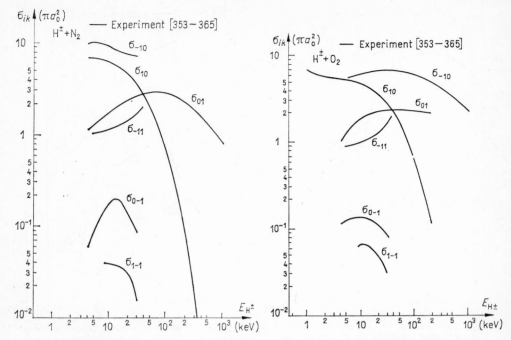

Fig. 1.58 Values of charge transfer cross section σ_{ik} for $H^{\pm} + N_2$ collisions as a function of hydrogen ion energy

Fig. 1.59 Values of charge transfer cross section σ_{ik} for $H^{\pm} + O_2$ collisions as a function of hydrogen ion energy

The values of charge transfer cross sections and the components $N_{i\infty}$ determined by the methods discussed above [354–363, 386–388] are plotted as a function of impact energy in Figs 1.56–1.63 for H^+ ions bombarding helium, neon, argon, hydrogen or oxygen gas with Be, Al, SiO, Ca, Ag or Au foil targets.

It can be seen from Figs 1.56–1.61 that the value of σ_{ik} depends on the atomic weight of the target element.

As a rule, it can be said that the charge transfer cross section is usually larger with heavier target elements. Figures 1.62 and 1.63 with the $N_{i\infty}$ values show that the component $N_{0\infty}$ has a maximum value at low impact energies both for gas and solid targets and that this maximum can be as high as 80–90 per cent of the total $N_{i\infty}$.

The component $N_{-1\infty}$ has its maximum, similarly to $N_{0\infty}$, at low impact energies. The highest values are obtained with alkali metal vapours (Fig. 1.64).

$N_{1\infty}$ increases with increasing impact energies and goes up to 90–100 per cent at energies $E_p \geq 200$ keV. This means that at these energies the beam consists almost entirely of positive hydrogen ions after its passage across the target.

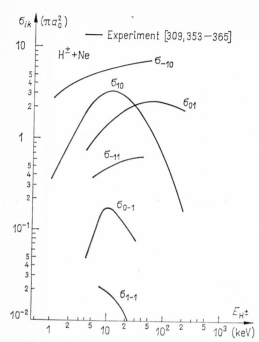

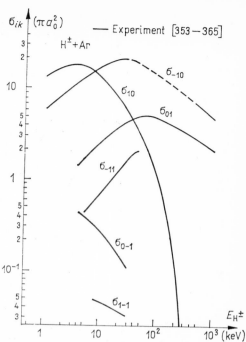

Fig. 1.60 Values of charge transfer cross section σ_{ik} for $H^{\pm} + Ne$ collisions as a function of hydrogen ion energy

Fig. 1.61 Values of charge transfer cross section σ_{ik} for $H^{\pm} + Ar$ collisions as a function of hydrogen ion energy

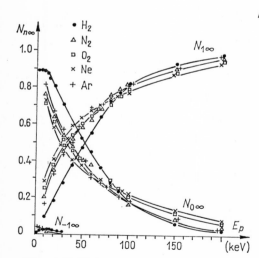

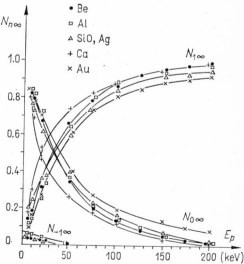

Fig. 1.62 Values of $N_{i\infty}$ beam components measured for hydrogen, nitrogen, oxygen neon or argon thick gas target as a function of hydrogen ion energy

Fig. 1.63 Values of $N_{i\infty}$ beam components measured for solid Be, Al, SiO, Ag, Ca or Au thin target as a function of hydrogen ion energy

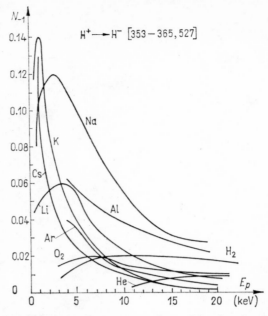

Fig. 1.64 Variation of N_{-1} beam components formed in the charge transfer process $H^+ \rightarrow H^-$ as a function of hydrogen ion energy in the case of hydrogen, helium, oxygen, argon, lithium, sodium, potassium, caesium or aluminium target

1.5 RECOMBINATION

Ions can transform to neutral atoms upon incorporation or loss of an electron. This process is called recombination. Recombination can occur following the inelastic collision between positive and negative ions or positive ions and electrons.

Ion-ion recombination can take place in the forms

$$A^+ + B^- \rightarrow A \rightarrow B \qquad (1.73)$$

$$A^+ + B^- \rightarrow A^* + B^* \qquad (1.74)$$

$$A^+ + B^- \rightarrow AB + h\nu \qquad (1.75)$$

$$A^+ + B^- + e \rightarrow AB + e \qquad (1.76)$$

$$A^+ + B^- + e \rightarrow A + B + e \qquad (1.77)$$

while the processes for electron–ion recombination can be described as

$$e + A^+ \rightarrow A + h\nu \qquad (1.78)$$

$$e + A^+ \rightarrow A^* + h\nu \qquad (1.79)$$

$$e + A^+ \rightarrow A^*(n) - A^*(n') + h\nu \qquad (1.80)$$

$$e + AB^+ \rightarrow AB^* \rightarrow A^* + B^* \tag{1.81}$$

$$e + e + A^+ \rightarrow A^* + e \rightarrow A + e + h\nu \tag{1.82}$$

$$e + A^+ + B \rightarrow A + B \tag{1.83}$$

The above recombination processes can be subdivided into double and triple collision processes.

Recombination in double collision of the forms (1.73)–(1.75) or (1.78)–(1.81) can take place only if the excess energy can transform to kinetic, excitation or radiative energy. Accordingly, the processes of type (1.73) are called collision, those of type (1.74) excitation and those of types (1.75), (1.78), (1.79) radiative recombinations. The process of type (1.80) is called dielectric recombination where the recombination occurs as a result of an autoionization state and the decay of this state is determined by the radiation. In case (1.81) the electrons collide with a molecular ion and the excess energy is expended on excitation and dissociation of the molecule. This is known as dissociation recombination.

In the triple collision recombinations of the type (1.76), (1.77), (1.82) or (1.83) the excess energy is generally consumed by the increase in the energy of the third particle or by its excitation.

In the recombination processes the colliding particles are always oppositely charged and this results in a contribution from the attractive Coulomb force to the interactions.

Recombination processes are generally observed in experiments with gases, vapours or gas discharges where the variations of the negative and positive charge concentrations are measured. The charge density variation can be characterized by the recombination coefficient α which determines the recombination cross section σ_r and is given by the relation [373]

$$\frac{dN^+}{dt} = \frac{dN^-}{dt} = -\alpha N^+ N^- \tag{1.84}$$

On the assumption that the positive charge density N^+ is nearly equal to the negative charge density N^-, (1.84) transforms to

$$\frac{dN}{dt} = -\alpha N^2 \tag{1.85}$$

The recombination coefficient α is related to the time taken by the ions to become neutral, that is, to the lifetime τ_i of the ions by the formula [373, 382]

$$\tau_i = \frac{1}{\alpha N_i} \tag{1.86}$$

In the ion–ion recombinations by double collision described by expressions (1.73)–(1.75), positive and negative ions are recombined so that the negative ion transfers an electron to the positive ion producing in this way either two neutral atoms in the ground state or in an excited state, or a molecule. In these processes, since the affinity energy S of the electron in the negative ion is usually lower than the ionization energy E_i of the atom, we get an excess energy given as $E_i - S = \Delta E > 0$. This excess energy can be transformed in the recombination to kinetic or excitation energies of the colliding particles or emitted as radiation. If one or both of the ions taking part in the recombination are molecular ions, then apart from being expended on excitation the energy ΔE can also be expended on the dissociation of the molecules.

For the calculation of the cross section value for recombination of two oppositely charged ions in an inelastic collision the process can be treated as a charge exchange between positive and negative ions and thus approximated in terms of the model of the quasi-molecule by use of the Landau–Zener formula [68, 69].

It follows that for $\Delta E > 0$, the recombination cross section of positive and negative ions is given as [369]

$$\sigma_r = 4\pi R_0^2 J(\eta) \qquad (1.87)$$

where R_0 is the distance between the nuclei at which there are the crossing points of the quasi-molecular levels, given in this case as $R_0 = 27.2/\Delta E$, while the function

$$J(\eta) = \int\limits_1^\infty \frac{e^{-\eta x}}{x^3} (1 - e^{-\eta x}) dx$$

has one maximum with a value of ~ 0.1 [378].

For double collision the cross section value was theoretically calculated for $H^- + H^+$ recombination [372]. The cross section values were predicted as $\sigma_r \approx 7 \cdot 10^{-14}$ cm^2 at an energy of 10^{-1} eV and $\sigma_r \approx 3 \cdot 10^{-16}$ cm^2 at 10^3 eV, while the recombination coefficient was estimated as $\alpha \approx 1.3 \cdot 10^{-7}$ cm^3/sec at room temperature.

The experimental investigation of double collision recombination is most conveniently performed at gas pressures $p < 100$ mmHg, since in this range the probability of triple collision recombination is lower than that of double collision recombination between ions.

Only a few experiments have been reported on double collision recombinations [369–371]. The measurements were carried out using iodine and bromine ions. The recombination coefficient for negative and positive ions was determined at different pressures and temperatures. It was found that at pressures from $5 \cdot 10^{-2}$ to 1 mmHg and at room temperature $\alpha \approx 10^{-7}$ cm^3/sec, which corresponds to $\sigma_r \approx 10^{-13}$ cm^2, and that it is practically independent of pressure. However, the value of α decreases with increasing temperatures from 0° to 100°C.

In triple collision recombinations — as has been already mentioned — an ion, before interacting with an oppositely charged ion, collides first with and imparts a large part of its kinetic energy to a neutral particle in its environment. Consequently, the relative velocities of the interacting ions will be so low that the recombination probability is high. Thus, triple collision ion recombinations will occur if a third particle is present at the collision of oppositely charged ions. This requirement can be met at gas or vapour pressures $p > 100$ mmHg where the recombination probability by triple collision is much higher than the double collision recombination probability. Theoretical studies of triple collision ion–ion recombinations were carried out by Thomson [373] at pressures $p < 760$ mmHg and by Langevin [376, 377] at pressures $p > 760$ mmHg. Both of them used classical assumptions.

Thomson's theory, applicable to the range of low pressures, permits the process of triple collision recombination to be interpreted up to pressures $p \leq 760$ mmHg. It is assumed in this theory that the mean energy of ions and molecules in gas is $3\,kT/2$ and that recombination occurs only if this energy decreases after the collision with a third particle to a value lower than that of the energy of electrostatic interaction between the ion pair, that is

$$\frac{3kT}{2} < \frac{e^2}{r}$$

where r is the distance between the colliding particles.

To satisfy this condition, the distance between the colliding ions must, at the time of the collision with the third particle, be less than

$$r_{\max} \leq \frac{2}{3}\,\frac{e^2}{kT}$$

Under this condition the recombination coefficient of the triple collision recombination of negative and positive ions of identical masses is given if $\lambda_i \gg r_{\max}$ by

$$\alpha_i = \frac{8\pi}{27}\,\frac{e^6}{\lambda_i(kT)^3}\,\sqrt{\frac{6kT}{M}} \qquad (1.88)$$

If $T \sim T_i$ and we know that $\lambda_i \sim p^{-1}$, the recombination coefficient can be expressed as

$$\alpha_i = C\,\frac{p}{T^{5/2}} \qquad \left[\frac{\text{cm}^3}{\text{sec}}\right]$$

where C is given for air as $1.5 \cdot 10^{-2}$, while p and T are given in mmHg and °K units, respectively.

For the pressure range $p > 760$ mmHg, the theoretical considerations of triple collision recombinations have been worked out by Langevin [376, 377]. These calculations permit the recombination coefficient to be predicted at these higher pressures. It is assumed in this calculation that all ions are at the same distance (r) from one another and that they are exposed to the action of the electric field $\dfrac{e}{r^2}$ which causes the ions to move and collide with one another. Under these conditions the ions approach one another with velocities equal to their drift velocity in the electric field given by

$$v_d = (\varkappa_+ + \varkappa_-)\left(\frac{e}{r^2}\right)$$

where $\varkappa_+$ and $\varkappa_-$ are the mobilities of the positive and negative ions in the gas, respectively. If the positive ion is taken to be at rest, the number

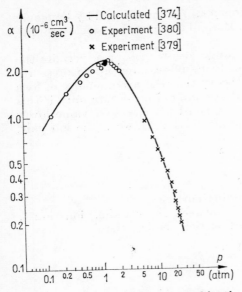

Fig. 1.65 Comparison of recombination coefficients as calculated from Natanson's formula with those obtained from experimental data

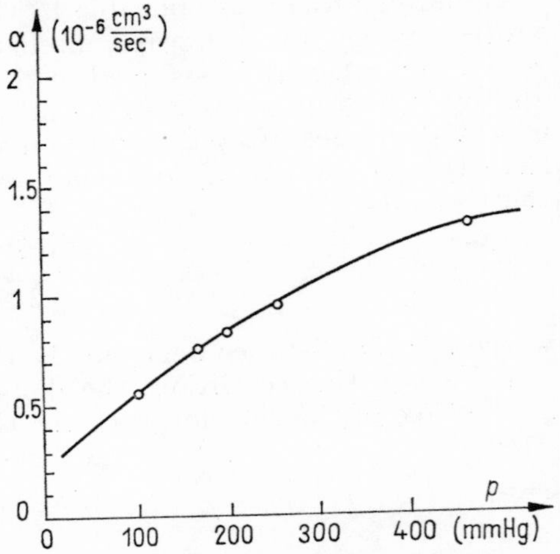

Fig. 1.66 Comparison of recombination coefficients as calculated from Nathanson's formula with those obtained from experimental data

of negative ions colliding with it per unit time is equal to $4\pi r^2 v_d N_i$. Hence, the recombination coefficient can be expressed by use of the formula for drift velocity as

$$\alpha_i = 4\pi e(\varkappa_+ + \varkappa_-) \tag{1.89}$$

Natanson [374] formulated a triple collision ion–ion recombination theory which applies to both of the above specified ranges of pressure. For the recombination coefficient, if $\lambda_i \gg r_{max}$ (i.e. at the lower pressures where $p < 760$ mmHg), this formulation yields a value similar to that obtained from eq. (1.88) and if $\lambda_i \ll r$ (i.e. at the higher pressures $p > 760$ mmHg), a value in good agreement with that obtained from eq. (1.89). In the derivation of the expression for the triple collision recombination coefficient, it was postulated that the energy of the ions is composed of the kinetic contribution $3kT/2$ and the contribution from the energy $\dfrac{e^2}{r}$ of the electrical interaction between the ions. Thus, the critical radius of the environment which must contain the third particle is given as $r = \dfrac{5}{12}\dfrac{e^2}{kT}$ and this value is lower than that of Thomson's r_{max}. Natanson's formula for α_i reads

$$\alpha_i = \frac{17\pi v_0 \omega \lambda^2 x e^{2x}}{1 + \dfrac{17 v_0 kT}{20 e^2} \dfrac{\lambda^2 x^2}{D} \omega (e^{2x} - 1)} \tag{1.90}$$

where v_0 is the mean thermal velocity of the ion, ω is the probability of the collision of ions with a third particle and given as $\omega = 2\omega_1 - \omega_1^2$, with

$$\omega_1 = \left\{ 1 - \frac{1}{2x^2} [1 - e^{-2x}(1 + 2x)] \right\} \text{ and for } x \gg 1, \text{ we have } \omega = 1 \text{ and}$$

for $x \ll 1$, we have $\omega = \dfrac{8}{3}x$; where $x = \dfrac{r}{\lambda}$ and $D = D_+ + D_-$, D_+ and D_- being the diffusion coefficients of the positive and negative ions, respectively.

The values calculated from eq. (1.90) are compared with the experimental data in Figs 1.65 and 1.66. It can be seen that the values obtained by taking $v_0 = 6.2 \cdot 10^4$ cm/sec and $\lambda/D = \text{constant} = 3.54 \cdot 10^{-5}$ are in good agreement with the experimental results [379, 380, 383].

For ion–electron recombination the collision energy between the particles must be such that it allows the electron to be captured by the ion. This condition is satisfied only if the kinetic energy of the electrons can be converted by the collision to radiative, excitation or dissociation energy, or if its energy is transferred to a third particle in a triple collision.

Photorecombination (see processes (1.78) and (1.79)) — when the excess energy is removed by the emission of a photon — occurs only if no other two-particle processes are available for recombination or if the triple collision recombination becomes negligible because of low gas density.

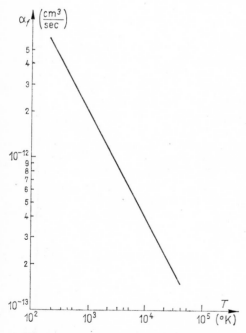

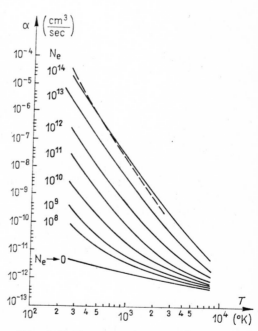

Fig. 1.67 Dependence of photorecombination coefficient on electron temperature T, as evaluated from formula (1.91)

Fig. 1.68 Dependence of radiative collision recombination coefficient on impact electron temperature T for different values of the electron density N_e

The coefficient of photorecombination at low electron temperatures is given by the expression

$$\alpha_f \approx 1.6 \cdot 10^2 \left(\frac{e^3}{\hbar C} \right) T^{-0.7} \tag{1.91}$$

The photorecombination coefficients calculated from this formula at temperatures $T = 10^2$ to $4 \cdot 10^4 °K$ are given in Fig. 1.67. It is apparent from the figure that the value of α_f decreases as the temperature increases and that it does not depend on the electron density N_e.

Recombination can lead to photon emission also in the case of triple collision. In this case the free electron captured in the triple collision is not in the ground state but in an excited state and the photon is emitted by the transition to the ground state. This process is called radiative collision recombination (see process (1.82)).

The coefficient of radiative collision recombination has been most precisely calculated theoretically with the third particle taken to be an electron [393, 394]. The values calculated as a function of electron temperature for various electron densities N_e are presented in Fig. 1.68. The predicted [393] and experimental [404–407] values are in reasonably good agreement.

The values of the radiative collision recombination coefficient calculated for hydrogen differ only by a few percent from those calculated for the alkali metal ions analogous to hydrogen [393]. This small difference, obtained at high electron densities and low temperatures, is negligible in many cases so that the values specified in Fig. 1.68 apply in both cases in the range mentioned above.

If less precise values of α can be used than those obtainable by the methods referred to above [393, 394], it is possible to work at low temperatures and high electron densities with the approximation [389]

$$\alpha \approx 3.3 \cdot 10^{-2} \, \frac{m^{1/2} e^8}{\hbar^2 (kT)^{5/2}} \, \frac{e^6 N_e}{(kT)^3} \tag{192}$$

In Fig. 1.68 the values calculated by use of the approximation (1.92) with $N_e = 10^{14}$ at the temperatures $T = 2.5 \cdot 10^2$ to $2.5 \cdot 10^3 °K$ are shown by the broken line. It can be seen that the difference between the two curves is not great, thus formula (1.92) should be satisfactory in a number of cases.

If the third particle is an atom (see process (1.83)), the triple collision electron–ion recombination coefficient can be estimated from the formula [392]

$$\alpha = \frac{32 \sqrt{2\pi}}{3} \, \frac{m_e^{1/2} e^6 N_a}{m_a (kT)^{5/2}} \, \sigma_{\text{elast}} \tag{1.93}$$

where N_a is the density of atoms of mass m_a and σ_{elast} is the cross section for elastic electron–ion scattering.

The ions colliding with free electrons are excited at the expense of the kinetic energy of electrons involved in the recombination process (1.80), thus the electron can be captured by an atom and this leads to the formation of an atom in the state of autoionization. The autoionized atom can subsequently decay to the initial ion and free electron. If this decay does not take place, the atom can return to a stable bound state either by emitting radiation or under the action of an impact.

It follows that the recombination via autoionization depends on the probabilities of these two processes. If the autoionized atom transforms to a stable bound state by radiation, the recombination process is called dielectric, if the stabilization occurs upon collision, it is called collision stabilization.

The coefficient of recombination through autoionization of the atom can be evaluated from the formula [408]:

$$\alpha = \frac{(2\pi)^{3/2}}{2} \left(\frac{\hbar^2}{m_e kT} \right)^{3/2} \omega_a \frac{g_a}{g_i} e^{-\frac{E_i - E_{ag}}{KT_e}} \qquad (1.94)$$

where ω_a is the transition probability from the autoionization to the stable bound state, T_e is the temperature of the free electron, g_a and g_i are the statistical weights of the autoionized atoms and the ions, respectively, E_i and E_{ag} are the excitation energies of the ionization and autoionization states.

The estimations of the dielectric recombination coefficient show that its value is generally lower than that of the photorecombination coefficient [409, 410].

Electron–molecular ion recombination processes of type (1.81) occur if the kinetic energy of the free electron is imparted to the core of a molecule. For this reason, this process usually has a higher probability than the electron–atomic ion recombination. If the thus imparted energy is higher than the molecular dissociation energy (see Appendix, Table 12), the molecule dissociates and the process is called dissociation recombination. The coefficient of dissociation recombination was measured as a function of temperature on Ne^+ [395, 396, 398], Ar_2^+ [397], N_2^+ [400, 401], O_2^+ [401, 402], NO^+ [399] gases.

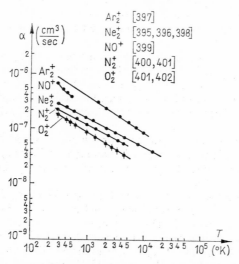

Fig. 1.69 Dissociation recombination coefficient versus electron temperature T curves for the molecular ions N_2^+, O_2^+, Ne_2^+, NO^+ and Ar_2^+

The measured values are presented in Fig. 1.69. The dissociation recombination coefficients at temperature $T \sim 300 \; °K$ evaluated for some molecules are

$$\alpha_{He_2^+} \sim 6 \cdot 10^{-9} \frac{cm^3}{sec} \;, \qquad\qquad \alpha_{CO_2} \sim 3.8 \cdot 10^{-7} \frac{cm^3}{sec} \;;$$

$$\alpha_{Ar_2^+} \sim 7.5 \cdot 10^{-7} \frac{cm^3}{sec} \; [401, 403]; \qquad \alpha_{Kr_2^+} \sim 1.2 \cdot 10^{-6} \frac{cm^3}{sec} \; [403];$$

$$\alpha_{Xe_2^+} \sim 1.4 \cdot 10^{-6} \frac{cm^3}{sec} \; [403]; \qquad \alpha_{H_2^+} \sim 5.6 \cdot 10^{-6} \frac{cm^3}{sec} \; [413];$$

$$\alpha_{N_2^+} \sim 2.5 \cdot 10^{-7} \frac{cm^3}{sec} \; [401, 402]; \qquad \alpha_{NO^+} \sim 4.1 \cdot 10^{-7} \frac{cm^3}{sec} \; [399];$$

$$\alpha_{Ne_2^+} \sim 1.8 \cdot 10^{-7} \frac{cm^3}{sec} \; [395, 396, 398].$$

1.6 DISSOCIATION

Dissociation means the process by which the molecule decomposes to its atoms due to the energy imparted to the molecule. The amount of energy required for inducing molecular dissociation always depends on the internal energy state of the molecule which at the time of the impact can be in various electronic, vibrational or rotational states. The energy needed for the dissociation of a molecule in the ground state is called dissociation energy and is symbolized by D.

The dissociation energies for a set of molecules are listed in Appendix, Table 11. The energy needed for dissociation can be imparted to the molecule by single or multiple collisions with electrons, atoms, molecules or photons.

The electronic, vibrational and rotational energy states of the hydrogen molecule are given by the solutions of the wave equation for the hydrogen molecule [3]. The potential energy curves for the electron states of the hydrogen molecule and the vibrational and rotational energy level structures characteristic of diatomic molecules are shown in Figs 1.10 and 1.11 respectively.

The wells on the potential energy curves for excited electron states are less deep than the well in the curve for the ground state, consequently, the molecule can be dissociated by less imparted energy in an excited state than in the ground state.

The potential curves for the lowest single $1^1\Sigma_g$ and the triplet $1^3\Sigma_u$ electron states in hydrogen molecules as a function of the distance between the nuclei of the two atoms in the molecule are shown in Fig. 1.70 together with the vibrational levels assigned to the state $1^1\Sigma_g$. As is apparent from the figure, the singlet state $1^1\Sigma_g$ is stable since the potential energy curve has a minimum, while the triplet state $1^3\Sigma_u$ does not allow the formation of a stable molecule because the potential energy decreases monotonically

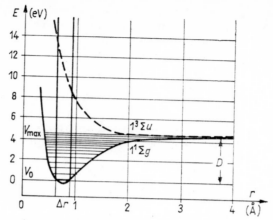

Fig. 1.70 Potential curves for the singlet $1^1\Sigma_g$ and for the triplet $1^3\Sigma_u$ electron states with vibrational levels belonging to the $1^1\Sigma_g$ state in hydrogen molecules

with increasing distance between the atomic nuclei. Dissociation in the case of vibrational excitation can be readily understood on inspection of the figure. It can be seen that the molecule can decompose to its atoms if the vibrational energy of the molecule exceeds the value of the dissociation energy. The vibrational energy can be increased due to single or multiple collisions with other particles. Molecular vibrational levels can be excited and lead to molecular dissociation in gases and vapours if the thermal energy kT of the atomic and molecular particles increases to a value higher than $\Delta V = V_n - V_m$ which is the difference between the vibrational energy levels.

The transition from one vibrational level to another is allowed in practice only between adjacent levels in the case of low lying discrete vibrational levels and the transition probability rapidly increases with increasing vibrational quantum numbers. At the higher energy range of the vibrational spectrum, transitions between non-adjacent levels are also allowed. At levels close to the dissociation energy (active band) the probability of transition to the continuous spectrum increases to a degree at which dissociation becomes possible. This type of vibration excitation is called thermal excitation and the resulting dissociation is known as thermal dissociation.

The rate of thermal dissociation is determined by the rate of the thermal excitation of the vibrational levels lying in the active band.

The dissociation rate constant of diatomic molecules for the process which, in the case of thermal equilibrium, can be described as

$$AB + C \rightarrow A + B + C \qquad (1.95)$$

is expressed by the equation [414–417]

$$K_d = \frac{g_{ABC}}{g_{AB}g_C} \frac{kT}{2\pi\hbar} e^{-\frac{D}{kT}} \qquad (1.96)$$

where g_{ABC} stands for the intermediate complex while g_{AB} and g_C are the statistical weights of the components AB and C respectively.

However, the dissociation processes has to compete with the simultaneously occurring association process of the form

$$A + B + C \rightarrow AB + C \qquad (1.97)$$

It follows that the concentration of the dissociated particles depends on the relative rates of the two processes at a given temperature T. The ratio of the dissociation to the association rate at a given temperature T is called the equilibrium constant K_e and it can be expressed in the form [414–417]

$$\frac{K_d}{K_a} = K_e = \frac{g_A g_B}{g_{AB}} e^{-\frac{D}{kT}} \qquad (1.98a)$$

Using the expressions for the statistical weight of the molecule, the equilibrium constant of a diatomic molecule can be written as

$$K_e = \left(\frac{2\pi m_A m_B kT}{(m_A + m_B)h^2} \right)^{3/2} 2k\theta_r \left(1 - e^{-\frac{\theta_v}{T}} \right) \frac{g_{0A} g_{0B}}{g_{AB}} e^{-\frac{D_0}{kT}}$$

where m_A and m_B are the atomic masses of the diatomic molecule; $\theta_r = \dfrac{h^2}{8\pi^2 kI}$ is the temperature characteristic of the rotation with $I = \mu r_0^2$ being the inertia momentum of the molecule relative to the axis of rotation in the centre of mass normal to the straight line connecting the atoms of the molecule, $\mu = \dfrac{m_A m_B}{m_A + m_B}$; r_0 is the half value of the equilibrium distance between the atoms in the molecule; $\theta_v = \dfrac{h\nu}{k}$ is the characteristic temperature of vibration; g_{0i} is the statistical weight of the electron state; and D_0 is the dissociation energy in the ground state of the molecule. For a molecule consisting of identical atoms the equilibrium constant can be given in the form

$$K_e = \left(\frac{\pi m_A kT}{h^2} \right)^{3/2} 2k\theta_r \left(1 - e^{-\frac{\theta_v}{T}} \right) \frac{(g_{0A})^2}{g_A} e^{-\frac{D_0}{RT}} \qquad (1.98b)$$

For thermal dissociation in hydrogen gas, when the equilibrium constant K_e depends only on the absolute temperature T, formula (1.98) reduces to [419, 420]

$$K_e(T) = \frac{(g_H)^2}{g_{H_2}} e^{-\frac{D}{kT}} = \frac{(N_H)^2}{N_{H_2}} \qquad (1.99)$$

where N_H and N_{H_2} are the equilibrium numbers of atomic and molecular hydrogen particles in the given volume, respectively. Substituting now the partial pressures for the numbers of the components, we get

$$\frac{(N_H)^2}{N_{H_2}} = \frac{(P_H)^2}{P_{H_2}} = K_e(T) \qquad (1.100)$$

The partial pressure of the dissociated atomic hydrogen can be written as

$$P_H = XP \tag{1.101}$$

where X can have any value between 0 and 1 and stands for the degree of dissociation, and P is the total pressure which equals the sum of the partial pressures, that is

$$P = P_H + P_{H_2} \tag{1.102}$$

The degree of dissociation X can be expressed from eqs (1.100), (1.101) and (1.102) [420] as

$$X = \frac{-K_e(T)}{P} + \frac{1}{2} \sqrt{\left(\frac{K_e(T)}{P}\right)^2 + 4\,\frac{K_e(T)}{P}} \tag{1.103}$$

It can be seen that the degree of dissociation can be evaluated if the equilibrium constant $K_e(T)$ and the total pressure are known.

In the case of hydrogen the value of the equilibrium constant has been determined [418] as

$$\log K_e(T) = -\frac{21\,200}{T} + 1.765 \cdot \log T - 9.85 \cdot 10^{-5}\,T - 0.265$$

Using the values determined by this expression the values of X obtained

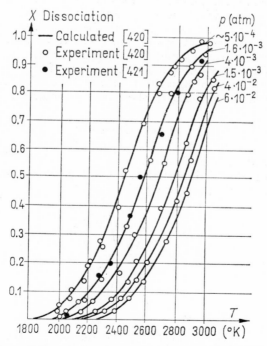

Fig. 1.71 Comparison of dissociation degree X against dissociator temperature curves as calculated from formula (1.103) with those obtained from experimental data

from eq. (1.103) for different values of P [420] are shown as a function of temperature in Fig. 1.71 together with the experimental curve [420, 421]. It can be seen that the calculated and experimental values are in good agreement.

It follows from expression (1.98) that the value of K_e and through it that of the degree of dissociation depends at a given temperature T on the dissociation energy D. At the same temperature T the molecule with a lower dissociation energy shows a higher degree of dissociation than the molecule with a higher value of D. Accordingly, a dissociation degree of 100 per cent can be reached in hydrogen, which has a dissociation energy $D = 4.48$ eV, only at 3000°K, while with iodine for which $D = 1.54$ eV, 100 per cent dissociation can already be achieved at 920°K at pressures in the range $P = 10^{-5}$ to 10^{-4} atm.

If the equilibrium constant $K_e(T)$ is known, the degree of dissociation can also be evaluated for other molecules using the method applied to hydrogen.

1.7 DIFFUSION AND MOBILITY OF IONS AND ELECTRONS

The motion of ions and electrons in gases are determined by diffusion and by the action of applied fields.

In the absence of an applied field the motion of ions and electrons can be treated in terms of the diffusion theory applied to gas kinematics. Diffusion is known to be the process which leads to the levelling of different particle concentrations in the gas volume due to thermal motion. This means that the ions or electrons move at a higher rate in the direction of lower concentration than in the opposite direction. Owing to this fact, a particle current is generated in the direction opposite to that of the concentration gradient. This current (j) can be related to the particle concentration n by the formula

$$j = - D \operatorname{grad} n \tag{1.104}$$

where D is the diffusion coefficient.

In the case of identical particles the diffusion coefficient can be well approximated [528, 529] in the form

$$D = \frac{3}{16N\bar{\sigma}} \sqrt{\frac{2\pi kT}{\mu}} \tag{1.105}$$

where

$$\bar{\sigma} = \int\limits_0^\infty \sigma(x) e^{-x} x^2 dx;$$

$x = \dfrac{\mu v_r^2}{2kT}$; $\mu = \dfrac{m_i m}{m_i + m}$; v_r is the relative velocity of the ions and gas particles, N is the number of particles per unit volume, $\sigma(x)$ is the diffusion cross section which corresponds to the inelastic scattering cross section

$$\sigma(x) = \frac{4\pi}{K^2} \sum_{l=0}^{\infty} (l + 1) \sin^2 (\delta_l - \delta_{l+1})$$

with $K = \dfrac{\mu v_r}{\hbar}$, and δ_l stands for the phase shift.

If the gas system is composed of two types of particles with masses m_1, m_2 and concentrations n_1, n_2, the diffusion coefficient is given as [528, 529]

$$D = \frac{3}{16(n_1 + n_2)\bar{\sigma}} \sqrt{\frac{2\pi kT}{\mu}} \qquad (1.106)$$

In the presence of an applied field the ion mobility ($\varkappa$) means generally the velocity at which the ions move in the direction of unit applied field.

The mobility of electrons or ions can be related also to the diffusion constant in the absence of an applied field, or if the applied field is so weak that the increase in particle velocity is much less than the mean velocity $v_n = \sqrt{\dfrac{3kT}{m}}$. In this case we can write

$$\varkappa = \frac{De}{kT} \qquad (1.107)$$

Using now formula (1.105), the mobility of ions or electrons can be expressed as

$$\varkappa = \frac{3}{8} \frac{e}{\bar{\sigma} N \sqrt{2kT\mu}} \qquad (1.108)$$

Formula (1.108) can be utilized to express the drift velocity of ions or electrons in the form

$$v_d = \varkappa F = \frac{3}{8} \frac{eF}{\bar{\sigma} N \sqrt{2kT\mu}} \qquad (1.109)$$

If not only an electric, but also a magnetic field H is applied and the direction of the former is normal to that of the latter, the drift velocity in the direction of the electric field becomes

$$v_d = \frac{\varkappa F}{1 + \dfrac{\varkappa^2}{v_k^2} H^2} \qquad (1.110a)$$

and the diffusion coefficient takes the form

$$D_H = \frac{D_{H=0}}{1 + \dfrac{\varkappa^2}{v_k^2} H^2} \qquad (1.110b)$$

It is apparent from the formula that the drift velocity and the diffusion coefficient of charged particles decrease in the proportion $1/H^2$ as the magnetic field increases.

Usually, in most of the cases the particles are displaced due both to diffusion and to the applied electric field. Therefore, the particle current can be described by the equation

$$j = -D \operatorname{grad} n + \varkappa F n \qquad (1.111)$$

In the case of ambipolar diffusion, that is, when the currents of the positive (j^+) and the negative (j^-) particles are equal, we have $(n^+ - n^-) \ll n^-$. In this case the net particle current is given by

$$j = -D_a \operatorname{grad} n$$

where D_a is the ambipolar diffusion coefficient, defined as

$$D_a = \frac{D_+ \varkappa_- + D_- \varkappa_+}{\varkappa_+ + \varkappa_-} \qquad (1.112)$$

since in the case of equilibrium

$$\frac{\varkappa_+}{D_+} = \frac{\varkappa_-}{D_-} = \frac{e}{kT}$$

and with $D_- \varkappa_+ = D_+ \varkappa_-$, we have

$$D_a = \frac{2 D_+ \varkappa_-}{\varkappa_+ + \varkappa_-} \qquad (1.113)$$

Now, if ambipolar diffusion takes place in a plasma, where the negatively charged particles are electrons, $D_a \sim 2D_+$, since $\varkappa_- \gg \varkappa_+$. Thus, the ion mobility in a plasma is given in the case of ambipolar diffusion by

$$\varkappa = \frac{e D_a}{2kT} \qquad (1.114)$$

At higher applied fields in which the mean velocity of ions and electrons parallel to the direction of the field appreciably exceeds the thermal particle velocity, the drift velocity is given by the equation [530]

$$v_d = \left(\frac{2kT}{m} \right)^{5/2} \frac{3\sqrt{\pi}}{8} \frac{eF}{\mu N \int\limits_0^\infty e^{-\frac{mv_r^2}{2kT}} v_k^5 \sigma(v_r) dr} \qquad (1.115)$$

If the applied field is high, the particle mobility $\varkappa$ can be described by the series [530]

$$\varkappa = \varkappa_0 + \varkappa_2 F^2 + \varkappa_4 F^4 + \cdots$$

where $\varkappa_i$ is a function of the collision integral. Calculations have been already made on the values of $\varkappa_i$ [531].

If the main contribution to the elastic scattering cross section, which determines the ion mobility, arises from the ion–atom collisions brought about by the polarization interaction with a large collision parameter, the drift velocity and the mobility can be expressed as

$$v_d = \frac{eF}{\mu N v_r \sigma_p(v_r)} \tag{1.116}$$

and

$$\varkappa = \frac{e}{\mu N v_r \sigma_p(v_r)} \tag{1.117}$$

Since the diffusion scattering cross section $\sigma_p(v_r)$ contains simultaneous contributions from polarization capture and small angle scattering, it takes the form [532]

$$\sigma_p(v_r) = 2.2\pi \sqrt{\frac{\beta}{\mu v_r^2}} \tag{1.118}$$

where β is the polarization of the atom. Expression (1.118) can be utilized for rewriting eqs (1.116) and (1.117) as

$$v_d = \frac{eF}{2.2\pi N \sqrt{\mu\beta}} \tag{1.119}$$

and

$$\varkappa = \frac{e}{2.2\pi N \sqrt{\mu\beta}} \tag{1.120}$$

The ion mobilities predicted from eq. (1.120) for Li^+, Na^+, K^+, Rb^+ and Cs^+ in argon, krypton, xenon, hydrogen and nitrogen gases agree with the experimental data to within an error of less than 10 per cent [533, 534]. The measurements in helium and neon gases show larger deviations from the predictions. The difference is attributed to the contribution from short range exchange interactions between the atoms and ions.

If the ions move in their own gas, their mobility and drift velocity are determined by resonance charge transfer processes of atoms and therefore the resonance charge transfer cross section σ_p (see expression (1.87)) is substituted for $2\sigma_r$ in expressions (1.116) and (1.117) [536].

1.8 SECONDARY ELECTRON EMISSION

The electrons or ions impinging on the surface of solids are either scattered back from or penetrate into the solids. The electrons or ions which penetrate into the solid lose a large fraction of their kinetic energy by imparting it to the electrons of the solid. If this imparted energy is large enough to enable the energized electrons to reach the solid surface with an energy higher than that which is required by the work function of the given solid, the electrons can escape from the surface. The so emitted electrons are called secondary electrons. The parameter value of secondary electron emission is the ratio of the escaping electrons to the primary particles incident on the surface, that is

$$\delta = \frac{n_{es}}{n_p} \qquad (1.121)$$

The quantity δ is called secondary electron emission coefficient.

The value of coefficient δ depends for a given surface on the primary particle energy E_{ep}, the angle θ of the direction of incident particles to the direction normal to the surface, and on the nature of the surface.

If the incident particles are electrons the experimental data on the function $\delta_e(E_{ep})$ [827–872] yield for primary electrons incident normal to the surface ($\theta = 0$) a curve of the form shown in Fig. 1.72. As can be seen from the figure, the initially steeply rising function exhibits a broad flat maximum which is reached at an energy E_{ep} of a few hundred electron volts (Appendix, Table 16).

The angular distribution of the secondary electrons emerging from the solid surface can be described by the cosine law [878] if the reflected primary electrons are ignored. This means that most of the secondary electrons emerge in a direction normal to the surface.

The energy distribution of secondary electrons can be described by the curve to be seen in Fig. 1.73 [879–881]. In the figure, the lower energy peak which contains the major fraction of the emerging electrons corresponds to the secondary electrons while the higher energy, sharp peak is produced

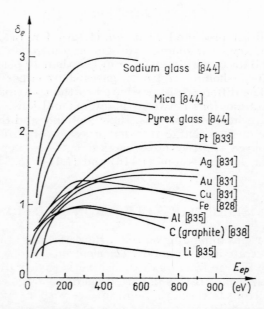

Fig. 1.72 Secondary electron emission coefficient versus impact electron energy curves plotted for impact on surfaces of different materials

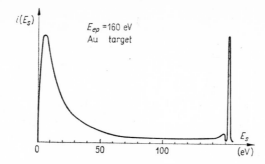

Fig. 1.73 Energy distribution of secondary electrons

by the reflected primary electrons which have kept almost entirely their initial energy. The small peaks near the high energy peak are due to the reflected primary electrons which have lost a fraction of their initial energy to the generation of soft bremsstrahlung radiation.

According to experimental studies [879–884] the most probable energy of secondary electrons lies in the range of a few electron volts and it is independent of the primary energy. The variance within this range of values is due to the difference between the work functions of different materials and surfaces [884].

At low primary energies E_{ep} the dependence of the secondary electron emission coefficient on the angle of incidence θ of the primary electron can be expressed as [876]

$$\delta_e(\theta) = \delta_{e_0} e^{\alpha_a \bar{x}(1-\cos\theta)} \qquad (1.122)$$

where $\delta_{e_0} = \delta_e(\theta = 0^0)$ is the value of the secondary electron emission coefficient determined for primary electrons incident in the direction normal to the surface, α_a is the absorption coefficient, $\bar{x}$ is the mean depth from which the secondary electrons are emitted.

The data obtained from measurements [827, 873–877] are consistent with the shape of the curve predicted from eq. (1.122) since the measured values also show that at lower values of E_{ep} the angular dependence of δ_e becomes of little importance due to the low mean value of the penetration depth. The values of the function $\delta_e(\theta)$ as measured for various metals are shown in Fig. 1.74 [875].

The dependence of secondary electron emission on the nature of the surface arises from several factors.

The value of δ_e is sensitive to the smoothness, purity and crystal structure which eventually determine the work function of the surface.

Rough and porous surfaces reduce the value of δ_e because a fraction of secondary electrons produced in the holes cannot escape from the solid [838]. The purity of the surface also has a considerable effect on δ_e which increases if gas atoms or molecules (impurities) are adsorbed on the solid surface [827, 849]. A thin film deposited on the surface by which the value of the work function is changed also leads to an alteration of δ_e. As regards the dependence of the secondary electron emission coefficient on the value of

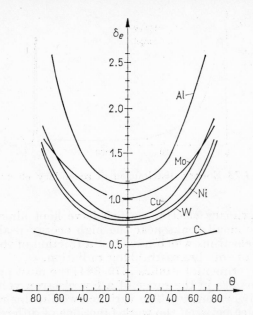

Fig. 1.74 Secondary electron emission
coefficient plotted as a function of the
angle of primary electron incidence for
different metal surfaces

the work function φ, it can be seen on inspection of Appendix, Tables 15 and 16 that, in general, the solids with a low value of the work function show low values of δ_e while those with high values of φ show high values of δ_e. On the other hand, it has been observed that the value of δ_e increases with the deposition of thin layers on the surface which decrease the value of the work function and that this increase in δ_e continues as long as the work function can be decreased by increasing the thickness of the deposited layer [885–889]. The maximum value, δ_{em}, can be obtained for the layer thickness associated with the minimum value of φ. If the layer thickness is further increased above this value with δ_{em}, the emission coefficient starts to decrease and settles eventually at a value corresponding to the work function of the deposited material.

It can be seen in Appendix, Table 16 that the secondary electron emission induced by electron impact has the lowest coefficient δ_{em} in the case of alkali metals ($\delta_{em} < 1$) while for the other metals we generally find $\delta_{em} = 1$–1.8. The highest values, varying between $\delta_{em} = 10$–25, are found with oxide and salt crystals.

On the incidence of positive ions in the solid surface we can observe the same processes as on the incidence of electrons, i.e. secondary electron emission and backscattered primary particles. However, the backscattered particles in the former case may not only be positive ions but may also be neutral particles in ground or excited states.

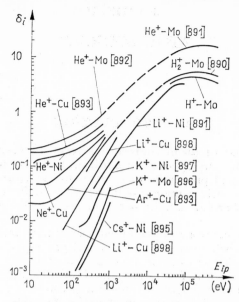

Fig. 1.75 Secondary electron emission coefficients in the case of ion impact on different metal surfaces

The condition of secondary electron emission in the case of electron impact is that the energy of the primary electron be higher than the energy determined by the work function of the surface.

In the case of positive ion impact secondary electron emission can be induced only if the sum of the potential energy E and the kinetic energy E_{ip} is higher than twice the value of the work function, i.e.

$$E_{ip} + E > 2\varphi \qquad (1.123)$$

The reason for this condition is that the positive ion must eject two electrons from the solid in order to induce the emission of a secondary electron since one of the electrons is needed for the neutralization of the positive ion.

On the analogy of expression (1.121) the secondary electron emission coefficient in the case of ion impact is symbolized by δ_i.

The values of δ_i as a function of the kinetic energy E_{ip} are shown for some metals and ions in Fig. 1.75. The values of $\delta_i(E_{ip})$ are usually in the range $\delta_i = 10^{-4}$–15. The highest values are reached, as a rule, at energies $E_{ip} = 10^5$–10^6 eV.

Experiments have shown that δ_i increases with increasing angles of ion incidence [894]. The coefficient δ_i increases considerably in the presence of an adsorbed gas layer on the solid surface like in the case of electron impact [897]. Similarly to the electron impact, the energy distribution of the secondary electrons emitted on ion impact is nearly Maxwellian and the most probable values are in the range of 2 to 5 eV [897].

CHAPTER II

ATOM SOURCES

2.1 INTRODUCTION

The application field of atom sources has recently been expanded. Such sources are now also employed — in addition to the experimental investigations of the physical properties of atomic collisions, atomic or molecular states and their magnetic moments — in experiments for the study of chemical processes at thermal and higher energies in solid state, nuclear and plasma physics.

Atom and molecular beams have already been successfully used for more than 50 years in investigations of the physical properties of atomic particles [2, 422, 426, 1415]. Their extensive application for the study of chemical behaviour started only a few years ago [1359–1368]. The utilization of neutral atomic or molecular beams made it possible to develop more effective methods and to obtain direct information in chemical research. These methods permit the mechanism and kinetics of chemical reactions to be observed both at thermal and at higher energies.

Atomic and molecular beams have also proved to be useful in solid state physics, particularly for obtaining data on the structure of solid surfaces, their impurities and on the physical and chemical effects brought about by particle beams colliding with the surface [1369–1377].

In nuclear physics research neutral atom beams are used partly for the production of nuclear spin polarized particle beams, partly in electrostatic particle accelerators utilizing charge transfer (see Chapter IV). Most recently, fast intense atom beams have been increasingly used in controlled thermonuclear fusion experiments for heating the plasma [1378–1397] and for plasma diagnostics [1296, 1301, 1339, 1347, 1349, 1398–1409].

According to the requirements of the above mentioned application fields the investigations devoted to the production of atomic and molecular beams have lately covered not only the sources of beams with thermal energy but also those capable of yielding well-collimated low and high intensity atom or molecular beams of a small spread in energies required for chemical (0.5–100 eV) and plasma physics (10–100 keV) research.

The present chapter deals with problems of atom sources yielding atom beams of thermal energies, with the extraction and collimation of these beams, and with the conditions under which fast atom beams can be produced.

2.2 EXTRACTION AND COLLIMATION OF ATOM BEAMS

The atom beam source is usually a small chamber (oven) in which the chosen material can be kept at a definite pressure and temperature. The atom beam pours out from the source into the adjacent vacuum chamber on passing through a slit shaped as required by the physical measurement. The vacuum chamber pressure has to be maintained optimally at a value at which the scattering of the atom beam does not occur or it is negligible in order to minimize the loss of atoms before their reaching the collimating slit in the wall separating the first vacuum chamber from the measuring chamber.

Atoms leave the source either by effusion or by aerodynamical flow.

2.2.1 EFFUSION ATOM BEAM

Atoms will leave the source by effusion if the source pressure and the slit dimensions are chosen such that the spatial and velocity distributions of atoms remain unchanged while passing through the slit, that is, if they cross the slit without collisions [422–426].

In the case of thin slits, when the slit thickness $l \sim 0$, the condition under which effusion can take place can be expressed as

$$d \ll \lambda_A \tag{2.1}$$

where d is the width (or in the case of a circular slit the diameter) of the slit and λ_A is the collision mean free path of atoms in the source, which in terms of gas kinetical theory is given by the formula

$$\lambda_A = \frac{1}{\sqrt{2}\,\sigma_k n} \tag{2.2}$$

where n is the number of atoms per unit volume at the source pressure p', σ_k is the collision cross section of atoms defined as $\sigma_k = \pi\delta^2/4$, and δ is the atomic diameter. Using the relation which holds for an ideal gas and has the form

$$p' = nkT \tag{2.3}$$

where p' is the source pressure in atmospheric units, T is the absolute temperature in °K and k is the Boltzmann constant, expression (2.2) can be transformed to

$$\lambda_A = \frac{kT}{\sqrt{2}\,p'\sigma_k} = 7.321 \cdot 10^{-20} \frac{T}{p\sigma_k} \tag{2.4}$$

where p is the source pressure in mmHg units. The values of the collision cross section σ_k and the atomic diameter δ are given in Appendix, Table 12.

If the beam of atoms is emitted by effusion from the source across a slit of width d and thickness $l \sim 0$, then the number of atoms in the solid angle $d\omega$ flying in the direction of angle Θ is given by the formula [423, 424]

$$dN = \frac{d\omega}{4\pi} X n \bar{v} A_s \cos \Theta \tag{2.5}$$

where $X = \dfrac{n_a}{n}$ is the ratio of the number of atoms per unit volume to the total number of particles per unit volume (the parameter of the degree of dissociation in the case of molecular gases) and A_s is the area of the cross section of the slit, $\bar{v}$ is the mean particle velocity which can be expressed as

$$\bar{v} = \sqrt{\frac{8kT}{\pi m}} \tag{2.6}$$

taking the particle mass to be m and assuming Maxwellian or nearly Maxwellian particle velocity distribution in the source.

The number of particles flying from the source in all directions is obtained by integrating eq. (2.5) over the solid angle 2π, thus

$$N = \frac{1}{4} Xn\bar{v}A_s \tag{2.7}$$

It can be seen from (2.7) that the number of atoms emitted from the source can be increased by increasing the pressure, the particle velocity and the size of the slit. However, the pressure and slit size are limited by condition (2.1) while the increase in velocity is subject to restrictions imposed by the methods applied in physical measurements. For this reason, it is often preferable to increase the source efficiency by applying instead of the strict condition (2.1) for effusion, the condition of the form

$$d \leq \lambda_A \tag{2.8}$$

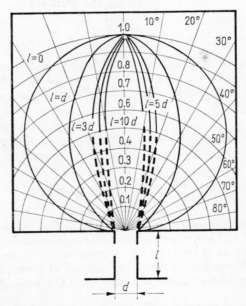

Fig. 2.1 Angular distributions of atom beams emitted from atom source through a slit of diameter d and of different lengths l

As can be seen, conditions (2.1) and (2.8) impose limits only on the width or the diameter of the slit but not on its height. Thus, if the slit is high enough, its area substantially increases and with the slits of this shape higher atom beam intensities can be obtained.

2.2.1.1 Effusion across long channels

The effusion of atoms from the source across a thin slit ($l \sim 0$) shows an angular distribution of cosine type, as it can be seen from formula (2.5) [427–430] (see Fig. 2.1: curve for $l \sim 0$). With the usual geometry of the measurements only a fraction of the atom beam falling into a small solid angle can be utilized — as apparent from the angular distribution shown in Fig. 2.1 — if a thin slit is used for the exit. If the slit is replaced for the good collimation of the beam by a channel of finite length, the angular distribution of the atoms flying out from the source will change according to the ratio l/d [431–434, 525] as is also apparent from Fig. 2.1.

The change in the angular distribution can be explained by the fact that the atoms flying into the channel at a large angle collide with the channel wall and have, therefore, less chance of flying out from the source. To prevent the collisions between atoms in a channel, the pressure has to be chosen such that

$$l \leq \lambda_A \qquad (2.9)$$

The number of atoms flying out from the channel decreases because of the above considered factors compared with that emitted from the source and therefore the number of outflying atoms can be calculated from eq. (2.7) only upon the introduction of a reduction coefficient symbolized by ξ in the form [432]

$$N = \frac{1}{4}\,\xi X n \bar{v} A_s \qquad (2.10)$$

The reduction coefficient ξ has been already evaluated for slits of different shapes [431, 432, 435].

If d stands for the slit width (or the diameter of circular slits), h for the height and l for the length of the slit, then a circular tube with a uniform diameter d and length l, gives for $d \ll l$, the reduction coefficient

$$\xi = \frac{4d}{3l} \qquad (2.11)$$

For a rectangular channel with a uniform cross section, we get with $l \gg d$ and $l \gg h$ the reduction coefficient

$$\xi = \frac{1}{ldh}\left\{ d^2h \ln\left(\frac{h}{d} + \sqrt{1 + \left(\frac{h}{d}\right)^2}\right) + \right.$$

$$\left. + dh \ln\left(\frac{d}{h} + \sqrt{1 + \left(\frac{d}{h}\right)^2}\right) - \frac{(l^2 + d^2)^{3/2}}{3} + \frac{l^3 + d^3}{3}\right\} \qquad (2.12)$$

If $l \gg d$ and $l \ll h$, we have

$$\xi = \frac{d}{l} \ln \left(\frac{l}{d} \right) \tag{2.13}$$

while for $l \gg d$, $l \gg h$ and $h \gg d$, we find

$$\xi = \frac{d}{2l} \left[1 + 2 \ln \left(\frac{2h}{l} \right) \right] \tag{2.14}$$

If the cross section of the quadratic ($h = d$) channel is given by $d \times d$ and $l \gg d$, expression (2.11) transforms to

$$\xi = \frac{2d}{l} \ln (1 + \sqrt{2}) = 1.76 \frac{d}{l} \tag{2.15}$$

In the case of a thin slit ($l \sim 0$) of any shape we have

$$\xi = 1 \tag{2.16}$$

Working with an exit slit of finite length and uniform circular cross section, the number of atoms flying out from the source can be calculated by the use of eqs (2.10) and (2.11) and by recalling that $A_s = \left(\frac{d}{2} \right)^2 \pi$, from the equation

$$N = \frac{\pi d^3}{12l} X n \bar{v} \tag{2.17}$$

The value of the half angle, $\theta_{1/2}$, associated with the half intensity has been calculated [436] as

$$\theta_{1/2} = \frac{0.84d}{l} \tag{2.18}$$

The application of slits with finite length presents great advantages if one wants to extract well-collimated atom beams from atom sources. In fact, the ratio l/d — as apparent also from the angular distribution diagram in Fig. 2.1 — can be chosen such that it corresponds to the angle $\theta_{1/2}$ in the given geometry of the apparatus. This increases the useful fraction of the emitted beam and thus also the utilization coefficient of the material (a fact of particular importance if costly or radioactive materials are consumed). Moreover, the pumping rate of the vacuum pumps can be reduced while still maintaining the required pressure. In the choice of the channel dimensions, conditions (2.8) and (2.9) have to be considered. The maximum value of the length l is restricted by the pressure in the source through the relation (2.9), therefore it is the most convenient to set the appropriate ratio l/d by changing the value of d. If, for example, this pressure is ~ 0.5 mmHg, then for hydrogen $\lambda_H \sim 10^{-1}$ cm. In order to obtain $\theta \sim 2.5°$, the choice can be $d \sim 5 \cdot 10^{-3}$ cm. At this pressure and channel size, assuming that $X = 0.8$ and $\bar{v} = 2 \cdot 10^5$ cm/sec, eq. (2.16) gives for the number of atoms flying out from the source

$$N = 3.5 \cdot 10^{14} \left[\frac{\text{atoms}}{\text{sec}} \right]$$

2.2.1.2 *Effusion through multi-collimator*

It has been shown in the foregoing that the strong collimation of the beam leads to the reduction of the beam current because of conditions (2.8) and (2.9) of the effusion. It is for this reason that experimentalists apply several parallel channels of the proper ratio l/d at the output slit of the atom source if well-collimated high intensity, very slow beams at thermal energies are needed for the experimental study [433, 434, 438–445, 525, 1413, 1424].

This arrangement permits well-collimated beams of higher intensities to be obtained, while the conditions of effusion are fulfilled.

Several methods are available for the arrangement of such parallel channels, the so-called multi-collimator. The multi-collimator depicted in Fig. 2.2 has been prepared from a ~0.03 mm thick corrugated nickel foil. The cross sections of channels thus obtained are trapezoidal with an average size of ~0.2 mm and a transparency of ~65–70 per cent [439].

Among the multi-collimators prepared from metal [438–440], the highest transparency, i.e. 85 per cent, has been achieved with an assembly prepared from thin-walled copper capillaries. The multi-collimators used for nuclear spin polarized particle sources are composed of sets of thin-walled glass capillaries [441–445]. The diameter and length of the glass capillary varies as $d = 30~\mu$ to 1 mm and $l = 300~\mu$–10 mm, depending on the presssure

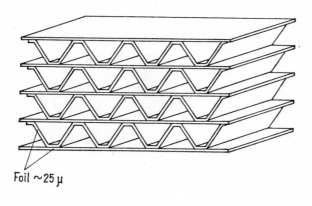

Foil ~25 μ

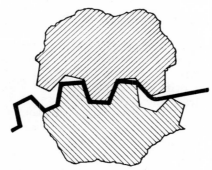

Fig. 2.2 Multi-collimator prepared from nickel foil

Fig. 2.3 Multi-collimator made of glass capillaries

applied in the dissociator which is employed as atom source. A glass multi-collimator of this type (for detailed description see ref. [446]) is shown in Fig. 2.3. With this type of output slit, the particle current obtained from the source can be increased proportionally to the number of capillaries applied. If the number of multi-collimator channels is symbolized by N_{ch}, the number of outflying atoms is given by use of expression (2.17) as

$$N = N_{ch} \frac{\pi d^3}{12l} X n \bar{v} \qquad (2.19)$$

Higher atom beam currents can be generated under the conditions of effusion if the mean velocity $\bar{v}$ of the atoms in the source can be further increased.

Since $\bar{v} \sim \sqrt{T}$, as is apparent from formula (2.6), the mean velocity of atoms can be increased by raising the temperature of the source. However, this method is not very effective in view of the $\sqrt{T}$ proportionality; moreover, the increase in temperature is limited by the melting point of the source material.

2.2.2. SUPERSONIC ATOM BEAMS

An alternative method for the production of intense atom and molecular beams has been proposed by Kantrowitz and Grey [447]. They suggested the use of a supersonic stream generated by the application of a miniature Laval nozzle as the first orifice of the source (see Fig. 2.4a).

The stream of atoms from the Laval nozzle is passed through a so-called "skimmer" orifice which permits the passage to the collimator slit only to atoms forming the core of the beam. This core consists of atoms flying with adequate velocities in adequate directions while the rest of the beam is deflected and pumped off by high speed pumps. The possibility of producing supersonic beams with a Laval nozzle follows from the general laws of gas dynamics which are valid for aerodynamical flow [455–458].

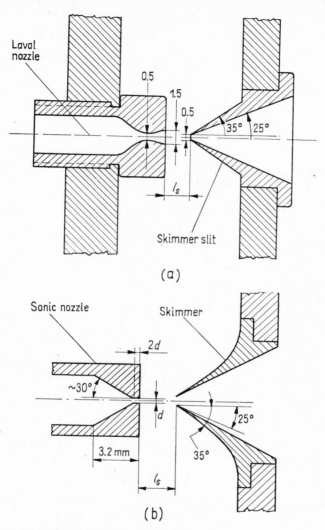

Fig. 2.4 Atom beam collimator system consisting of Laval nozzle
and skimmer slit

It has been shown [447–458, 1412, 1428, 1431] that gas passed through
a Laval or supersonic nozzle is expanded and at the same time accelerated
when leaving the nozzle. The measure of acceleration, that is the Mach
number obtainable with a Laval nozzle (the Mach number $M = u/a$, where
u is the flow rate and $a = \sqrt{\dfrac{\gamma p_0}{\varrho_0}}$ is the speed of sound) is determined theoretically by the ratio of the inlet cross section to the minimum cross section
of the nozzle since this value determines the expansion ratio p_0/p_1 at the
Laval nozzle (p_0 is the pressure in the source chamber and p_1 is the pressure

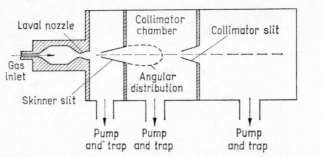

Fig. 2.5 Angular distribution of atoms emitted from source through Laval nozzle and skimmer slit collimator system

at the outlet orifice of the nozzle). In addition, the passage through the nozzle also causes the random thermal motion of atoms to change from any possible direction to a motion oriented preponderantly in the direction of the outgoing beam. Thus, the angular distribution of atoms leaving the supersonic nozzle shows an elongation (see Fig. 2.5) in the direction of beam propagation as compared with the distribution for a thin slit ($l \sim 0$) to be seen in Fig. 2.1.

It can be seen from the above considerations that the intensity of supersonic beams increases because of the following two effects. The passage through the thin nozzle increases the atomic velocity and correspondingly the intensity of the beam. The intensity is further increased because of the strong cooling due to adiabatic expansion which considerably decreases the velocity component arising from the thermal motion of atoms. Consequently, the number of atoms having velocities close to the mean atomic velocity of the beam increases in a measure depending on the Mach number [1431] (Fig. 2.8a). This self-collimation of the supersonic beam substantially increases the beam intensity.

The quantity of gas passing through supersonic nozzles can be evaluated from the equation [459]

$$j_0 = \varrho^* u^* A^* \tag{2.20}$$

where ϱ^* is the gas density, u^* is the flow rate at the critical diameter of the nozzle (where $M = 1$), and A^* is the cross sectional area at the critical diameter. It has been shown [449] that for a miniature supersonic nozzle the size of A^* corresponds to the minimum cross section of the nozzle. If it is also considered that

$$\varrho^* = \varrho_0 \left(\frac{2}{\gamma + 1} \right)^{\frac{1}{\gamma - 1}} \tag{2.21}$$

and that

$$\mu^* = a^* = a_0 \sqrt{\frac{2}{\gamma + 1}} = \sqrt{\frac{\gamma p_0}{\varrho_0}} \sqrt{\frac{2}{\gamma + 1}} = \sqrt{\frac{\gamma R T_0}{m}} \sqrt{\frac{2}{\gamma + 1}}$$

where a^* is the sound velocity in cm/sec units at the critical cross section and a_0 is the sound velocity in cm/sec in the chamber of the nozzle at temperature T, $\gamma = \dfrac{C_p}{C_v}$ is the heat capacity ratio (given as $\gamma = \dfrac{5}{3} = 1.667$ for monoatomic and as $\gamma = 1.4$ for diatomic gas), ϱ_0 is the gas density in the nozzle chamber at pressure p_0, m is the weight of gas per unit volume at pressure p_0, then eq. (2.20) can be written in the form

$$j_0 = \sqrt{\frac{2\gamma p_0}{\varrho_0(\gamma + 1)}} \left(\frac{2}{\gamma + 1}\right)^{\frac{1}{\gamma-1}} \varrho_0 A^* \qquad (2.22)$$

If $T_0 = 400°K$, $p_0 = 30$ mmHg and the minimum nozzle diameter is 0.6 mm, then the quantity of gas, that is, the corresponding number of H_1 atoms passing through the nozzle is given as

$$j_0 \approx 5.2 \cdot 10^{20} \text{ atoms/sec}$$

For an atomic density

$$n_b = n_0 \left[1 + \frac{\gamma - 1}{2} M_s^2\right]^{-\frac{1}{\gamma-1}}$$

and an atomic velocity

$$v_b = u M_s = a_0 \frac{M_s}{\left[1 + \dfrac{\gamma - 1}{2} M_s^2\right]^{1/2}}$$

of the beam before the skimmer orifice, the fraction of the particle current from the supersonic nozzle passing through the skimmer can be evaluated [452] by use of the equation

$$j_1 = n_b v_b s_1 = n_0 a_0 s_1 \frac{M_s}{\left[1 + \dfrac{\gamma - 1}{2} M_s^2\right]^{\frac{1}{2} + \frac{1}{\gamma-1}}} \qquad (2.23)$$

where n_0 is the atomic density in the nozzle chamber, s_1 is the area of the skimmer cross section, and M is the Mach number. If the skimmer diameter $d_1 = 1.1$ mm and $M = 5$, the number of atoms passing through the skimmer is given as

$$j_1 \approx 2 \cdot 10^{18} \text{ atoms/sec}$$

The current obtainable in the measuring chamber after the collimator slit can be evaluated from the equation

$$j_2 = n_b v_b s_1 \frac{s_2}{\pi l_d^2} \left(\frac{3}{2} + \frac{\gamma}{2} M_s^2\right) =$$

$$= n_0 a_0 s_1 s_2 \frac{1}{\pi l_d^2} \frac{M_s \left(\dfrac{3}{2} + \dfrac{\gamma}{2} M_s^2\right)}{\left[1 + \dfrac{\gamma - 1}{2} M_s^2\right]^{\frac{1}{2} + \frac{1}{\gamma-1}}} \qquad (2.24)$$

where s_2 is the area of the collimator slit cross section and l_d is the distance from the skimmer to the collimator slit. For a collimator slit diameter $d_2 = 1.4$ mm and $l_d = 25$ mm, we get

$$j_2 \approx 1\text{-}2 \cdot 10^{16} \text{ atoms/sec}$$

Results in good agreement with expression (2.24) have also been obtained by other authors [454, 1428–1431].

It follows from eqs (2.23) and (2.24) that the intensity of supersonic beams is a function of the values of p_0, a_0, s_1, s_2, l and M. The dimensions of the supersonic nozzle are taken into consideration only through the Mach number. The value of the Mach number was calculated [454, 1360, 1414] in terms of the ratio of the distance l_s between the supersonic nozzle inlet orifice and the skimmer to the diameter of the former and it was found that the function $M = M\left(\dfrac{l_s}{d}\right)$ (shown in Fig. 2.6) can be expressed for a circular inlet orifice as

$$M \approx b\left(\frac{l_s}{d}\right)^{\gamma-1} \sim b\sqrt{\frac{l_s}{d}} \tag{2.25}$$

It is apparent from Fig. 2.6 that the values calculated from formula (2.25) for $b = 2.75$ are in good agreement with the experimental values [454, 1414]. Consequently, the values of M in the axis of the beam can be calculated for given values of l_s/d by use of the formula (2.25) if $M < M_T$, where M_T is the maximum value of the Mach number in the beam axis. M_T can be evaluated from the expression [1428, 1431]

$$M_T = A K_{n_0}^{-B} \tag{2.26}$$

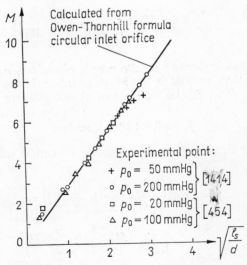

Fig. 2.6 Dependence of Mach number on the ratio of distance l_s between nozzle and skimmer to nozzle inlet diameter d

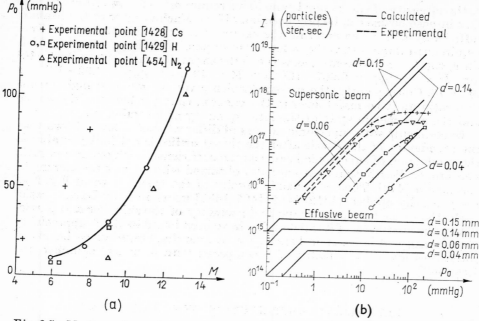

Fig. 2.7a Mach number values associated with maximum atom or molecular beam intensities for given pressure p_0

Fig. 2.7b Atom or molecular beam intensity as a function pressure p_0 for effusion and supersonic beams [1428]

where $K_{n_0} = \dfrac{\lambda_0}{d}$ is the Knudsen number, that is the ratio of the mean free path λ_0 of atoms or molecules in the nozzle chamber at temperature T_0 and at pressure p_0 to the nozzle throat diameter d, A and B are constants with values depending on the nozzle throat diameter and temperature; e.g. for argon gas with $d = 0.33$ mm and $T_0 = 297°$K, $A = 1.17$, $B = 0.4$ [1431], for rubidium gas with $d = 0.14$ mm and $T_0 + 100°$K, $A = 0.33$, $B = 0.70$, while with $T_0 + 350°$K we get $A = 0.94$, $B = 0.50$ [1428].

The beam intensity is expected to increase linearly with increasing pressure. However, as the value of p_0 increases, the effect of atomic collisions becomes more and more important in the region before the skimmer and a deviation from linearity is to be expected (Fig. 2.7b).

Intensity measurements made on supersonic nozzles for hydrogen and nitrogen gases and rubidium and caesium alkali metal vapours at different values of p_0 [1428, 1429, 1454] have shown that the Mach number value obtained for the maximum intensity associated with the given pressure p_0 varies with pressure in the same way as shown in Fig. 2.7a. The monochromaticity of the beam improves as the value of the Mach number increases, that is, the value of relative atomic velocity decreases.

It is pointed out in any report dealing with the study and application of supersonic nozzles [447–454, 1412, 1414, 1425–1432] that the use of high performance vacuum pumps is imperative for obtaining an adequate increase in the intensity of supersonic beams. This requirement follows from the fact that intense beams can be produced only if the Mach number of the beam is sufficiently high when entering the skimmer after leaving the nozzle. This requirement implies that the value of the expansion ratio p_0/p must be high since, e.g. in a monoatomic gas we need the expansion ratio $p_0/p = 2.66 \cdot 10^2$ for $M = 5$, $p_0/p = 6.907 \cdot 10^3$ for $M = 10$ and $p_0/p = 6.93 \cdot 10^4$ for $M = 16$. Consequently, the pressure which gives the required high expansion ratio must be maintained in the vacuum chamber between the nozzle and the skimmer in spite of the large quantity of incoming gas.

Experiments with supersonic nozzles of different shapes have shown that the broadening of the nozzle after the throat orifice is not indispensable to the production of supersonic beams as the sufficiently high expansion ratio and proper distribution shape are also obtained without any broadening of the nozzle [449]. Thus, supersonic nozzles of the type shown in Fig. 2.4b are generally applied [1412, 1414, 1425–1431] instead of the Laval nozzle.

On the other hand, the shape and geometry of the skimmer and particularly the sharpness of its inlet orifice have been found to affect appreciably the intensity of supersonic beams; e.g. a hole diaphragm gives beams of intensities lower by a factor of 5 compared with a conical skimmer orifice [449].

2.2.3 VELOCITY DISTRIBUTION OF ATOMIC BEAMS

The atomic velocity distribution in the atom source is usually of Maxwellian type. Taking dN to be the number of atoms with velocities in the interval $(v, v + dv)$, while the total number of atoms is N and the atomic mass is m, then the velocity distribution can be described by the equation

$$\frac{dN}{dv_s} = Nf(v_s) = N \frac{4}{\sqrt{\pi}} \frac{v_s^2}{\alpha^3} e^{-\frac{v_s^2}{\alpha^2}} \qquad (2.27)$$

where

$$v_s = \alpha = \sqrt{\frac{2kT_0}{m}} \qquad (2.28)$$

is the most probable velocity of atoms in the source because of $\dfrac{d^2N}{dv^2} = 0$.

The atomic velocity distribution in the beam obtained by effusion differs from that described by eq. (2.27) since the effusion probability of atoms through the outlet aperture of the source is proportional to the atomic velocity. The velocity and intensity of atoms are related by expression (2.5).

Consequently, the atomic velocity distribution of the outcoming beam is proportional to the expression (2.27) as multiplied by v. The proportionality coefficient I_0 can be evaluated by normalization to the total beam intensity. If the intensity of atoms with velocities in the interval $(v, v + dv)$ is given by $I(v)dv$, then

$$\frac{dI(v)_e}{dv} = I_0 \frac{v^3}{\alpha^3} e^{-\frac{v^2}{\alpha^2}} \qquad (2.29)$$

where we find that with $\dfrac{d^2 I(v)_e}{dv^2} = 0$, the most probable velocity is given in the form

$$v_b = \sqrt{\frac{3}{2}\,\alpha} \qquad (2.30)$$

It is apparent from eqs (2.27) and (2.29) that the atomic velocity distribution in the source differs from that in the beam, thus, the most probable source and beam velocities are expected to differ from each other, as shown by eqs (2.28) and (2.30).

Similarly, it is possible to evaluate for atoms or molecules the mean velocity in the source as

$$\bar{v}_s = \int\limits_0^\infty v f(v)\,dv = \sqrt{\frac{8kT_0}{\pi m}} = \frac{2}{\sqrt{\pi}}\,\alpha \qquad (2.31)$$

and the mean velocity in the beam as

$$v_b = \frac{3\sqrt{\pi}}{4}\,\alpha \qquad (2.32)$$

The velocity distribution of atoms or molecules in the beam axis is given [1431] as

$$\frac{dI(v)_{su}}{dv} = I_0\,\frac{v^3}{\alpha^3}\,e^{-\frac{(v-u)^2}{\alpha^2}} \qquad (2.33)$$

where u is the velocity of atoms flying out from the supersonic nozzle and v is the velocity corresponding to the nozzle temperature T.

It can be seen that formula (2.33) transforms to the effusion beam formula (2.29) if $u = 0$.

These values of velocity distribution calculated from formula (2.33) for an atomic argon $(\gamma = \dfrac{5}{3})$ beam are shown in Fig. 2.8a for various values of M. It shows that the velocity spread of atoms in the beam decreases with increasing Mach numbers. A good agreement with the theoretical velocity distribution was obtained in the measurement on an argon atom beam for $M = 16$ [1431] and, as to be seen in Fig. 2.8b, also for Rb atom and Rb_2 molecular beams [1428].

Comparing the parameters of effusion beams with those of supersonic beams obtainable from atom or molecule sources by use of the different methods described above, we find the following.

The intensity given for beams obtained by the effusion of atoms or molecules by expression (2.17) can be determined by the appropriate choice of p_0, d and v. Condition (2.9) also determines the effect of the increase in p_0 for a given value of d. Thus, higher intensities of effusion beams are obtainable above all by use of multi-collimators. The intensity of beams

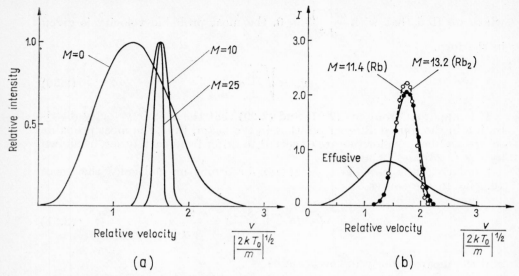

Fig. 2.8 Velocity distribution in supersonic atom and molecular beams

(a) for atomic gas $\gamma = \dfrac{5}{3}$ [1431]

(b) for rubidium atom and molecular beam [1428]

from effusion sources with a single outlet aperture has been evaluated as $I_{\text{effus}} \approx 10^{14}$–$10^{15}$ atoms/ster.sec. This value can be increased by several orders of magnitude by the use of a multi-collimator since the increase in intensity depends on the number N_{ch} of multi-collimator channels.

Taking, for example a multi-collimator of 3 mm in diameter with $d = (3\text{–}5) \cdot 10^{-2}$ mm channels, $N_{ch} = (1\text{–}2) \cdot 10^3$, thus for $p_0 = 0.5$ mmHg, $d = 5 \cdot 10^{-2}$ mm, $v = 2 \cdot 10^5$ cm/sec and $N_{ch} = 10^3$ well-collimated beams can attain intensities of the order

$$I_{\text{effus}} \approx N_{ch}(3\text{–}6) \cdot 10^{14} = (3\text{–}6) \cdot 10^{17} \frac{\text{atoms}}{\text{ster.sec}} \tag{2.34}$$

In the case of supersonic beams, the intensity of atoms or molecules, as calculable from expression (2.24), can be determined by the choice of p_0, d and M. A substantial increase in intensity can be achieved by increasing the pressure p_0 which is necessarily higher than the pressure applicable in the case of effusion beams (Fig. 2.7b).

Taking, for example $d = 6 \cdot 10^{-2}$ mm, $p_0 = 10^2$ mmHg, $T_0 \sim 900°$K, we get for well-collimated beams

$$I_{\text{sup}} \approx (2\text{–}5) \cdot 10^{17} \frac{\text{atoms}}{\text{ster.sec}} \tag{2.35}$$

The angular distribution in both effusion and supersonic beams can be the same if the value of l/d for the effusion beam and that of l_s/d for the supersonic beam are properly chosen.

The essential difference between effusion and supersonic beams is due to their different velocity distributions. As is apparent from Fig. 2.8, the energy spread decreases in the supersonic atom or molecular beams as the Mach number increases. It follows that for experiments requiring particle beams of a small spread in energy, supersonic beams have to be used.

2.3 ATOMIC GAS SOURCES

The most simple atom sources can be obtained with non-condensing atomic gases, particularly with gases effusing from sources which can be kept at the ambient temperature. In this case the source assembly can be composed of a small container made of glass or metal connected to a gas inlet tube and having an outlet aperture of the type required by the measurements to be performed. However, even in this case the source geometry must be adjusted by a special device to direct the outcoming beam in the direction of the detector mounted into the apparatus.

If the temperature of the non-condensing atomic gases (that is, the velocity of the outflying atoms, according to eq. (2 6)) must be lower or higher than the ambient temperature, the source assembly has to be built such that the source chamber can be appropriately cooled or heated. A source assembly of this type is shown in Fig. 2.9.

The source chamber and the double-walled tube, together with the gas inlet tube should be prepared preferably from stainless steel because this permits the components to be welded under protecting gas, furthermore the heat conduction of the stainless steel tube is conveniently low. The source chamber can be cooled to the temperature required by the experiments using liquid nitrogen, hydrogen or helium which is introduced into the long thin-walled tube in direct contact with the walls of the chamber and the gas inlet tube to bring the gas to the required temperature. If atoms with higher mean velocities are needed for the experiment a heater can be placed in the tube instead of the coolant liquid and thus the source chamber with the gas can be heated to the required temperature.

The dimensions of the chamber slit should be chosen with respect to the performance of the pumps of the differential vacuum system, while the shape of the slit must be formed according to the measurement as described in Paragraph 2.2.

The source geometry may be set with an adjusting nut and screws. The push valve enables one to replace the source without letting atmospheric pressure get into the entire vacuum system.

If non-condensing gases are applied, the evacuation of each chamber in the vacuum system requires pumps with higher pumping power than for substances of low vapour pressure at room temperature. This difference is particularly important if a Laval nozzle is used since the equipment provided with a supersonic atom beam source must be operated with pumps working at a rate of several thousand litres/sec [447–454].

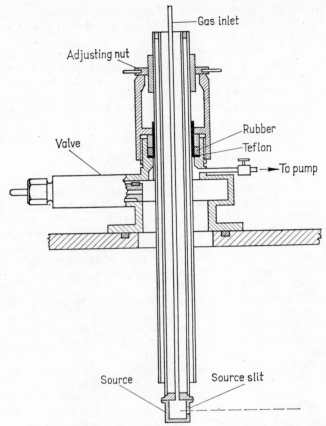

Fig. 2.9 Source of atoms from non-condensing gases

Atomic beams are produced from non-condensing gases in the molecular state with the aid of dissociators in which the molecular gases are decomposed by one or another dissociation method. These methods will be discussed in detail in Paragraph 2.5.

2.4 THERMAL ATOM SOURCES
[460–485, 1367, 1374, 1375, 1400, 1410, 1411]

In order to produce atom beams of elements in the solid state at room temperature the atom source chamber and slit must be heated. The material placed into the source chamber has to be heated to a temperature at which the vapour pressure of the material attains the value required by the conditions (2.8) and (2.9) of effusion.

The chamber of thermal atom sources must be prepared from a material which does not melt, alloy or react chemically with the evaporating material at the temperature (see Appendix, Table 13) required for the necessary vapour pressure. Another serious problem is the appropriate heat insulation and the necessity to prevent the particles transmitted through the slit from adhering to the border of the slit. Poor thermal insulation causes the ambient temperature to rise and impairs the conditions of vacuum generation while the source particles adhering to the outlet slit modify the slit dimensions and this leads eventually to the total obstruction of the exit. For rather low experimental temperatures the thermal source chambers are usually prepared from iron, stainless steel, cooper, nickel, Monel metal and (rarely) also from gold or silver. At higher experimental temperatures molybdenum, tantalum, tungsten, carbon and the high melting aluminium oxide, titanium oxide, etc. (see Appendix, Table 14) are used. Apart from the materials listed above, a number of materials can be employed which are useful in cases when conventional materials are unsuitable because of chemical or physical reactions. For the melting of hydroxides, for example, chambers prepared from silver were found to be the most adequate since iron reacted appreciably with the hydroxides [460, 461]. Iridium and gallium also alloy with iron, thus a source in a molybdenum chamber at 1000°C can yield iridium atom beams [462–463]. On the other hand a beam of gallium atoms can be obtained only with a graphite lined [464] or pure graphite [465] source chamber and slits. For the production of aluminium atom beams it proved useful to place an Al_2O_3 crucible in a graphite chamber separating the crucible from the graphite with a tantalum foil to prevent the crucible material from reacting with the graphite [466]. For the production of a praseodymium atom beam a ThO_2 crucible was placed in the molybdenum chamber [467] and this minimized the probability of slit obstruction.

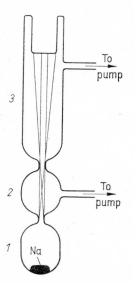

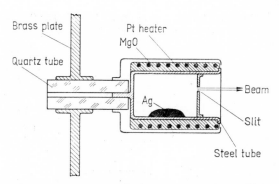

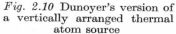

Fig. 2.10 Dunoyer's version of a vertically arranged thermal atom source

Fig. 2.11 Stern–Gerlach-type horizontally arranged thermal atom source

The few examples cited in the foregoing are sufficient to show that the preparation of thermal atom sources may raise a number of problems depending on the type of measurement to be performed.

The first tentative construction of an atom (molecule) source was reported by Dunoyer [468, 469]. With the equipment made of glass (Fig. 2.10) Dunoyer generated a beam of Na_2 molecules using as thermal source the spherical glass chamber (1) and the small glass tube between chambers (1) and (2) acted as outlet slit. The Na vapour condensed on the cooled surface of chamber (3) and showed a pattern from which it could be inferred that Na_2 molecules moved in a straight direction in the evacuated equipment. This simple arrangement was useful for the demonstration of this effect but the later experiments called for more sophisticated thermal sources.

The source shown in Fig. 2.11 is suitable for the production of silver atom beams (Gerlach, Stern [471]). The chamber was prepared from an iron tube, the outlet slit of circular shape 1 mm in diameter, 1 mm in length. The chamber can be heated with a helical heater made of 0.3 mm platinum wire up to temperatures of $\sim 1300\,°C$. An iron cylinder is used as a thermal reflector. The copper disc by which the quartz tube supporting the chamber is kept in a fixed position is water-cooled. The heating filament is embedded in a paste prepared from quartz powder, magnesium oxide, kaolin and water-glass which serves at the same time as an insulator and fixing medium.

Figure 2.12 also shows a horizontally arranged atom source (Leu [472]). With this source of more advantageous properties than the former, atom beams of high stability can be generated. The copper block is heated with a Pt wire wound on a quartz tube introduced into a constantan sleeve. The metal vapours are brought to the temperature of the copper block during the relatively long path downwards in the copper block. Thus the vapour temperature is uniform even at the extracting slit. The sufficiently high heat capacity of the copper block stabilizes the temperature of the atomic beam so that the slit temperature was found to differ from that of the evaporation chamber by not more than $3.5\,°C$ at $\sim 650\,°C$ owing to the good heat conductivity of copper. This high temperature of the slit prevents the metal vapours from condensation at the slit and the obstruction of the exit. Using iron instead of copper, the temperature difference was measured as $30\,°C$.

In the sources to be seen in Figs 2.13, 2.14 and 2.15 the outlet slit is kept at a higher temperature than the other parts of the source to avoid obstruc-

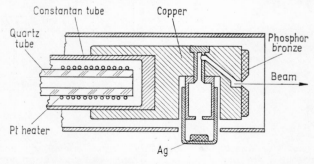

Fig. 2.12 Leu's version of a horizontally arranged thermal atom source

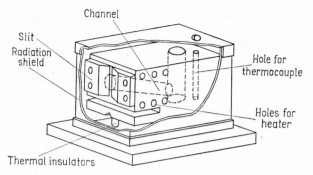

Fig. 2.13 Horizontally arranged thermal atom source with thermal shielding [470]

tion. This is achieved by placing the heater close to the slit. In the source of Fig. 2.13 the cut below the slit is designed to diminish the cooling in the environment of the slit (Kusch [470]).

The inlet aperture on the upper part of the block is closed with a conical metal plug after the introduction of the material to be evaporated. The plug is pressed into the block. The inlet can be opened by extracting the plug or by boring a hole into the plug. The pair of plates forming the slit are usually fixed by screws to the block. The plates are made, as a rule, of the

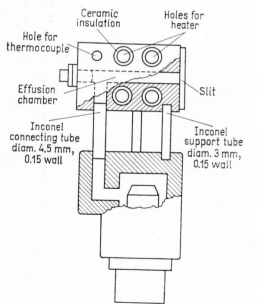

Fig. 2.14 Thermal atom source with separate control of effusion and of evaporator chamber temperatures [473]

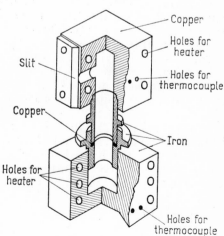

Fig. 2.15 Thermal atom source with separate control of effusion and of evaporator chamber temperatures [473]

same material as the block. The diameter of the channel connecting the evaporator to the slit is somewhat smaller than the height of the slit. The block temperature is measured by use of a thermocouple fastened into a bore close to the evaporation chamber. The heaters are spirals, 2 mm in diameter, prepared from a 50 cm long, 0.25 mm diameter tungsten wire, which are introduced into holes drilled into the block. Thin-walled ceramic tubes inside the holes serve for the insulation of the spirals. The spirals are connected to one another with screws on nickel support or by spot welding with nickel. The temperature can be varied by setting the heating voltage from 2 to 20 V. The heating current is usually a variac-regulated, ferro-resonance-stabilized a.c. current. To avoid leakage to earth it is expedient to use isolating transformers. If one wants to use the spirals several times, they should be prepared from tantalum which does not get brittle after use.

If the source is prepared from molybdenum, its channel is lined with graphite and if the slit is also graphite, the source can be used for the production of gallium atom beams [461].

The sources shown in Figs 2.14 and 2.15 are suitable for the production of atomic beams with improved temperature stability. In both cases the evaporator and effusion chambers are located in separate blocks.

In the source to be seen in Fig. 2.14 both blocks are prepared from copper (McFee, Marcus [473]). Thin (0.15 mm thick) tubes are mounted between the two parts. The tube of 4.5 mm diameter connects the evaporator to the effusion chamber, while the other tubes, which are 3 mm in diameter, simply connect the two blocks mechanically. Since the thin-walled "Inconel" tube is of a low thermal conductivity, the copper block comprising the effusion chamber can be kept at the desired temperature in a given range independently of that of the other block. A source built in this way lends itself well to the study of velocity distribution in atomic beams at different temperatures.

The construction of the source in Fig. 2.15 is somewhat different from that of the former. The block containing the evaporator is made of iron, while the block of the effusion chamber is made of copper. The two blocks can be heated separately. The connecting tube is made of iron which can be screwed off in order to introduce the material to be evaporated. The inlet to the block is fitted with a copper gasket ring.

In vertically arranged thermal atom sources, particularly, if atomic beams of elements with low melting point (e.g. alkali metals) have to be produced, quite simple constructions are practicable.

A thermal atom source of this type is shown in Fig. 2.16 (Eisinger, Bederson, Feld [475, 476]). The source chamber is 63 mm in height and 19 mm in diameter. Two reflector spacers are mounted inside the chamber which prevent the melted material from sputtering and thus obstructing the slit. The chamber is heated by a tungsten spiral to the required temperature. This spiral is mounted close to the slit in the upper part of the chamber. This slit is milled with thin discs into one side of the chamber to a width of 0.15 mm.

A small, vertically arranged source of caesium atoms, which is mounted on the same base plate as another atom source of an element with higher melting point, is shown in Fig. 2.19. The caesium atom source chamber is of rectangular cross section and made of iron in a size of $6.3 \times 6.3 \times 22$ mm^3.

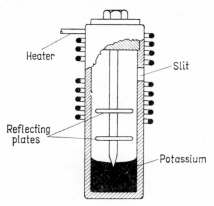

Fig. 2.16 Eisinger's version of a vertically arranged thermal atom source

The slit, which is 3 mm in length and 0.1 mm in width, is located at one side of the chamber. The evaporator chamber is placed into the bore of an iron block, somewhat larger in size than the evaporator. The iron block is heated with molybdenum spirals. The source is filled with metallic sodium and caesium chloride. Stable caesium atom beams are obtained at a chamber temperature of 200°C (Lew [467]).

If small quantities are available of the material used, for reasons of economy, a design should be chosen which permits high utilization coefficients to be achieved.

In one such design reported in the literature (Davis [478]), a long slit with a small cross section is employed in accordance with the requirements (2.10) and (2.11) to collimate the atom beam and to reduce thereby the number of atoms flying out from the chamber in directions not covered by the measurements. To decrease the adherence of the experimental material to the chamber, the evaporator size is less than 25 mm³. This source was used for the production of a ^{22}Na atom beam activated in a cyclotron. It was for this reason that the material consumption had to be kept at a minimum.

The source was heated with Al_2O_3 coated molybdenum spirals. Al_2O_3 serves also as the electrical isolator of the molybdenum from the Monel metal components. The source block was coupled to a horizontal axle by the aid of which the source could be moved and the beam appropriately adjusted. This was of great importance in the given case, since the angle at which the strongly collimated beam is at half intensity was found to be 1.5°, i.e. the same as the solid angle of the detector covering the source. The Monel metal block containing the collimating channel and the evaporator chamber consist of two parts attached to each other by grinding to a precision of 0.00025 mm. The $0.25 \times 0.25 \times 28$ mm³ sized channel was milled into the lower half of the upper part of the block while the lower part of the block housed the evaporator of 3 mm in diameter and 4.5 mm in height. The pressure of the chamber was kept at $3 \cdot 10^{-3}$ mmHg so that the mean free path of the ^{22}Na atoms should be longer than the length of the collimator channel.

The utilization coefficient of the material can be further increased if the collimation of the beam is further improved with the use of a slit made of parallelly arranged capillary tubes or a finely corrugated foil. Such types of collimators are discussed in detail under Paragraph 2.2.1.2.

A number of materials (e.g. potassium, rubidium, caesium, etc.) yielding atomic beams of experimental interest may cause an explosion if they react with water or air. For this reason, precautions must be taken if the evaporator of the source needs to be opened while it still contains some remainder of this type of material. In the case of sodium, the metallic sodium can be immersed — after cleaning — in a less volatile liquid before its introduction to the evaporator and the vacuum can be generated during the evaporation of this protective liquid. This method cannot be applied to metallic rubidium or caesium. These metals are first introduced by vacuum evaporation into small glass capsules which are subsequently closed and placed in the source evaporator filled with nitrogen or some other non-active gas. While the closing plug is being inserted into the evaporator, the glass capsules are crushed. An alternative method for the production of sodium or caesium atom beams is to use some given compounds of these metals. Thus, for example, a stable caesium atom beam can be obtained from a mixture of metallic sodium and caesium chloride heated to 200°C in the evaporator of the atom source (Lew [467]). The successful application of a potassium and caesium chloride mixture has been also reported [480]. It is possible to avoid the use of metallic alkali metals if one employs caesiumazide (CsN_3) (Vályi [481]; Császár, Rózsa [482]) which does not react with air or water at room temperature but decomposes at 390°C to become free caesium and nitrogen. Similarly, it is possible to obtain a Na atom beam by heating NaN_3 (sodiumazide) to 320°C. A sodium atom beam can also be generated by heating a mixture of sodium chloride and fresh calcium splinters (Ramsey [426]).

Alkali metal (sodium, potassium, etc.) atom beams are usually contaminated by sodium and other molecules to an extent which can be as high as 0.5 per cent. The contamination can be reduced by increasing the temperature. Moreover, the application of sodium chloride or sodium fluoride salts can lead to the so-called "polymerization", that is, to the presence of dimers of the form $NaCl_2$, in the beam [483–485].

2.4.1 ATOM SOURCES OF HIGH MELTING ELEMENTS

There is a very limited number of materials with high melting point which can be utilized in a thermal source for the production of atom beams. Such high melting materials are, as already mentioned, molybdenum, tantalum, tungsten, carbon and the oxides Al_2O_3, ThO_2, etc. (see Appendix, Tables 13 and 14). Sources prepared from molybdenum metal or graphite-lined molybdenum heated by tungsten wire can be employed from 1500° to 1600°K [462, 465]. Higher temperatures, particularly values above 2300°K, present serious difficulties. In this case the available choice of source material and type of heating are equally restricted. Temperatures above 2300°K can be obtained only by direct heating or by electron bombardment.

Figure 2.17 shows a thermal atom source with direct heating. The source chamber can be heated up to 2400°K with the high current intensities flowing through the chamber wall. The thin-walled sleeve, serving as resistor, can be made of molybdenum, tantalum or graphite as required by the experiment. The electric connector to the thin-walled cylindrical source chamber is inserted into a copper block cooled by water. The cylindrical thermal shield is also water-cooled. The dimensions of the generally used graphite sleeve are 9 mm in outside diameter, 6.5 mm in inside diameter, 80 mm in total length and to reach a temperature of 2300°K a current of 400 A is needed at a voltage of 11.5 V (Lew [466]).

In Fig. 2.18 an atom source arrangement with graphite chamber can be seen which has been used for the production of an aluminium atom beam (Lew [466]). Here a crucible prepared from Al_2O_3 destined to contain the liquid aluminium is placed in a thin-walled graphite cylinder. This arrangement is necessary because molybdenum, tungsten and tantalum are soluble in liquid aluminium, while graphite forms tungsten carbide crystals with tungsten (Lew [467]). The Al_2O_3 aluminium oxide crucible is isolated from the graphite by a tantalum foil. Aluminium wire to be evaporated is placed in the crucible and the evaporation temperature is obtained by the a.c. current flowing through the graphite sleeve. Owing to the water cooling of the thermal shield surrounding the evaporator chamber, the required temperature of 1670°K can be obtained with a power consumption of 800 W. An amount of 0.17 g of aluminium in the crucible is sufficient for a run of ∼6 hours.

Sources of similar arrangement to those shown in Figs 2.17 and 2.18 were used to produce silver and gold atom beams (Wessel, Lew [487]). In these cases a tantalum crucible is employed in the graphite chamber.

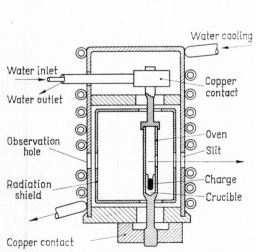

Fig. 2.17 Lew's version of a thermal atom source with direct heating and of high temperature

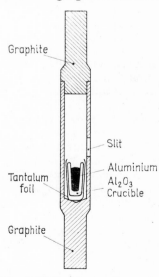

Fig. 2.18 Lew's version of a directly heated high temperature evaporator chamber made of graphite

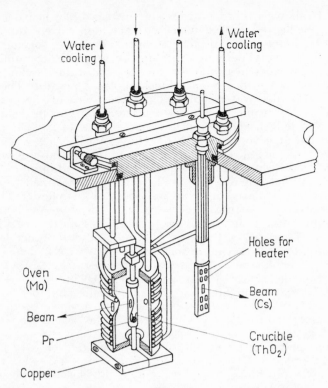

Fig. 2.19 An indirectly heated low temperature and a directly heated high temperature thermal atom source mounted on common base plate [467]

No insert crucible is needed for the production of copper and boron atom beams (Ting, Lew [488]; Lew, Title [489]). For the production of yttrium and scandium beams, a tantalum crucible was placed in the chamber made of tantalum (Friche, Kopferman, Penselin [490, 491]). If the sources are directly heated, it is important to ensure good electrical contact of the connections, i.e. to keep the contact resistance to a minimum.

The maximum temperature obtainable in these sources is determined by the value of the thermal gradient in the thermal shielding system.

Another variant of the directly heated sources is shown in Fig. 2.19. It is placed together with another type of atom source on the same rotatable base plate. The arrangement permits alternative measurements to be performed with two different types of beams without opening the vacuum system. The directly heated source shown in the figure, was used for the generation of praseodymium and caesium atom beams (Lew [467]). The source chamber for praseodymium evaporation is a thin-walled molybdenum cylinder with a slit on the cylinder wall located parallel to its axis. The cylinder is made of a molybdenum rod, 9.5 mm in diameter, formed into a chamber of 6.3 mm inside diameter and 8 mm outside diameter. The slit was 0.1 mm in width and 3 mm in height. The lower end of the cylinder was inserted into the thick copper connectors of the electrical heating in contact with the water-cooled cylindrical thermal shield of the chamber. The open upper end of the molybdenum sleeve is connected to a copper rod which closes the open end of the evaporator and serves as the electrical connection. This copper rod passes through the opening of the thermal shield and is connected in this way to an electrically insulated, water-cooled copper block. The praseodymium to be evaporated is placed into a ThO_2 crucible to prevent direct contact between praseodymium and molybdenum. This arrangement substantially reduces the obstruction probability of the slit. The evaporator can be heated to 2000°K with a consumption of 240 A at a voltage of 10 V.

An alternative method of heating is applied in the source of cobalt atom beams to be seen in Fig. 2.20 (Ehrenstein [486]). The electron bombardment

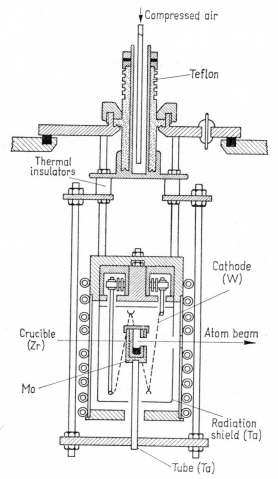

Fig. 2.20 Ehrenstein's version of a high temperature atom source heated by electron bombardment

applied here can raise the chamber temperature to 2300°K if a tantalum crucible is employed. The electron beam is led to the crucible at a positive voltage used as anode through a tungsten wire mounted in a zig-zag.

The energy of the electrons is 1000 eV and the current intensity is 0.8 A. The evaporator can be heated to 2300°K with a power consumption of 800 W. The stabilization of the electron current can also stabilize the chamber temperature. If an electron beam is used for heating, the atoms will always be ionized to a small extent. The thus produced ions strike the cathode and induce a sputtering of the cathode material and thus reduce the lifetime of the cathode. A small fraction of the ions can escape through the slit along with the atoms. These ions can, however, be deflected from the atom beam with a weak applied magnetic field. This method can be used also if the ion sources of high melting elements of the types described in Paragraph 3.9 have to be converted for some reason to a thermal beam source by some slight modification of the construction.

2.5 MOLECULAR DISSOCIATION

Atom beams of gases which are in a molecular state under normal conditions and atom beams of some elements can be produced by dissociation technique. As already mentioned (see Paragraph 1.6), molecular dissociation can be induced by imparting a given amount of energy to the molecules. This amount of energy is determined partly by the species of the molecule, partly by the internal energy states of the given molecule.

Different methods are available for imparting this energy to the molecules. Molecular dissociation is generally induced by the use of dissociators in which thermal, low frequency, radiofrequency or arc-discharge energy transfer takes place.

2.5.1 THERMAL DISSOCIATORS

Thermal dissociation is one of the methods by which molecules can be dissociated. The thermal dissociator is a source chamber made of tungsten [419–421, 493, 495, 1336]. The chamber is heated to a given temperature at which the molecular gas of a required pressure passing through the chamber dissociates and leaves the dissociator preponderantly in the form of atomic gas. The temperature dependence of molecular dissociation has been discussed in detail in Paragraph 1.6.

The temperature and pressure dependence of hydrogen dissociation, for which the required dissociation energy is 4.476 eV, is formulated by eq. (1.103) while the temperature curves for the degree of dissociation at different pressures are shown in Fig. 1.70.

The variations of the degree of hydrogen dissociation with temperature and pressure have been investigated in measurements at 1 mmHg pressure at different temperatures (Hendrie [421]) and in measurements made at pressures from $6 \cdot 10^{-1}$ to 200 mmHg (Vályi [420]). In both cases the measured values were found to be in good agreement with the predictions

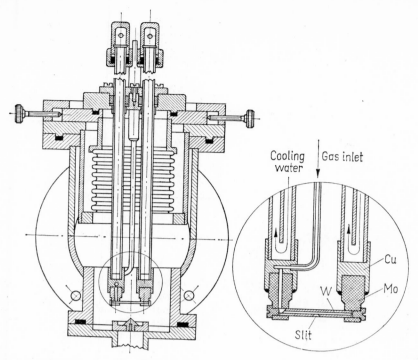

Fig. 2.21 Vályi's version of a dissociator with directly heated tungsten chamber

from eq. (1.103). In the above and in some other reported measurements (Lamb, Retherford [419, 493]; Vályi [495]) the thermal dissociator arrangement shown in Fig. 2.21 was applied. The dissociator chamber for heating the gas to the required temperature was made of tungsten in the measurements reported by [419, 493]. It was a 0.1 mm thick plate formed to a tube of ∼1.7 mm inside diameter with a longitudinal slit, 0.2×1.5 mm² in size, running parallel to the tube axis. In the reported experiments (Vályi [420, 495]) the dissociator chamber was also a tungsten tube formed from a tungsten rod by spark machining. The extracting slit was a 0.3 mm diameter aperture (Fig. 2.21). In both cases the tungsten chamber was fixed in a molybdenum support soldered to a water-cooled copper block. This arrangement also served as electrical connection. To heat the tungsten chamber to temperatures from 2500° to 2800°K, voltages from 1.2 to 2 V and currents between 80 and 180 A are needed. The degree of dissociation $X = 0.65$–0.9 at a pressure of $6 \cdot 10^{-1}$ mmHg. At a temperature of 3000°K the hydrogen can be dissociated to 98 per cent at a pressure of $6 \cdot 10^{-1}$ mmHg.

Tungsten chambers can also be heated by electron bombardment. This method is used in the dissociator arrangement shown in Fig. 2.22 (Heberle, Reich, Kusch [494]). The region of the tungsten chamber in the vicinity of the extracting slit is bombarded by an electron beam of 2000 eV and 10 mA

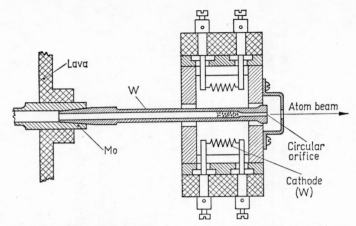

Fig. 2.22 Heberle–Reich–Kusch's version of a dissociator with electron
bombardment-heated tungsten chamber

which raises the chamber temperature to the required 2500–2600°K. The
tungsten chamber is at earth potential and has a polished connection to
a molybdenum sleeve. The packing of thin crumpled tungsten wires placed
into the chamber before the circular outlet slit improves the heat transfer
from the chamber wall to the gas flowing through the chamber.

The dissociator to be seen in Fig. 2.23 shows a similar arrangement for
the dissociation of bismuth molecules (Leu [496]). The chamber is heated
with the electron current from a tungsten spiral. This type of dissociator
was used by other workers (Lindgen, Johannsson [497]) for the production
of a radioactive bismuth isotope atom beam by use of a molybdenum crucible. In this case the nozzle temperature can be from 1300° to 1500°C and
the chamber temperature from 600° to 700°C as the dissociation energy of
the bismuth molecule is only 1.7 eV.

Radioactive iodine molecules have been also dissociated using a similar
arrangement. In this case the chamber was made of gold and since the dissociation energy of these molecules is even lower — not more than 1.54 eV,
~100 per cent dissociation could be achieved at 650°C. The gold chamber
was chosen in order to reduce the corrosion occurring at high temperatures.

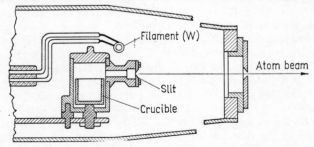

Fig. 2.23 Leu's version of a dissociator with tungsten chamber heated by electron
bombardment

Thermal dissociators are suitable for the production of beams containing atoms in the pure ground state. They are therefore very useful for the study of atoms in excited states since in this case there is no interference from the ultraviolet radiation emitted in the transition of short-lived excited states to the ground state nor from excited metastable states. The investigations can thus be performed without these disturbing effects.

Thermal dissociators are not suitable for the production of such high intensity atom beams as obtainable with high frequency dissociators. This is due to the fact that in thermal dissociators the higher degree of dissociation is associated with a lower pressure, thus, the high temperature does not permit multi-collimators to be used as an extracting slit of the dissociator.

2.5.2 LOW FREQUENCY DISSOCIATORS (WOOD TUBE)

The dissociation of molecular gases in a gas discharge at low frequencies was first observed by Wood [499, 500]. This observation led to the idea of utilizing the tubes in which low frequency gas discharge takes place as a source of atom beams. The so-called Wood tubes have been used by many experimentalists (Clausnitzer [501]; Sherwood [504]; Kellogg, Rabi, Zacharias [505]; Clausnitzer, Fleischmann, Schopper [514]) as a tool for the dissociation of molecular gas (hydrogen, oxygen, nitrogen and the gaseous compounds of halogen atoms). The usually applied variant of the Wood tube type atom source is shown in Fig. 2.24. The gas discharge is initiated by applying a voltage of 10 to 20 kV at 50 to 60 Hz frequency to the alu-

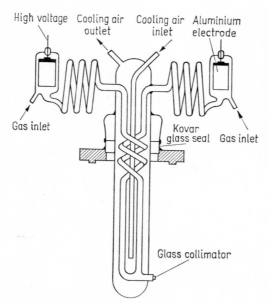

Fig. 2.24 Low and high frequency dissociator with internal electrode (Wood tube)

minium electrodes in the broadened part at the end of the glass tube of
2 to 5 m length. The discharge current varies from 50 to 200 mA. The
pressure of 0.5 to 1 mmHg needed for maintaining the discharge can be set
by a control valve located between the discharge tube and the gas container.
The gas inlet is close to the electrodes. The outlet slit, which can be a longitudinal or circular hole or a multi-collimator is at about the middle of the
discharge tube (see Paragraph 2.2.1). The middle section of the discharge
tube can be cooled by different types of coolant (air, water, liquid, nitrogen,
etc.) depending on the atomic velocities needed for the experiments. With
the use of hydrogen a gas discharge current of 50 mA can be obtained with
an applied voltage of 10 kV, at a pressure of 1 mmHg. At a sufficiently
high degree of dissociation the discharge is ruby coloured; if the colour turns
bluish this means that the degree of dissociation decreases because of a
higher rate of recombination due to contamination. With this type of dissociator a dissociation degree $X = 0.8$ to 0.9 can be achieved, that is, the
beam extracted from the dissociator contains from 80–90 per cent atomic
hydrogen. The main problem arising in the application of Wood tube type
dissociators is the contamination of the gas from the sputtering of electrodes, oil diffusion pumps and rubber sealing. The presence of these impurities enhances recombination and diminishes thereby the intensity of the
atomic component of the beam. This effect can be reduced if the electrodes
are cooled [494] and placed at an appropriate distance from the outlet slit
and if the oil diffusion pumps and the rubber sealing of the vacuum system
are replaced by ion getter pumps and metal sealings [433, 445, 495] while
the discharge tube is lined with phosphoric acid or potassium borate.

2.5.3 HIGH FREQUENCY DISSOCIATORS

Molecules can be dissociated in high frequency discharges of various types
at appropriate values of gas pressure, frequency and power.

It has been shown (Prodeil, Kusch [506]; Gravin, Geren, Lipworth [507])
that the Wood tube can be applied as a dissociator also if the applied field
is not of the low but of the high frequency type. A dissociator to which
a high frequency field is applied is suitable also for the production of the
high intensity (10^{16} to 10^{17} atoms/sec) atom beams needed for proton or
deuteron beams polarized with respect to nuclear spin (Adyasevits, Antonenko, Polunin, Fomenko [442]). In the high frequency dissociator employed
to generate spin polarized beams a multi-collimator is used as the outlet
slit to obtain a well collimated beam and to decrease the amount of gas
to be pumped off.

High frequency discharge with alternating electric or magnetic field and
external electrodes (Thomson [508]) presents a number of advantages over
the Wood tube with internal electrodes. The physical properties of the former
will be discussed in more detail in Chapter III.

With the use of external electrodes the lifetime of the dissociator increases
since in contrast with the Wood tube sputtering of the electrodes does not
contaminate gas. A further advantage is the small size of the discharge tube
and the simple construction, particularly in the case of discharge tubes
operating in alternating magnetic field.

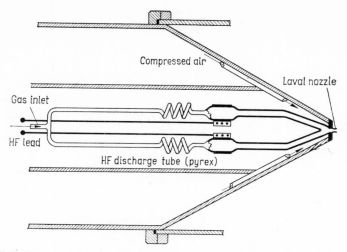

Fig. 2.25 High frequency dissociator with external electrode, capacitive coupling
and output slit equipped with Laval nozzle [512]

A high frequency discharge dissociator set up in which an alternating
electric field is applied can be seen in Fig. 2.25 (Thirion, Beurtey, Papineau
[512]). Dissociators of similar construction were also used in other reported
experiments [502, 509, 549, 550]. The discharge tube is made of pyrex glass,
10 mm in diameter, and of U shape with both branches 25 mm long. The
gas is introduced through a glass tube, 5 mm in diameter, coupled to the
broadened end of the U-shaped discharge tube. The two electrodes mounted
on the end of the discharge tube provide for capacitive coupling. The
discharge is generated by means of a 20 MHz frequency voltage generator
with a power output of 1 kW. The beam is obtained from the dissociator
through a Laval nozzle (see Paragraph 2.2.2). The cross section of the
atomic beam is either circular (Clausnitzer [502]; Thirion, Beurtey, Papineau
[512]) or ring form (Keller, Dick, Fedecaro [509]). The discharge tube is
cooled with a compressed air current. The maximum possible dissociation
degree $X = 0.95$.

Another variant of the high frequency dissociator with alternating electric
field, which has been also used for the production of nuclear spin polarized
beams, is shown in Fig. 2.26 (Vályi [433, 445, 495]). Here the discharge
tube is 8 mm in diameter and the section of the U-shaped tube between the
two electrodes is 50 cm long. The high frequency discharge is generated
by means of an oscillator of variable frequency and output power. A frequency of 23 MHz and a power of 360 W are applied to the dissociator. The
U-shaped part of the assembly has a forced air cooling with compressed air.
The dissociator geometry can be set by a control system using sylphon
coupling. A multi-collimator is employed at the outlet slit of the discharge
tube. Relative intensity measurements made with capillaries of different
lengths ($l = 4 \times 10^{-2}$; 6×10^{-2}; 1×10^{-1}; 2×10^{-1} cm) and the same diameter $d = 8 \cdot 10^{-3}$ cm, forming a multi-collimator covering the entire
aperture of 4 mm diameter, showed that the optimum pressure for hydrogen

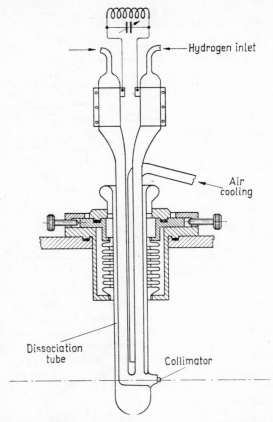

Fig. 2.26 High frequency dissociator with external electrode, capacitive coupling and multi-collimator output slit [433, 445, 495]

lies in the range of 0.2 to 0.4 mmHg (see Fig. 2.27). The atomic beam current obtainable with this dissociator is $2 \cdot 10^{16}$ atoms/sec in a solid angle of 6° (Vályi [433]). The vacuum system equipped with titanium cathode ion getter pumps can be operated for 80–100 hours without appreciable change in atom beam intensity.

The high frequency dissociator with alternating magnetic field is of the most simple design and of the smallest size among the dissociators. This type of dissociator has been used by a number of experimentalists [434, 441, 444, 503, 510, 511, 513, 544, 550–552] to generate the high intensity beams utilized as polarized beam sources. The dissociator in question is shown in Fig. 2.28. The high frequency discharge is generated by inductive coupling from a voltage generator of 20 MHz frequency and 400 W power output [434, 510]. The atomic beam current from this dissociator can be as high as 10^{16} atoms/sec. Usually, multi-collimators are applied as outlet slit and the useful life of the apparatus amounts to 100–150 hours. The discharge tube is made of quartz or ceramic and cooled by a ventilator. The optimum

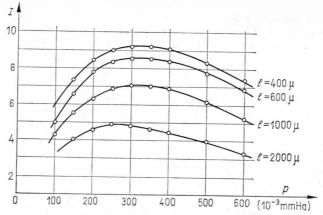

Fig. 2.27 Beam current as a function of the pressure in the dissociator of Fig. 2.26 for atom beams emitted through a multi-collimator consisting of glass tubes of different lengths l and of $d = 8 \cdot 10^{-3}$ cm in diameter and filling the outlet diaphragm, 4 mm in diameter [433, 495]

pressure of the discharge tube varies between 0.2 and 0.4 mmHg. The degree of dissociation can be as high as $X = 0.8$ at an appropriate pressure and power, provided the discharge tube is well kept (washed with a 3 per cent phosphoric acid solution and rinsed with potassium tetraborate or metaphosphorous solution).

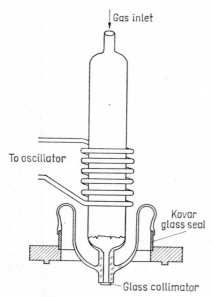

Fig. 2.28 High frequency dissociator with external electrode, inductive coupling and multi-collimator output aperture

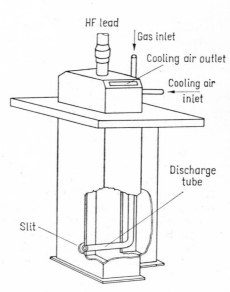

Fig. 2.29 Dissociator with microwave-generated discharge [425]

Microwaves can also induce a discharge capable of dissociating molecules. Microwave-operated dissociators have been used for the production of hydrogen and halogen atom beams [479, 518, 520, 521, 1421]. A source of this type can be seen in Fig. 2.29. The L-shaped tube is mounted into the microwave cavity resonator where the amplitude of the standing wave is at a maximum. The discharge is maintained by an input power of 50 W at a frequency of 300 MHz to the cavity resonator. With this input current the optimum dissociation was found to be 90 per cent in the case of chlorides (King, Zacharias [425]).

The external surface of the glass tube is cooled by forced air across the cavity resonator. The gas consumption rate of the dissociator was measured with a Burdon tube as 1 cm³/min at normal pressure and room temperature. The discharge tube is 7 mm in diameter and made of 707-type glass. The outlet slit is 0.076 mm wide, 4.5 mm high and ~ 0.25 mm long. In microwave dissociators, similarly to those with external electrodes, there is no possible contact between the atomic gas and the metallic parts of the equipment. This factor allows one to obtain the appropriate atom to molecule ratio in the outflying beam. In the case of high frequency dissociators with external electrodes the atomic fraction is higher because with the microwave type an atomic fraction of only 60 per cent could be achieved.

Another type of microwave dissociator with coaxial arrangement has been also successfully used for atom beam production (Christener, Benewitz, Hamilton, Reyholds, Stroke [520]).

2.5.4 ARC DISCHARGE DISSOCIATORS

If molecular gas is led through an arc discharge the high temperature of the discharge causes the gas molecules to dissociate. This principle has been utilized in a dissociator built for the generation of high intensity atomic beams (Fleismann [515]; Clausnitzer [516]; Fleismann [518]). The drawing of this dissociator is to be seen in Fig. 2.30. The molecular gas is dissociated in the high temperature arc discharge occurring between a tungsten rod and the border of the hole in the middle of the tungsten disc shown in the figure. Both supports of the tungsten electrodes are water cooled. The supporting tungsten rod is movable to permit the spacing of the two electrodes to be set equal to the diameter, usually $d = 0.6$ mm, of the tungsten disc aperture. The discharge is generated with a current of 5 A at 500 V. The arc discharge takes place, depending on the aperture in the tungsten disc at pressures from 100 to 600 mmHg. At pressures below 10 mmHg the arc discharge turns to glow discharge. Because of the high temperature (3000° to 5000°K) of the arc, the maximum lifetime of the electrodes is not more than 50 hours. The degree of dissociation depends on the distance of the electrodes and the maximum obtainable dissociation was found to be 50 per cent. The beam current was measured as 10^{18} atoms/sec at 8 cm from the dissociator output.

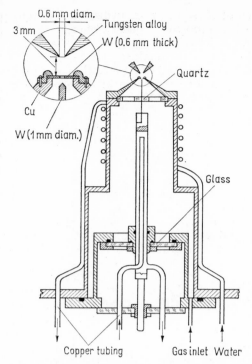

Fig. 2.30 Dissociator with arc-discharge and tungsten electrode [516]

2.6 FAST ATOM BEAM SOURCES

Besides atom or molecular beams of thermal energy ($E_a < 0.5$ eV) considered in the foregoing sections, the applications of these beams in the study of chemical processes, solid state, nuclear and plasma physics, require also the use of fast neutral particle beams with energies higher than the thermal range (0.5 eV to 10^6 eV).

Most problems are involved in the production of intense, well-collimated atom or molecular beams of small energy spread in the range of 0.5 to 10 eV. Neutral particle beams with energies in the range from 10 to 10^6 eV can be generated by applying the fairly well known charge transfer technique to ion beams accelerated to sufficiently high energies.

2.6.1 SOURCES OF ATOM OR MOLECULAR BEAMS WITH ENERGIES FROM 0.5 TO 20 eV

In recent years neutral particle beams with energies of a few electron volts have been extensively applied in chemical as well as solid state physical investigations and in technological processes. Several methods are available for the production of atom or molecular beams in this range of low energies. Each of these methods utilizes a different possibility, namely, gas dynamics (with a mixture of gases), charge transfer, sputtering, shock waves and laser radiation.

The gas dynamical method enables one to accelerate the heavy component of a gas mixture composed of light and heavy particles and to produce in this way atom or molecular beams of energies in the range from 0.5 to 20 eV [450, 451, 1412, 1425–1444].

This method is based on the fact that both components of the outgoing gas, which is a mixture of light and heavy atoms or molecules, will have about the same velocity if there is a sufficiently high number of collisions between the two types of components in the expansion range characteristic of the supersonic nozzle. Depending on the percentual fraction of the heavy component, the about identical velocity of the two components can be brought close to the mean velocity obtainable in the supersonic beam from a pure source of light particles [451, 1431, 1438, 1444].

If this method is used, the particle energy of the atom or molecular beam can be determined by the choice of the percentual fraction of the heavy component, the temperature T_0 in the nozzle chamber and the pressure ratio p_0/p, that is, the measure of the expansion.

Thus, e.g. Ar + H_2 gas mixture $M = (10–20)$, Ar = $(1.09–5.13)$ per cent, $T_0 = (300–1296)°K$, the argon atom energies can be varied in the range from 0.62 to 4.35 eV by setting the parameters to the appropriate values [1436].

Neutral particle beams with energies ranging from 5 to 20 eV can be produced from gas mixtures ignited by arc discharge, choosing the value of p_0/p to be sufficiently high and the heavy component concentration to be less than 10 per cent [1370, 1435–1438].

The light component of the gas mixture is usually H_1, H_2 or He.

The low concentration of the heavy component of gas mixtures does not mean that the percentual fraction of heavy particles will be similarly low in the beam crossing the skimmer. It has been experimentally shown [450, 1431, 1438, 1444] that heavy particles have a stronger tendency to focus on the beam axis than the lighter species, consequently, the heavy component concentration increases in the beam crossing the skimmer. It follows that gas dynamical technique lends itself to the production of atom or molecular beams with energies from 0.5 to 20 eV and enriched in heavy particle content.

Neutral particle beams in the energy range from 5 to 20 eV can also be generated by a special version of the extensively applied charge transfer method [1438, 1452]. This method (see Chapter IV) requires the generation of an ion beam from an ion source, then the acceleration of this beam to the required energy before letting the beam pass through the charge transfer target. Collisions in the charge transfer target transform the fraction of ions determined by the cross section $\sigma_{10}(E_i)$ (see Paragraph 1.4) to neutral atoms or molecules.

Intense atom or molecular beams with low energy spread can be produced by the above method as follows: The high intensity supersonic atom or molecular beam is led through a high frequency (20 MHz–200 MHz) ionizer and a fraction of the beam is ionized in this process. The ions leaving the ionizer are deflected by a magnetic field to be introduced into the accelerator tube and accelerated to the required energy. The accelerated ion beam is then neutralized by a supersonic atom or molecular beam crossing the path of the ion beam in a direction normal to its direction of propagation and acting as a charge transfer target. The non-neutralized ions remaining in the beam are deflected by a magnetic field so that the beam reaching the measuring chamber consists entirely of neutral particles.

Further methods for the production of atom or molecular beams in the energy range from 1 to 10 eV are known as the sputtering, the shock wave and the laser radiation methods.

The sputtering method utilizes the fact that the atoms of the material sputtering from monocrystals show an anisotropic angular distribution with most atoms flying in the direction of the closest packing if the crystal is bombarded with fast ions [1445–1449] and that the energy of the sputtered atoms can reach 10 eV [1447, 1448]. This method makes it possible to generate any metal atom beam up to intensities of 10^{17} atoms/cm^2 sec.

Intense neutral beam pulses with energies from 1 to 10 eV can be produced by use of a shock wave tube as the nozzle chamber of the source emitting the beam pulses. The atom or molecular beam is passed through the nozzle mounted at the end of the shock wave tube [1450, 1451] and the outgoing ions of the beam are deflected from the beam transmitted by the nozzle by a magnetic field so that the beam entering the measuring chamber consists of neutral particles.

Low intensity beam pulses with energies in the range of 1 to 10 eV can be generated by focusing a laser beam with an energy density, which can be from 10 to 100 Joule/cm^2, on a thin foil. Depending on the applied energy density, atom beam pulses with energies from 1 to 10 eV are released from the foil heated by the absorbed energy. The atomic species is determined by the material of the foil [1416].

2.6.2 SOURCES OF ATOM BEAMS WITH ENERGIES IN THE RANGE
10 TO 10^6 eV

Neutral particle beams in this higher range of energies are currently applied for various purposes in solid state, nuclear and plasma physics. In the former two fields and in plasma diagnostics mainly low beam currents ($I < 10^{-1}$ A equivalent currents) are used while for controlled thermonuclear fusion experiments high neutral beam currents ($I \sim 1$ to 200 A equivalent currents) are required to heat the plasma.

Neutral atom beams with energies from 10 to 10^6 eV can be generated, as already mentioned, by neutralizing ion beams obtainable from different types of ion sources. Neutralization takes place during the passage of the ions through the charge transfer target (see Paragraph 1.4 and [1363, 1364, 1378–1383, 1452–1458]). Practically, any type of the ion sources described in Chapter III can be utilized for the production of positive ion beams with the required energy. The positive ions are neutralized due to charge transfer when the beam is passed through the target which can be some gas, vapour or metal foil. The charge transfer takes place in a measure depending on the value of the cross section $\sigma_{10}(E)$ in accordance with the considerations in

Paragraph 1.4 and with the data shown in Figs 1.56–1.63. The positive ion beam passing through the charge transfer target is transformed to a beam composed of neutral, positive and negative charged particles. The neutral component of the beam can be as important as 70 to 90 per cent of the beam, for example H_1^+ ions having energies of a few keV are passed through gas or metal foil targets. Even with 30 keV H_1^+ ion energy the neutral component can reach 50 to 70 per cent: as to be seen in Figs 1.62 and 1.63. It has been found that the neutral component of the beam rapidly decreases with increasing ion energies so that for 200 keV H_1^+ ion energy the neutral fraction decreases to 1–9 per cent and for 1 MeV to 0.01–1 per cent. Consequently, intense fast atom beams are more efficiently produced from low than from high energy positive ion beams. Any type of the negative ion sources described in Chapter IV can be utilized for the production of fast atom beams either in the original arrangement or with some slight modification.

The high energy particle beams needed for the study of nuclear reactions are generated by the use of electrostatic accelerators with multiple charge exchange. Here the intense neutral beam is usually obtained from a positive ion source with a special gas target [522–524].

Figure 2.31 shows a neutral beam source (Rose [522]) consisting of a duoplasmatron and a gas target. The positive ion beam is focused by a unipotential lens on the input of the charge transfer channel. The extracting electrode as well as the charge transfer channel are cooled. The percentual neutralization achievable with different gas targets used for the production of neutral hydrogen atom beams can be seen from the diagram in Fig. 2.32 (Rose [522]). Maximum neutralization (87 per cent) is obtainable with a hydrogen molecule gas target and 10 keV ion energy. The charged particles are deflected by a magnet from the direction of the neutral beam.

Fast atom beams with currents equivalent to 0.1–200 A, which are suitable for controlled fusion experiments, are usually generated by leading the positive ion beam extracted from the source through a gas target consisting of the same gas as that in the source. In this target the positive ions are neutralized due to resonance charge transfer (see Paragraph 1.4) [1296, 1301, 1339, 1347–1349, 1363, 1364, 1378–1397, 1452–1458].

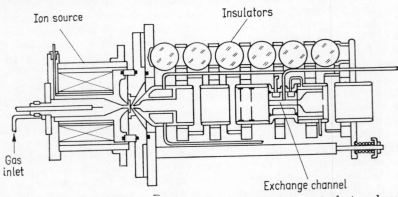

Ion source   Insulators

Gas inlet

Exchange channel

Fig. 2.31 Rose's version of ion source for production of fast neutral atom beams by charge transfer

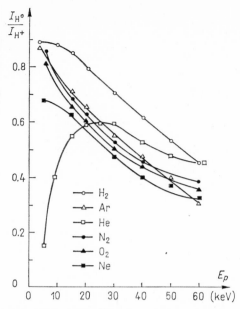

Fig. 2.32 Ratios of outgoing neutral atom beam current I_H° to the ingoing hydrogen ion current I_H^{+} of the charge transfer target for different gas targets (H_2, Ne, He, N_2, O_2 or Ar)

One of the source types which can generate such fast atom beams with ~ 1 A equivalent current is shown in Fig. 2.33 (Fumelli [1339]).

The ion source of the apparatus generating the fast atom beam is seen to be a combination of a duoplasmatron with a Penning-type discharge ion source, which is termed by the author a duopigatron, similarly to the assembly reported in [1296].

This ion source has two oxide cathodes prepared from oxide coated nickel ribbons, mounted in the intermediate electrode and shielded from the magnetic field. The gas pressure in the discharge chamber is 2×10^{-2} torr. The maximum arc-current of the discharge is 200 A for a voltage of 130 V.

The extracting system of the ion source is an assembly of three multi-aperture electrodes. The electrodes are copper discs, 90 mm in diameter and 2 mm thick, each with 199 bores having a radius of 2 mm. The transparency of the multi-aperture is 39 per cent. The distance from the first electrode to the intermediate, accelerating electrode is 4.5 mm while that of the latter to the decelerating electrode is 2 mm. The first electrode mounted at the plasma boundary is grounded through a resistor of 20 Ohm. The accelerating voltage $-V_a$ is variable from 8 to 20 kV. The decelerating electrode is coupled to the neutralizer tube and connected through a variable resistor to the high voltage supply. Thus, for an input ion current I_n to the neutralizer tube, the decelerating electrode potential is more positive by $V_n = RI_n$ than the accelerating electrode potential V_a. Consequently the ion beam enters the neutralizer with energy $E_i = e(V_a - V_n)$. The neutralizer is a 400 mm long copper tube, 100 mm in diameter, and it also

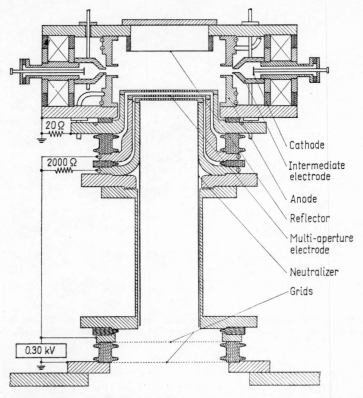

Fig. 2.33 Schematic drawing of fast atom beam source with 1 A equivalent current
output [Fumelli 1339]

serves as a beam collimator. The two grids prepared from 0.3 mm diameter
molybdenum filament are mounted at the neutralizer output aperture to
retain the charged particles. The transparency of the grids is 93 per cent.
The distance between them is 25 mm.

The fast neutral hydrogen atom beam output from the neutralizer has
been measured to have 18 keV energy, 1 A equivalent current and an angular
divergence of 1°.

Much higher, (10–500) A equivalent neutral beam currents can be obtained
from fast atom sources with a heated cathode arc-discharge ion source and
resonance charge transfer on the gas of the source [1347, 1348, 1386–1389,
1394, 1456, 1457]. A specimen of this type of source is shown in Fig. 2.34a
(Ehlers, Baker, Berkner *et al.* [1456]).

This fast atom beam source operates with a heated cathode, pulsed arc-
discharge ion source. The cathode consists of 20 circularly arranged U-shaped
heated tungsten wires of 0.5 mm in diameter which give a total emitter
surface of ~34 cm². The parallel connected hot cathodes are heated with
a current of ~500 A. Under these conditions the operational temperature
of the cathodes is ~3200°K during the pulse.

The anode is a copper ring surrounding the cathodes. This ring forms at the same time a part of the ion source container. The arc-discharge current I_a can be as high as 1000 A for values of the arc-voltage V_a between 25–70 V.

The plasma density distribution of the ion source during arc-discharge is — as to be seen in Fig. 2.34b — nearly homogeneous over the total area of the cross section formed by the cathodes within the ion source if the param

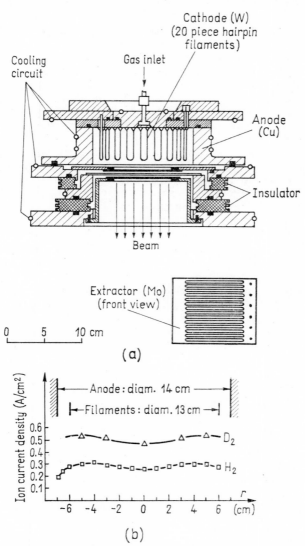

Fig. 2.34 (a) Schematic drawing of a fast atom beam source with large emitter surface arc-discharge ion source
(b) Plasma density distribution in the source [Ehlers 1456]

$(I_{a_{D_2}} = 1000$ A, $V_{a_{D_2}} = 48$ V; $I_{a_{H_2}} = 660$ A, $V_{a_{H_2}} = 40$ V$)$

eters are $I_a = 1000$ A, $V_a = 48$ V for D_2 gas, and $I_a = 660$ A, $V_a = 40$ V for H_2 gas.

The extracting system of the ion source consists of three multi-aperture electrodes made of copper (Fig. 2.34a). The upper electrode at the plasma has a multi-aperture composed of 21 slits which are 70 mm long and 2 mm wide. The distance between the slits is 1.3 mm, thus the transparency of the multi-aperture is $\sim$65 per cent. The apertures on the intermediate electrode are 1.6 mm, those on the lower electrode 1.25 mm wide. This distance from the first to the intermediate electrode is 2.2 mm and from the latter to the lower electrode 1.8 mm.

During the 30 msec period of the beam pulse the upper multi-aperture of the extracting system is at the anode potential $U_a = (10–20)$ kV, the intermediate electrode at $U = -(1–2)$ kV, and the lower at $U = 0$.

The ion current passing through the multi-aperture is 0.5 to 1 A/cm², the angular divergence of the beam is 2 to 3.5 degrees. The input gas of the ion source is also pulsed. Gas consumption of the source has been measured as 1.4 ml/sec for D^+ current of 15 A.

The ions leaving the ion source are passed through a charge transfer tube containing the same gas as the source where the ions are neutralized by electron capture in a measure determined by the value of the cross section $\sigma_{10}(E)$. The percentual neutralization values can be as high as 70 to 85 per cent for H_2 or D_2 gas target with H^+ or D^+ ion energies of 10 to 20 keV (Fig. 2.32).

A neutral beam of 20 keV energy with a 15 A equivalent current can be obtained with the source shown in Fig. 2.34a. With a heated cathode, arc-discharge ion source of this type or with one of its modified versions having a suitable large emitter surface and a large multi-aperture area, even 10^2 A equivalent beam currents can be generated [1394, 1457].

CHAPTER III

ION SOURCES

3.1 INTRODUCTION

Ion rays were first observed by Goldstein [553] while studying gas discharge at low gas pressure. Wien investigated the properties of the ion rays and concluded that the ion beam originated from the atoms of the gas in the discharge tube [554]. An ion source was built by Thomson [555, 556] using a long tubular electrode as cathode in order to produce the ion beam which permitted the isotopes ^{20}Ne and ^{22}Ne to be identified and the nature of the beam to be investigated. The so-shaped cathodes also serve for the collimation of the beam and for the production of differential pressures.

The ion beams from the Thomson-type sources operated at high applied voltage [556–567] showed a large spread of ion energy. The fraction of atomic ions is low and the stability of the gas discharge cannot be continuously maintained. These and other limitations considerably restricted the application field of this type of ion sources.

Attempts were started therefore, to develop new types of sources with less spread of ion energy, low power consumption, and suitable for the ionization of atoms in gases or solids.

This work was stimulated to a great extent by the start of mass spectrometric and nuclear physics research. Corresponding to the different objectives of research, ion sources with different properties have been worked out and investigated over the last few decades.

For mass spectrometric and nuclear physics experiments which require small ion currents, ion sources utilizing atom–electron collisions [111, 568–583, 1277–1280], thermal sources utilizing surface ionization [216, 584–600, 1284–1287] and sources with surface ionization by impact [220, 601–611, 1281–1283] which have low current intensities and a small spread of energy have been developed.

The aim of the development of ion sources utilizing gas discharge was the generation of high ion concentrations by the use of relatively low applied voltage. A potential source of this type seemed to be arc-discharge with a high intensity current at low voltage.

This principle has been applied in the different versions of the arc- and capillary arc-discharge ion sources from a hot cathode [612–626].

Another possibility was suggested by the observation that the ion concentration in the plasma of the gas discharge can be increased if the path of the ionizing electrons in the discharge can be lengthened. The ion concentration is increased by this method in the ion sources with hot or cold cathode in which the gas is being ionized by electrons oscillating in an applied homogeneous or inhomogeneous magnetic field [627–681, 805, 939–952, 1301–1315] and in the ion sources utilizing high frequency discharge in a magnetic field [682–740, 1298–1300].

The ion sources grouped into the different categories briefly specified above differ in physical properties and operational parameters. This always allows one to find the most appropriate ion source for a given purpose.

Ion sources are employed mostly in particle accelerators, mass separators, mass spectrometers and ion implantation equipment. These apparatuses are nowadays indispensable tools in scientific research, therapy, technical analyses and in a number of industrial fields.

3.2 ELECTRON IMPACT ION SOURCE

In the ion sources utilizing bombardment by electrons the atoms or molecules of the analysed element are ionized by inelastic collision with electrons. The number of atoms or molecules ionized in the source by electron impact depends on the following parameters: the ionization cross section σ_i, the current density I_e of the bombarding electron beam, the number of atoms or molecules to be ionized per unit volume n_a and the effective volume V of the ionization.

Thus, if all ions produced in the effective volume of ionization are able to leave the source, the ion current emitted from the source is given by

$$I_i = I_e V n_a \sigma_i \qquad (3.1)$$

The ionization cross section value can be determined as described in Paragraph 1.2.1. The electron energy is best chosen to be equal to the impact energy associated with the maximum value of the ionization cross section of the atom or molecule to be ionized. The number of atoms or molecules per unit volume can be calculated from the pressure and temperature of the gas or vapour in the ionization volume using the formula

$$n_a = \frac{p}{kT} \qquad (3.2)$$

In eq. (3.2) the pressure p is given in atmospheric units and the temperature T in °K units. The effective ionization volume is a part of the source in which the atoms or molecules are actually ionized. The impact of the electron beams or molecules can be brought about in several ways.

A possible method is to lead the ionizing electron beam through the ion source chamber in which the analysed material is being kept in the form of gas or vapour at a pressure at which ionization takes place (Fig. 3.1a). An alternative procedure is to let the electron beam collide with the beam of atoms or molecules. The collision may take place between the beams oriented in either normal or parallel directions with respect to each other (Fig. 3.1b, c).

The effective volume of ionization can be evaluated according to Figs 3.1a, b and c, as

$$V_{\text{eff}} = r^2 \pi l$$

or

$$V_{\text{eff}} = dhl$$

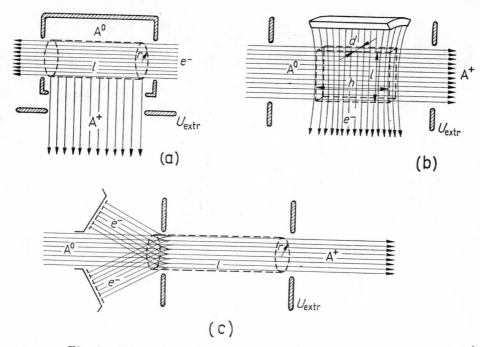

Fig. 3.1 Schematic drawing of ionizer with electron impact

The ions can be extracted from the ionization volume with an electrode at a potential U_{extr} of opposite polarity to that of the charge of the ions. The ions can be extracted in a direction either normal (Fig. 3.1a) to that of the atom and electron beams or parallel (Fig. 3.1b, c) to the direction of the atom beam. The beam extracted from the ion source can then be shaped according to the experimental aim by means of electric or magnetic lenses.

The ion sources originating from electron impact are generally suitable for experiments requiring a low intensity ion beam with a small energy spread. Ion sources of this type are preferably employed in mass spectrometric analyses [111, 568–583]. This method is also useful for the generation of ions destined to ionize nuclear spin polarized atom beams [422–445, 495, 502–518, 544–552].

Figure 3.2 shows an ion source arrangement of the electron impact type, suitable for the generation of ion beams from elements in the gas phase.

Here the atoms or molecules of the gas introduced into the ionization chamber of the source are ionized by inelastic collisions with the $\sim$100 eV energy electrons passed through the chamber. The ions thus formed are extracted from the ionizer by the $\sim$2 kV voltage applied to the extracting electrode and reach the ion beam analyser by passing through a focusing system. A pair of deflector plates placed between the two limiting slits permits the ion beam to be adjusted or interrupted without switching off the ion source.

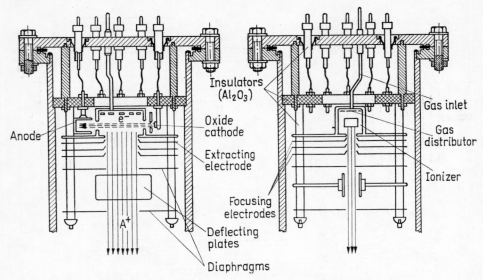

Insulators
(Al$_2$O$_3$)

Gas inlet

Gas
distributor

Anode

Oxide
cathode

Extracting
electrode

Ionizer

Focusing
electrodes

A$^+$

Deflecting
plates

Diaphragms

Fig. 3.2 Electron impact ion source for production of gas ions

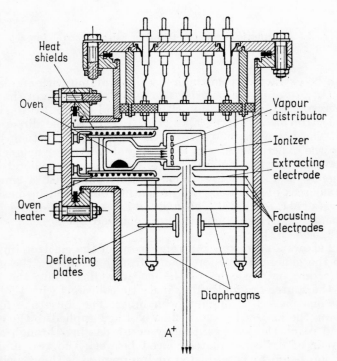

Heat
shields

Oven

Vapour
distributor

Ionizer

Extracting
electrode

Focusing
electrodes

Oven
heater

Deflecting
plates

Diaphragms

A$^+$

Fig. 3.3 Electron impact ions source for production of ions of solid elements

The ionizing electrons are emitted from an oxide cathode. The electrons are accelerated and the beam is shaped by use of the electrode mounted in front of the cathode. The electron energy is 100 eV. The electrode of the ion source containing the last limiting slit is at earth potential, thus the value of the voltage applied to the ionizer is $\sim +2$ kV. The electric lead and the gas inlet insulation are made of Al_2O_3, the connections are soldered by metal-ceramic technique. The base plate and the housing of the ion source are made of stainless steel welded under argon gas. The vacuum gasket is prepared from high purity copper.

If ion beams have to be obtained from solids, the arrangement of the ion source has to be modified in the manner shown in Fig. 3.3.

In this case the solid element has to be vaporized in an evaporator and introduced into the ionizer by a tube heated to and kept at a given temperature. The heater is a tantalum wire wound on a quartz tube surrounding the evaporator and the inlet tube. The quartz tube serves at the same time for the insulation of the evaporator from the heating wire as the chamber with the ionizer are at a potential of $\sim +2$ kV relative to the housing of the ion source.

Depending on the ionized elements, the ion current extractable from the ion source varies from 1 to 30 μA.

The ions obtained from the beam of atoms or molecules led through the ionizer are generally utilized for the production of nuclear spin polarized ion beams.

It can be seen from Fig. 3.1b and c how the collision leading to ionization can take place between the electron and atom beams flying in parallel directions or in directions normal to each other.

The specified parameter of the ionizer is its efficiency which can be defined as

$$\eta = \frac{\text{ions/sec}}{\text{atoms/sec}}$$

Using the ion current formula (3.1) and the expression $N_a = n_a \bar{v} A$ for the number of atoms crossing the ionizer per second (where $\bar{v}$ is the mean velocity of the atoms and A is the cross section of the atom beam), the ionizer efficiency can be written in the form

$$\eta = \frac{I_e V_{eff} n_a \sigma_i}{n_a \bar{v} A} = \frac{I_e V_{eff} \sigma_i}{\bar{v} A} \tag{3.3}$$

In the design of the ionizer the main objective is always to obtain the maximum possible efficiency. It follows from formula (3.3) that the efficiency increases if the electron density and the effective volume are increased or if the mean velocity of the atoms in the beam is decreased.

However, the increase or decrease in these values is limited.

The increase in electron beam density is limited by the space charge, that of the effective volume by the geometrical dimensions required for confining the extracted beam, while the mean velocity cannot be slowed down below the velocity of the thermal motion of atoms.

The ionizer efficiency can become as high as $\eta = (1-3)10^{-3}$ [444, 502, 743–746] if a multicathode or Pierce-type cathode [741, 742] is applied.

Figure 3.4 shows the schematic drawing of the multicathode ionizer (Clausnitzer [502]). The electron density at the atomic beam passing in the axis of the ionizer is ensured by the acceleration of the electrons emitted from 6 tungsten cathodes arranged concentrically around the axis due to the voltage applied to the grid.

The cylindrical screen, the extracting and focusing electrodes of the ionizer are made of antimagnetic stainless steel. The cathode and the grid are tungsten filaments of 0.4 mm diameter. The electrical connections are made of copper with quartz insulators. The effective ionizer volume is determined by the diameter of the atom beam ($2r = 9$ mm) and the length ($l = 100$ mm) of the electron emitter cathodes. The parallel connected cathodes are heated by a current from 60 to 70 A at 11 to 15 V. The grid current varies from 300 to 500 mA at a potential difference of $\sim$600 V between cathode and grid. The ions are taken out from the ionizer with an extracting electrode. The potential difference between the extracting electrode and the cathode is 30 V since it has been found that a potential gradient above this value does not change the value of the extracted ion current. After its passage across the limiting electrode the ion beam is shaped by an electrostatic quadrupole lens. The ionizer efficiency was measured as $\eta = (0.5–1.5)10^{-3}$.

The ionizer with Pierce-type oxide cathodes is shown with the electron and atom beams flying in directions normal to each other in Fig. 3.5 (Adyasevits, Antonenko, Polunin et al. [442]) and with parallel flying electron and atom beams in Fig. 3.6 (Slabospitskii, Kiselev, Karnaukhov et al. [444, 746, 747]).

In the arrangement shown in Fig. 3.5 the electrons are emitted from a curved oxide cathode surface area of 18×22 mm with a curvature radius

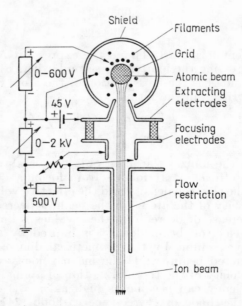

Fig. 3.4 Multi-cathode ionizer [502]

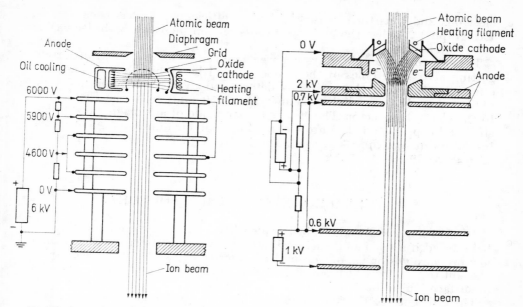

Fig. 3.5 Ionizer with oxide cathode of Pierce-type geometry [442]

Fig. 3.6 Ionizer with annular oxide cathode of Pierce-type geometry [444]

of 40 mm. The cathode surface is heated by the spiral placed behind it. The grid is at 2 mm from and parallel to the surface of the cathode. The potential difference of 550 V between cathode and grid is sufficient for the acceleration of the electrons to the required velocity.

The electron current emitted from the cathode is 400 mA. The electrons crossing the atom beam strike the oil-cooled anode. The ions formed in the ionizer are extracted and focused by a set of 6 electrodes to yield an ion beam with a diameter of 3 mm at 100 mm from the plane of the last electrode. The efficiency of this ionizer has been reported as $\eta = (0.5-1.0) \, 10^{-4}$ (Adyasevits, Antonenko, Polunin et al. [442]).

The ionizer shown in Fig. 3.6 also has a cathode of Pierce-type geometry. Here the effective volume of the ionizer is increased by the arrangement that the electron and atom beams are led in the same directions, parallel to each other (Slabospitskii, Kiselev, Karnaukhov et al. [444, 746, 747]).

The electrons are emitted from the oxide cathode ring, prepared from porous tungsten, shaped as can be seen in the figure. The cathode is 9.6 mm in outside diameter, while the internal hole is 5.5 mm in diameter, the solid angle of the emitter surface of the cathode is 130°. The heater spiral is located in the ring. The emitted electron current varies from 100 to 150 mA. The ionizer has two anodes, as is also apparent from the figure.

The potential difference between the first anode and the cathode is 2 kV. This relatively high potential gradient is necessary to obtain a reasonable electron current density (0.5 A/cm²). The potential difference between the second anode and the cathode is only 600 to 700 V in order to slow down the electrons and thus to increase the ionizer efficiency. (It has been seen in Paragraph 1.2.1 that the ionization cross section increases with decreasing collision energies.)

The electron optics of the system are chosen such that the electron current getting across the holes of the anodes should be as high as possible. The effective volume of the ionizer is determined by the diameter (5.5 mm) of the atom beam and the distance (1 = 150 mm) from the first anode to the electron collector. The electrodes are made of antimagnetic stainless steel and copper, and they are mounted on molybdenum rods. As electrical insulators quartz tubes are used. The ionizer efficiency has been reported as $\eta = 2 \cdot 10^{-3}$ [746, 747].

3.3 SURFACE IONIZATION ION SOURCES

It has been already shown in Paragraph 1.2.5 that surface ionization can be induced if a hot metal surface is struck either by atoms incident from a low pressure gas or vapour around the surface or by the atom beam of an element to be ionized. Either of these principles can be utilized in the design of surface ionization ion sources.

The efficiency of the ion sources of this type can be given in the absence of an external electric field by applying expression (1.44a) and formula

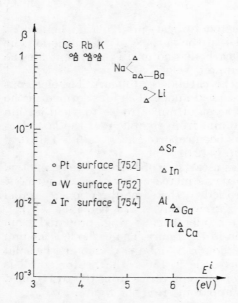

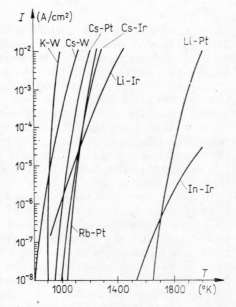

Fig. 3.7 Values of the surface ionization coefficient β at different energies of various metal atoms striking platinum, tungsten or iridium metal surface

Fig. 3.8 Ion current density I as a function of surface temperature in interval $T_c < T < T_m$. The first symbol on the curves stands for the striking atom, the second for the surface material

(1.45) defining the degree of ionization α_0^+, in the form

$$\beta = \frac{A}{A + \exp\left[\dfrac{z_i(E^i - \varphi)}{kT}\right]} \qquad (3.4)$$

This equation only holds, as has been already mentioned, for monocrystals with a homogeneous work function and incident particles of thermal energies. The efficiency for alkali metals with a surface having a work function satisfying the condition $E^i < \varphi$, can be as high as $\beta \approx 1$, however, there are elements for which we have $\beta \approx 5 \cdot 10^{-3}$ (Fig. 3.7).

Assuming a surface with uniform values of the work function, the ion current emitted from the surface can be calculated for an incident neutral atom beam of density n_0 and a surface temperature $T > T_m$ from the formula

$$I^+ = \frac{z_i n_0 A}{A + \exp\left[\dfrac{z_i(E^i - \varphi)}{kT}\right]} \qquad (3.5)$$

It is to be noted that for polycrystalline surface and external electric field formula (1.53) has to be used instead of (3.5).

It is obvious from (3.5) that with $E^i > \varphi$, the ion current density increases as the surface temperature increases and that the maximum current density $I^+ = z_i n_0$ is obtainable at $T = \infty$. Now, if $E^i < \varphi$, the ion current density decreases if the surface temperature continues to increase past a given value of $T_m < T$ (Fig. 1.48). The ion current density undergoes an abrupt change between zero and its maximum value in the surface temperature interval $T_c < T < T_m$ which is not predicted by the theoretical calculations (Figs 1.48 and 3.8).

It can be seen from expression (3.5) and Fig. 3.8 that at easily attainable surface temperatures practically speaking the highest ion current densities can be obtained if $E^i - \varphi < 0$. That is why it is the most reasonable to use for this type of ion sources surfaces with higher values of φ than that of the ionization potential E^i of the atom or molecule to be ionized.

Surfaces with a high value of the work function which can be used for surface ionization sources are the high melting metals, such as, Ta ($\varphi = 4.10$ eV; $t_m = 3273°$K), Mo ($\varphi = 4.24$ eV; $t_m = 2893°$K), Tc ($\varphi = 4.4$ eV; $t_m = 2413°$K), Ru ($\varphi = 4.6$ eV; $t_m = 2703°$K), W ($\varphi = 4.52$ eV; $t_m = 3653°$K), Os ($\varphi = 4.83$ eV; $t_m = 2973°$K), Re ($\varphi = 4.96$ eV; $t_m = 3440°$K), Ir ($\varphi = 5.27$ eV; $t_m = 2716°$K), Pt ($\varphi = 5.32$ eV; $t_m = 2042°$K), and oxides, e.g. NiO ($\varphi = 5.55$ eV) or metal surfaces coated with a molecular oxygen layer, like W + O_2 layer ($\varphi = 6.28$ eV), Pt + O_2 layer ($\varphi = 6.55$ eV).

Despite the fact that the oxide or oxygen coated surfaces have higher work functions than pure metal surfaces, the latter are preferably employed in practice. The reason for this is that the metal surfaces are easier to handle and that with a proper vacuum system they yield more stable ion currents than those obtained from oxidized or oxygen coated surfaces.

Surface ionization sources can be built so that the ionizer is incorporated in the evaporator of the element to be ionized. In this type of source the directly heated surface is a thin tungsten or tantalum filament [759–762] or ribbon [588, 589, 604]. A poorly evaporating compound of the element to be ionized is deposited on these filaments. If the ionizer is not directly heated the compound of the ionized element is deposited on a somewhat larger metal surface heated to the required temperature by a heater spiral or by electron bombardment.

A surface ionization source assembly with indirect heating of the ionizer is shown in Fig. 3.9 (Cornides, Roosz, Siegler [600]). In this source the lithium coating was deposited on the concave surface of the molybdenum crucible in the form of a suspension prepared from 50 to 60 μ glass powder containing 15.2 per cent Li_2O, 14.8 per cent Al_2O_3 and 70 per cent SiO_2 in amyl acetate and activated after the deposition. The ions emerging from the surface with thermal energies leave the surface under the action of the voltage applied to the extracting electrode and reach the accelerator in the form properly shaped by the focusing electrodes. The ion current intensity obtained from this source was measured as 100 μA/cm^2 at a power consumption not exceeding 100 W [600].

Higher current intensities are obtainable if the compound of the element to be ionized is mixed with tungsten powder and this mixture is put into the crucible [584, 585, 592, 594, 754]. The higher intensity is due to the thus substantially increased metal surface.

For the preparation of the above mixture the pure tungsten powder has to be added to a few volume percents of alkali metal chloride and heated at a slow rate to 800°C. If the heating is continued past 800°C the ions start to leave the surface. An ion source with this type of ionizer can be constructed in the same way as the assembly shown in Fig. 3.9.

Owing to an increase in surface area an ion current of higher density is obtained also if a porous tungsten ionizer surface is employed [763–766]. Experimental studies, performed in order to find ion source types which

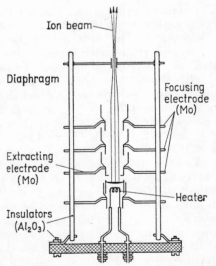

Fig. 3.9 Surface ionization ion source with indirect heating [600]

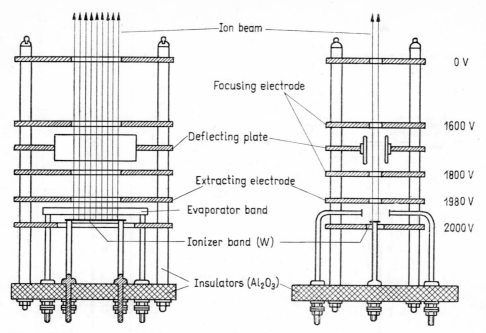

Fig. 3.10 Ion source with directly heated separated evaporator and ionizer surface
[604, 755—759]

can be used to fuel ion-drive mechanisms of rockets, have shown that ion
current densities of 10 mA/cm² can be attained. This value was measured
for a ~0.51 mm thick surface with a porosity of 10^6 holes/cm². The hole
diameter varied from 1.8 to 3.2 μ. The efficiency of this type of surface was
found to be $\beta - 1$ in the case of alkali metals (Husmann [766]).

One of the ion source types with separate evaporator and ionizer surface
shown in Fig. 3.10 is the so-called "triple band" ion source (Inghram,
Chupka [604]; Dietz [755, 756]; Hayden [758]; Langmuir [759]). In this
source the ionizer and the evaporator generating the atoms of the element
to be ionized are separate units. As it can be seen in the figure, a horizontally placed tungsten band functions as ionizer while two other tungsten
bands coated with the compound of the element to be ionized and located
symmetrically, each at one side of the ionizer, function as evaporator. This
arrangement permits the temperatures of the ionizer and the evaporator to
be set independently of each other, thus both can be heated to the optimum
temperature. Ion sources of this type can yield ion current densities exceeding by about an order of magnitude those obtainable from a "single band"
ion source where the compound of the ionized element is deposited directly
on the surface of the ionizer filament.

In the type of surface ionization ion source shown in Fig. 3.11a (Hintenberg, Lang [607]; Hintenberg, Voshage [608]) the atoms of the ionized
element are obtained from an evaporator and led to the ionizing surface
through a guide tube. The evaporator crucible and the tube are heated to

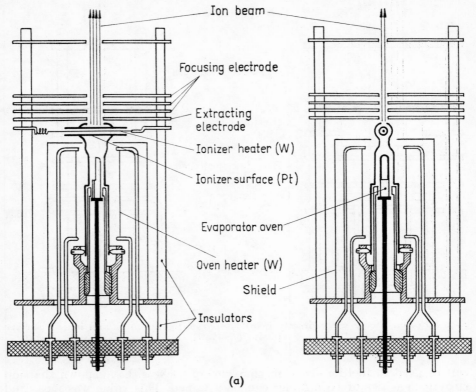

Ion beam

Focusing electrode

Extracting electrode

Ionizer heater (W)

Ionizer surface (Pt)

Evaporator oven

Oven heater (W)

Shield

Insulators

(a)

Fig. 3.11a Hintenberg–Lang–Voshage-type ion source with an evaporator and an ionizer surface heated by electron bombardment

the required temperature by the electron beam emitted from the tungsten filaments placed around the guide tube.

The ionizer surface, which consists of the surfaces of concentrically arranged platinum or platinum–iridium tubes, is heated by the electrons emitted from the tungsten filament passed across the inner tube.

The source types shown in Figs 3.10 and 3.11 are of poor efficiency from the point of view of material consumption since only a small fraction of the atoms of the ionized element is incident on the ionizer surface.

A more efficient material utilization can be achieved by use of an ionizer made of porous tungsten at the output slit of the source [1281, 1325].

If a porous tungsten ionizer is used, most of the atoms leaving the source will impinge on the surface of the tungsten ionizer. This porous tungsten ionizer disc is kept at the appropriate temperature.

The ionization efficiency for alkali metals can be as high as 70 per cent if such an ionizer is applied [1281].

A source assembly with a porous tungsten ionizer can be seen in Fig. 3.11b (Middleton, Adams [1325]).

In this source the alkali metal atoms are led from the stainless steel eva-

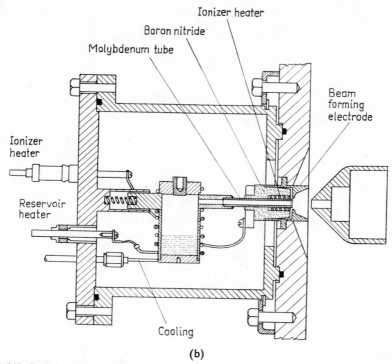

Fig. 3.11b Surface ionization ion source type with porous tungsten ionizer proposed by Middleton and Adams [1325]

porator through a vapour tube made of molybdenum with a 6.4 mm outside and a 4.8 mm inside diameter to the porous tungsten ionizer disc of 4.8 mm in diameter and 1 mm thick. The tungsten disc is welded by an electron beam to the molybdenum vapour conducting tube. The temperature of 1100° to 1200°C needed for the operation of the ionizer is produced by a heater made of tantalum wire of 1 mm in diameter with 7 windings. The surface ionization ion source shown in Fig. 3.11b yields alkali metal ion currents from 1 to 2 mA, formed by means of electrostatic gap lens.

A more advantageous material consumption can be obtained and the particle current density at the ionizer surface can be determined if a collimated atom beam is applied in the surface ionization source [610, 611, 752, 767–769].

A variant of this type of ion source can be seen in Fig. 3.12 (Wilson [610]; Jamba [611]). Here the collimated atom beam is generated in an atom source with a sapphire crucible. This crucible is 4 cm long and consists of a 2 cm long, 8 mm diameter cylinder of 1 mm wall thickness and closed at the bottom which is tightly fitted to a 2 cm long tube of greater wall thickness. The output slit of the crucible is 1.5 mm in diameter and 1.5 mm in length. The crucible is heated with a heater made of tungsten and surrounded by a thermal shield made of molybdenum. The atom beam emerging from the

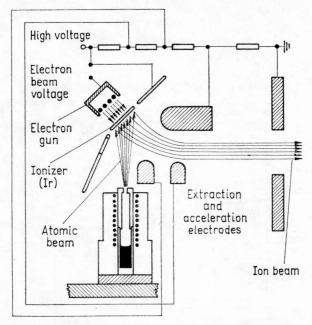

Fig. 3.12 Wilson–Jamba-type ion source with surface ionizer of curved ion trajectory

atom source strikes the ionizer surface at an angle of 45° to the direction of the beam axis. The ionizer is a disc of 1 cm in diameter with an iridium surface sintered onto 1.5 mm thick tungsten. The ionizer surface is heated with an electron beam to the appropriate temperature. The thin thermal shield surrounding the ionizer is made of tantalum. The ions emerging from the surface are led to the ion accelerator through the curved trajectory shown in the figure by a system composed of extracting, focusing and deflecting electrodes.

Owing to the curved trajectory, no neutral particles can reach the accelerator. This arrangement also permits the target to be observed through the aperture on the tantalum shield. The ion current density obtainable from this source is 10 μA/cm^2 for aluminium, gallium, indium and thallium with efficiencies $\beta = 5 \cdot 10^{-3}$ to $3 \cdot 10^{-2}$ while for the alkali metal ions with efficiencies $\beta = 1$, the ion current densities can vary from 1 to 5 mA/cm^2.

3.4 GAS DISCHARGE ION SOURCES

In the ion sources utilizing various types of gas discharge the ions are generated in the elementary physical processes taking place in the discharge. Gas discharges which occur at low gas pressures can be utilized in this type of source.

To obtain a continuous ion current from the source, the gas discharge by which the ions are produced must also be continuous. The gas discharge can be continuously maintained only if the electrons and ions lost in the process are supplanted by electrons and ions generated in the discharge, or by electrons from another source.

If the particles produced by the discharge processes compensate for the particle loss, the discharge is self-maintained. In self-maintained discharges different types of gas discharge can take place depending on the ion current intensity I_a flowing in the discharge. The discharge with $I_a < 10^{-6}$ A, that is with a dark current or with slight light radiation near the anode, is called Townsend-type discharge in which the space charge effect is negligible. The space charge effect, which increases with increasing discharge current, causes the distortion of the electric field, thus, the light phenomena extend in the direction of the cathode. This process continues with further increasing currents up to values of $I_a = 10^{-4}$ to 10^{-1} A when the discharge becomes of the type known as glow discharge. Now, if the discharge current continues to increase this leads eventually to the onset of arc-discharge at low gas pressure with discharge currents $I_a > 10^{-1}$ A.

No theory is available yet which could describe unambiguously the features of these different types of gas discharge and account for all the elementary processes which contribute to the gas discharge.

The theories proposed so far contain a set of simplifications and therefore they can describe only the main processes of the gas discharge. Nevertheless, the different features of the gas discharge can be interpreted and the processes involved in the discharge qualitatively approximated in terms of the simplified formulation.

3.4.1 TOWNSEND-TYPE GAS DISCHARGE

The self-maintained discharge which takes place at low gas pressures with discharge current intensities $I_a < 10^{-6}$ is called Townsend discharge. Because of the low current intensity the space charge effect can be ignored as the distortion of the electric field generated by the discharge is practically negligible.

In this type of discharge the field between the electrodes is dark and only some slight glow appears close to the anode. This phenomenon can be attributed to the fact that the electron avalanche formed in the discharge attains only in the vicinity of the anode the density at which the light effect produced by the transition of the electron-excited atoms to their ground state becomes of an appreciable intensity.

The condition of the formation of a self-maintained discharge with such low currents is given in terms of the Townsend–Rogowski avalanche theory [801, 802, 803, 804, 806–808] as follows.

Let us assume that the primary electron emerging from the cathode is driven toward the anode by the potential difference U_a applied between the cathode and the anode. Let α be the number of ions generated per electron per cm of its track, while β is the number of electrons generated per ion per cm of its track and d is the distance from the cathode to the anode. If the number of electrons generated by the ions is ignored ($\beta = 0$), the number of ions generated over the distance d is given by αd. However, simultaneously with the ions also electrons are generated in the same number as the ions. The applied electric field drives the electrons toward the anode and the ions toward the cathode. If the total energy E, consisting of the initial energies of the colliding and generated particles and of the energies imparted to them before the next collision, is higher than the ionization energy E_i, new ionizations can take place. As a result of these processes $e^{\alpha d}$ electrons will strike the anode due to the set of collisions initiated by a primary electron emerging from the cathode. Accordingly, the number of ions generated in the field between the electrodes and driven by the homogeneous applied field to the cathode, is given by $(e^{\alpha d} - 1)$. The number of secondary electrons generated by the ions striking the cathode is then obtained as

$$n_s = \gamma(e^{\alpha d} - 1) \tag{3.6}$$

where γ is the number of secondary electrons generated per ion incident on the cathode surface (see Paragraph 1.8).

The cycle described above can then start anew and the role of the primary electrons is played in the next cycle by the $n_s = \gamma(e^{\alpha d} - 1)$ secondary electrons and the cycles continue with an ever increasing number of electrons. The number of electrons striking the anode per primary electron in the successive cycles — ignoring the electrons generated by ions and assuming that the applied field between the cathode and anode is homogeneous — can be expressed as

$$n = \frac{e^{\alpha d}}{1 - \gamma(e^{\alpha d} - 1)} \tag{3.7}$$

Taking the initial current density to be i_0, formula (3.7) gives for the discharge current density in a homogeneous applied field

$$i = i_0 \frac{e^{\alpha d}}{1 - \gamma(e^{\alpha d} - 1)} \tag{3.8}$$

If the applied field between the cathode and the anode is inhomogeneous because α depends not only on the gas pressure but also on the electron energy, that is on the field intensity $F = \dfrac{U_a}{d}$, expression (3.8) can be replaced by the formula

$$i = i_0 \frac{e^{\int_0^d \alpha dx}}{1 - \gamma \left(e^{\int_0^d \alpha dx} - 1\right)} \tag{3.9}$$

It is apparent from expressions (3.8) and (3.9) that the value of the discharge current density can go to infinity if the denominator is set to zero. This can happen if

$$\gamma(e^{\alpha d} - 1) = 1 \tag{3.10}$$

It can be seen from expression (3.8) that if equation (3.10) holds, the discharge current density can reach high values independently of i_0, thus the discharge becomes self-maintained. It follows that equation (3.10) is taken to be the condition of self-maintaining discharge with the understanding that in this case the number of electrons generated by ions is negligible ($\beta = 0$) and that the applied field is homogeneous [801–804].

Townsend's theory has been seen to allow discharge currents of infinite value. The discharge currents of infinite value are restricted to high, but finite values in Rogowski's gas discharge theory. In this theory the condition of self-maintained discharge follows from the requirement that the ions generated by n_1 electrons leaving the cathode surface have to release $n_2 = n_1$ electrons from the cathode, that is, by using formula (3.6) for $\beta = 0$, with homogeneous applied field, this requirement reads

$$n_2 = n_1 \gamma (e^{\alpha d} - 1) \tag{3.11}$$

hence

$$\mu = \frac{n_2}{n_1} = \gamma (e^{\alpha d} - 1) \tag{3.12}$$

The condition obtained in terms of Rogowski's theory reproduces the condition (3.10) in the case of $\mu = 1$. The infinite value of the possible current density is limited to finite values by the following reasoning.

It is known that the electrons move faster in the discharge than the ions and consequently they leave more rapidly the discharge volume. The ions moving at a slower rate to the cathode generate a space charge. The effect of this space charge is negligible at low discharge current densities ($\mu < 1$) but if an increase in the potential gradient U_a causes the current density to become as high as $\mu \approx 1$, the space charge effect becomes appreciable. The field between the cathode and the anode is distorted by the space charge so that the anode appears to be closer to the cathode than it actually is and the distance between them seems to be not $x = d$, but $x < d$ (Fig. 3.13). Accordingly, relation (3.12) can be replaced by the equation

$$\mu = \gamma (e^{\alpha x} - 1) \tag{3.13}$$

The decrease in the distance x involves an increase in field intensity because of the relation $\dfrac{U_a}{x} = F$. Since the value of α increases with increasing F more rapidly than the value of x decreases [809–811], μ also increases according to the formula (3.13). The increase in μ entails further increase in the discharge current density accompanied by a comparable further distortion of the field. This process can continue until the value of x becomes less than that of the mean free path for collision, that is, $x < \lambda_i$, which leads to a decrease in μ.

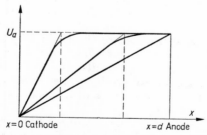

Fig. 3.13 Variation of cathode fall region with space charge: in absence of space charge, the uniform potential distribution between cathode and anode corresponds to a straight line; as space charge increases, the potential distribution and cathode fall region progressively decreases

The decrease in μ is mainly due to the fact that in the major part of the discharge field which is exposed to the effect of space charge no energy is imparted by the applied field to the colliding and generated particles after collision and they even continue to move in the direction of the recoil from the collision and can reach in this way the wall of the discharge tube. This state of affairs leads to a decrease in the number of ions and electrons, thus in the value of μ. The decrease in μ entails, in turn, a decrease in space charge and field distortion, thus the distance x becomes again $x \approx d$. At this stage the value of μ can start to increase again and initiate the cycle described above.

The Townsend–Rogowski theory of self-maintained discharge thus allows a continuous discharge to occur in which the current density may fluctuate to some extent and the discharge current has finite values.

The value of the ignition voltage U_b required for the ignition of the gas discharge, which, in fact, corresponds to the voltage at which the non self-maintained transforms to self-maintained discharge, depends on many factors. As apparent from expression (3.12), the ignition of the discharge depends on the value of α, γ and the electrode distance d. The value of α is, in turn, a function of the properties of the gas, the applied field intensity and of the gas pressure p. The value of γ depends on the material of the cathode, the energy of the ions and on the state of the cathode surface (see Paragraph 1.8). Consequently, the value of the ignition voltage U_b was calculated for different gases, especially, for cathode materials, as a function of pd [801–804, 806–822].

The investigations have shown that the curve for U_b has a minimum in the range of $pd = 0.1$ to 10 mmHg $\cdot$ cm. The value of pd at the minimum varies with the type of gas (Fig. 3.14).

The equation relating the ignition voltage U_b to the pressure p in the discharge field and to the distance d between the electrodes in the form

$$U_b = f(pd)$$

is called the Paschen law and the curves representing this relation are called Paschen curves.

The dependence of U_b on the cathode material arises from the different values of the secondary electron emission coefficient of different metals (see Paragraph 1.8). The minimum values of U_b were investigated in experiments on H_2 gas under identical conditions with cathodes prepared from different metals [817].

The results of these experiments are shown in Fig. 3.15.

To obtain very low ignition voltages, alkali metal coated cathode surfaces can be employed. The value of U_b can also be decreased by the utilization of thermal electron emission to produce the electrons in the discharge volume. The results of the investigations into this effect [826] are to be seen in Fig. 3.16.

It is apparent from this figure that by increasing the cathode temperatures, that is, by an enhanced contribution from thermal emission to the electron current density, U_b starts to decrease first slowly, then steeply down to a value determined by the distance between the electrodes.

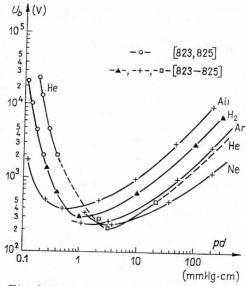

Fig. 3.14 Paschen curves for various gases. Dependence of ignition voltage U_b on the product of distance d from anode to cathode and the gas pressure p

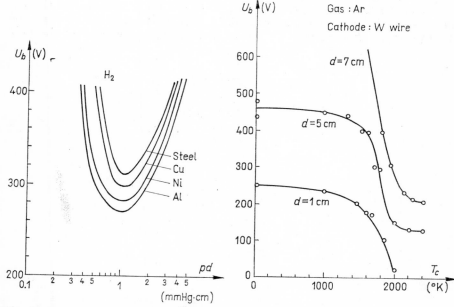

Fig. 3.15 Paschen curves for electrodes made of different metals

Fig. 3.16 Dependence of ignition voltage U_b on temperature of the electron emitter tungsten filament cathode in argon gas for various values of distance d from anode to cathode

3.4.2 GLOW DISCHARGE

Glow discharge occurs in the discharge tube at low gas pressures ($p = 10^{-1}$ to 1 mmHg) if the discharge current varies in the range $I_a = 10^{-4}$ to 10^{-1} A. The space charge effect cannot be ignored in the glow discharge as has been done in the case of Townsend discharge since in the former case the contribution from the space charge to the discharge volume causes an appreciable distortion of the electric field. It follows that the potential distribution between the cathode and the anode is not linear but it shows a distribution of the shape depicted in Fig. 3.17. The light phenomenon observed in the case of glow discharge can be divided into several zones within the tube.

As is apparent from Fig. 3.17 the so-called Aston's dark space [899, 900] occupies the closest environment of the cathode (*1*), next to it we find the luminous cathode zone (*2*), followed by the cathode dark space (*3*). The boundary of the luminous cathode zone is dimmed towards the dark cathode space which is separated by a sharp boundary line from the glow (*4*). The zone of the glow dims towards the Faraday dark space (*5*), which is followed in the direction of the anode by the luminous positive column of the glow discharge (*6*). This positive column is separated by the anode dark space (*7*), from the anode glow (*8*) next to the anode itself.

Aston's dark space can be attributed to the fact that the energy of the electrons emerging from the cathode is insufficient for the excitation of the gas atoms. Thus, the luminous zone can be observed only at that distance from the cathode surface at which the electrons are sufficiently energized by the electric field in the cathode fall zone for the excitation of the gas atoms responsible for the emission of light in their transition to the ground state. Consequently, the width of Aston's dark space depends on the field

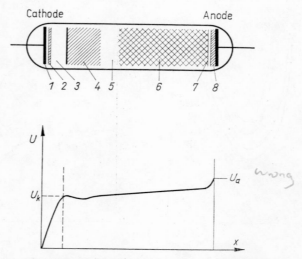

Fig. 3.17 Potential distribution in glow-discharge
(*1*) Aston's dark space, (*2*) Luminous cathode space, (*3*) Dark cathode space, (*4*) Glowlight, (*5*) Faraday's dark space, (*6*) Positive column, (*7*) Dark anode space, (*8*) Anode light

intensity in the zone of the cathode fall and on the excitation energy of the gas atoms. The field intensity depends, in turn, (see Paragraph 3.4.1) on the intensity of the discharge current because the length of the cathode fall zone is determined by the field distortion due to the effect of space charge. Aston's dark space does not appear if the cathode is struck by high velocity ions which are capable of exciting the atoms in the environment of the cathode.

The cathode dark space appears after the luminous zone because the energy imparted by the electric field to the electrons at this distance from the cathode is so high that the collisions lead predominantly to ionization and only in a small measure to the excitation of higher atomic levels. Moreover, the high electron energies do not allow electron–ion recombinations to occur which involve the emission of light, while the very low number of highly excited atoms in the cathode dark space does not lead to an appreciable emission of light by the transition to the ground state. In the cathode dark space electron energies are lost to inelastic collisions. However, the colliding electrons and those generated by ionization are sufficiently energized by the field in the cathode fall zone to induce further ionizations. This process is repeated as long as the collisions occur in the high intensity field of the cathode fall zone. As a result of this process the ion concentration increases in the cathode dark space towards the glow region.

The field strength in the glow region of the gas discharge is too low to impart, before the next collision, enough energy to all electrons for further ionization. Thus, the number of slow electrons increases, a large fraction of low excited atoms appears and also electron–ion recombination can occur.

Both of these processes involve the emission of light and the glow becomes visible. The length of the glow region increases towards the positive column if the gas pressure is reduced or if the cathode fall increases. Measurements with a probe have shown that a negative field gradient can be present in the glow region.

The potential distribution shown in Fig. 3.17 indicates that the field intensity, though much lower than in the cathode fall region, is not negligible in the glow, the Faraday dark space and the positive column regions. Thus, the energy of the electrons moving towards the anode increases as the consequence of the low field. This increase in electron energies is sufficient to prevent electron–ion recombination and the appearance of recombination light, thus the Faraday dark space sets in.

The Faraday dark space is followed by the positive column of glow discharge with a light contributed mainly by the transitions of excited atoms. In the positive column the space charge is nearly zero since the electron and ion concentrations are almost identical. The charged particle concentration shows the maximum value along the axis of the discharge and the minimum value at the wall of the discharge tube. In the positive column it is mainly the electrons that are responsible for the electrical conduction since the mobility and flow rate of the positive ions is extremely low. The positive column extends radially as the gas pressure increases and approaches the wall of the discharge tube. In this case the positive column of the glow discharge behaves in exactly the same way as the positive column in the arc-discharge at low gas pressure (see Paragraph 3.4.3).

The potential distribution in the glow discharge (Fig. 3.17) shows the greatest fall of the potential in the relatively short region of the discharge field. The fall in potential occurring near the cathode is called the cathode fall.

If the current flowing in the discharge increases without involving a change in current density and in the fall of the potential, the cathode fall is considered to be normal. This phenomenon can be explained by the smaller area of the luminous zone before the cathode than that of the cathode surface itself in the case of low discharge currents. As the discharge current increases, the luminous area also increases although the current density remains unchanged until the area of the luminous region becomes equal to that of the cathode surface.

If the discharge current continues to increase after this equality, both the current density and the value of the cathode fall start to increase and the normal glow discharge turns to anomalous glow discharge. The fall of the potential is called in this case anomalous cathode fall.

3.4.3 ARC-DISCHARGE AT LOW GAS PRESSURE

If a discharge tube is filled with a gas at low pressure, arc-discharge takes place at discharge currents $I_a > 10^{-1}$ A.

In this case three regions, which are characteristic of the low gas pressure arc-discharge, can be distinguished between the cathode and the anode (Fig. 3.18) in the discharge chamber.

The first is the region of the cathode fall; this is the zone with the highest fall in the potential during the discharge and the value of this fall is only slightly below that of the total potential difference in the discharge. The second is the region of the quasi-neutral plasma, the positive column of the arc-discharge which extends almost over the entire space between the electrodes. In this region the potential fall is very small. The third region, where the potential fall is still small, is called the anode fall.

The potential difference required to bring about the discharge as well as the value of U_n in the cathode fall depends mainly on the fact whether the electrons are obtained from the cathode by cold or by thermal emission.

In the case of cold emission, a potential difference of 10 to 50 kV has to be applied to the electrodes of the discharge tube in order to have as many electrons ejected from the cathode by the strong applied electric field as needed for inducing and maintaining the discharge (high voltage arc-discharge).

In the case of thermal electron emission the potential difference U_n in the cathode fall need not be higher than required for the ionization of the gas atoms (molecules) in the discharge field ($eU_k > E_i$) and high enough to enable the electron current to leave the cathode surface with a density necessary for the maintenance of the discharge.

If the gas pressure in the discharge tube is such that the length d of the cathode fall region is nearly equal to the free mean path λ_e of the electrons, the electrons and ions flow in this region against each other as in a vacuum.

Assuming that the electric field strength is zero at the cathode surface and at the cathode side plasma boundary, the electron and ion current densities can be evaluated in terms of the double layer theory from the expressions [798].

$$j_e = \frac{1.86}{9\pi} \sqrt{\frac{2e}{m_e}} \frac{U_k^{3/2}}{d^2} \qquad (3.14)$$

$$j_i = \frac{1.86}{9\pi} \sqrt{\frac{2e}{m_i}} \frac{U_k^{3/2}}{d^2} \qquad (3.15)$$

where d is the distance between the cathode surface and the plasma boundary and U_k is the potential difference in this region (Fig. 3.18).

By expressions (3.14) and (3.15), the ion and electron currents are related to each other as

$$\frac{j_i}{j_e} = \sqrt{\frac{m_e}{m_i}} \qquad (3.16)$$

It follows from this formula that the ion current density flowing through the boundary of the plasma on the cathode side is limited to $j_i < 4.27 \cdot 10^{-2} j_e \, \text{A/cm}^2$ in the arc discharge taking place at low gas pressure.

The electrons leaving the cathode surface are accelerated in the cathode fall region by the potential difference U_k and thus sufficient energy is imparted to them for the ionization of the gas atoms (molecules) in the discharge tube. After ionization the primary electrons and the secondaries produced by ionization continue to interact undergoing in this way several

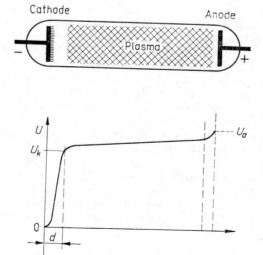

Fig. 3.18 Potential distribution in arc-discharge at low gas pressure

successive collisions with each other and the neutral and ionized particles. As a result of these interactions the primary electrons lose their initial energies and contribute to the electron gas in which the electrons move at random velocities. In the electron gas the velocity distribution of the electrons is nearly Maxwellian, thus their mean velocity $\bar{v} = \sqrt{\dfrac{8kT_e}{\pi m_e}}$, where T_e is the average electron temperature.

The electron gas together with the ions produced in the collisions form a quasi-neutral system, this is the plasma of the arc-discharge. The quasi-neutral plasma occupies almost the total volume of the discharge tube except for a thin layer along the tube wall and the cathode and anode fall regions which are not quasi-neutral. The thickness of the layer along the wall is approximately $2\lambda_D$, where

$$\lambda_D = \sqrt{\frac{kT}{4\pi n_0 e^2}} \tag{3.17}$$

is the Debye length. The formation of this layer in the plasma of the arc-discharge can be explained as follows [786–798].

In the plasma of arc-discharge the ion temperature T_i is much lower than the electron temperature T_e, whereas in the isothermal plasma the mean kinetic energy of the particles forming the plasma is the same. In the former case the ions lose much of their energy in ion–atom collisions because $m_i \approx m_a$, while very little energy is transferred to the ions in the electron–ion collisions owing to the difference $m_e \ll m_i$.

In the arc-discharge plasma the electron velocities are higher than those of the ions not only because of their much smaller mass but this difference increases also because of $T_e \gg T_i$. It follows that electrons can escape in a greater number from the arc discharge plasma to reach the wall of the discharge tube. This causes a negative charge of the walls resulting in the formation of a layer in which the ions are accelerated towards the wall while the electrons flying towards the wall are slowed down. This layer is responsible for the fact that not only electrons but also ions can escape from the plasma. Thus, for the subsistence of the discharge, the loss of electrons and ions due only to the neutralization at the tube wall — as the volume recombination at low gas pressure is negligible — must be compensated by the ions and electrons formed in electron–neutral gas atom collisions.

The numbers N_e and N_i of electrons and ions flying towards the wall of the discharge chamber are determined by the radial direction diffusion in the plasma and the electric field, F. This radial direction diffusion is generated by the radial direction concentration difference arising due to the neutralization on the wall, and the electric field F arises from the charge separation due to the different electron and ion velocities. Using equation (1.111), which holds under these conditions, the numbers of ions and electrons moving towards the tube wall per square centimetre of surface area are given [786–788] by

$$N_e = - D_e \frac{\partial n_e}{\partial r} - n_e \varkappa_e F \tag{3.18}$$

$$N_i = - D_i \frac{\partial n_e}{\partial r} + n_i \varkappa_i F \qquad (3.19)$$

where n_e and n_i are the concentrations, D_e and D_i the diffusion constants, while $\varkappa_e$ and $\varkappa_i$ are the mobilities of the electrons and the ions, respectively.

Since it is assumed that the charges are at equilibrium in the plasma of the arc-discharge, $n_e = n_i = n$, and the electron current is taken to be equal to the ion current at the wall, thus $N_e = N_i = N$, if we recall the considerations in Paragraph 1.7, we can use the ambipolar diffusion equation

$$N = - D_a \frac{\partial n_i}{\partial r} \qquad (3.20)$$

The ambipolar diffusion coefficient formula (1.113) can be written as

$$D_a = \frac{2 D_i \varkappa_e}{\varkappa_i + \varkappa_e}$$

Since in our case the electron mobility $\varkappa_e$ is much higher than the ion mobility $\varkappa_i$, we have

$$D_a \approx 2 D_i \qquad (3.21)$$

With expression (3.21) the relation (3.20) can be replaced by the expression

$$N_i = - 2 D_i \frac{\partial n_i}{\partial r} \qquad (3.22)$$

It is apparent from this relation that the number of ions reaching the discharge tube wall is proportional to the ion diffusion coefficient in the diffusion region and to the radial gradient of the ion concentration.

The ion distribution can be described on the assumption that the numbers of ions and electrons produced in the arc-discharge plasma are proportional to the electron concentration, i.e. $N_{ip} = \alpha n_e$, by using the equation [786–788]

$$n_i = n_i(r = 0) J_0(x) \qquad (3.23)$$

where $n_i(r = 0)$ is the ion concentration in the axis of the discharge tube, $J_0(x)$ is a Bessel function of zero order which for $J_0(0) = 1$, and $J_0(x = 2.4) = 0$. In this case the argument of the Bessel function is

$$x = r \sqrt{\frac{\alpha}{D_a}}$$

where $\alpha = n_a \sigma_i \bar{v}_e$ is the ionization per electron. Since it was assumed in the differential equation for the ion distribution [786–788] that $n_i \approx 0$ at the wall of the discharge tube ($r = R$) it can be written that

$$R\sqrt{\frac{\alpha}{D}} = 2.4 \tag{3.24}$$

This equation relates the radius of the discharge tube to the ionization and diffusion parameters of the gas applied in the discharge tube.

The discharge current is determined by the electron concentration n_e in the plasma, the electron and ion mobilities and the axial component F_{zp} of the applied field, that is

$$I_a = n_e e (\varkappa_i + \varkappa_e) F_{zp}$$

and since in our case $\varkappa_i \ll \varkappa_e$, we have

$$I_a = n_e e \varkappa_e F_{zp} \tag{3.25}$$

The power consumed in the discharge per unit length of the discharge tube is $I_a F_{zp}$. This power is consumed partly by the production of ions and electrons through inelastic collisions.

The power consumed in this process must be equal to the product of the number of ions generated per unit time per unit length of the tube and the ionization energy E_i of the gas in the discharge tube, thus

$$\eta I_a F_{zp} = N_{ip} E_i \tag{3.26}$$

Using expressions (3.25) and (3.26) together with the relation $N_{ip} = \alpha n_e$, we get for α the formula

$$\alpha = \frac{\eta \varkappa_e F_{zp}}{E_i} \tag{3.27}$$

Substituting this expression for α into formula (3.24), the axial field component can be related to the radius of the discharge tube and to the ionization and diffusion parameters of the gas applied in the discharge [786–788] as

$$F_{zp} = \frac{2.4}{R}\sqrt{\frac{D_a E_i}{e \eta \varkappa_e}} \tag{3.28}$$

This result has been derived from Schottky's diffusion theory. Let us now apply Langmuir's theory of the arc-discharge plasma [789, 790] under the following simplifying conditions: a) the ions reach the tube wall without undergoing any collision in the discharge; b) the discharge current is so low that the successive excitation and ionization processes can be ignored; c) the volume recombination of ions and electrons is negligible; d) the particle velocity distribution is Maxwellian and the electrons have a Boltzmann-type radial distribution. Under these conditions the axial field component within the plasma can be expressed [794–798] as

$$F_{zp} = 2kT_e \left\{ S_0 \frac{\dfrac{E_i}{kT_e} + \ln\left(\dfrac{m_i}{m_e}\right)^{1/2} + 2.7}{0.75\sqrt{\pi e^2 R \bar{\lambda}_e}} \right\}^{1/2} \tag{3.29}$$

where S_0 is a dimensionless constant [790] and $\bar{\lambda}_e$ is the mean free path of the electrons. The current density at the probe introduced into the interior of the discharge plasma is the sum of the ion and electron current densities [798–800] given by

$$j_s = j_i + j_e = \frac{1}{4} n_i \bar{v}_i e - \frac{1}{4} n_e \bar{v}_e e$$

Since a layer similar to that found at the tube wall is also formed around the probe introduced into the plasma of the arc-discharge, the ions are accelerated in the direction of the probe while part of the electrons is stopped or forced back by the negative field around the probe. Consequently, only the electrons with sufficiently high energies can cross the repulsive field and reach the surface of the probe.

On the assumptions that the particle velocity distribution is Maxwellian in the discharge plasma, the electron distribution in the environment of the probe obeys the Boltzmann law and that $n_e = n_i = n$, the current density striking the probe can be expressed as

$$j_s = n_e \sqrt{\frac{k}{2\pi}} \left(\sqrt{\frac{T_i}{m_i}} - \sqrt{\frac{T_e}{m_e}} e^{-\frac{eU}{kT_e}} \right)$$

If a negative potential higher than the value of $\dfrac{kT_e}{e} = U$ is applied to the probe as compared with the plasma, only ions can strike the probe surface with a maximum current density given by

$$j_s = j_i = n_i e \sqrt{\frac{kT_i}{2\pi m_i}} \tag{3.30}$$

3.4.4 HIGH FREQUENCY GAS DISCHARGES

Gas discharges in a high frequency field differ in many features from the gas discharges with a d.c. current discussed in the foregoing. In the high frequency gas discharge in an applied field with periodically alternating magnitude and direction the charged particles are not always capable of escaping from the discharge field. Consequently, the loss in electrons and ions is less important and their concentration increases sufficiently for the onset of self-consistent gas discharge.

The secondary processes occurring on the cathode also play a role but of smaller importance in the high frequency field than in the d.c. gas discharges. This makes it possible to apply the field by which the discharge is induced to electrodes mounted either inside or outside the discharge tube.

The considerations of high frequency discharges will be restricted to those processes taking place at low gas pressures since the gas or vapour pressure in the ion sources varies generally within the range $p = (10^{-3}\text{–}10^{-1})$ mmHg.

At low gas pressures, with $\lambda_e \geq r$ (where λ_e is the mean free path of electrons and r is the dimension of the discharge tube) and with a collision frequency ν between electrons and gas particles lower than that between electrons and the tube wall, the initiation of the discharge is markedly affected by the secondary processes at the discharge tube wall. The gas discharge can be induced only under the following conditions: the secondary emission coefficient of the wall must be $\delta > 1$, the electrons have to fly over the distance between the two opposite walls during the half period of the high frequency voltage cycle and over this time they take up from the field the energy needed for ionization and initiation of secondary emission.

If these conditions are satisfied, the electron concentration in the discharge tube can become high enough for an appreciable ionization to occur in spite of the low number of collisions and for the quasi-neutral plasma of the high frequency discharge to develop.

High frequency discharge can be induced by applying either an alternating electric or an alternating magnetic field.

The discharge generated by an alternating electric field applied to two electrodes located outsideor inside the discharge tube is called either capacitively coupled or linear high frequency discharge [902–910].

The type of discharge generated by the alternating magnetic field in a discharge tube placed inside the solenoid is named either inductively coupled or ring high frequency discharge [508, 911–916].

3.4.4.1 *Alternating electric field or linear high frequency discharge*
[902–910]

The discharge tube is prepared from an insulating material. The high frequency field is applied to the electrodes located internally or externally, one at each end of the tube. Under the action of the applied field the free electrons in the discharge field oscillate at the same frequency f as that of the applied field. If the electrons do not loss any energy during their oscillation, no energy is transferred to them from the field. If the electrons loss energy because of collisions energy transfer from the applied high frequency field takes place and if the transferred amount of energy is sufficient for inelastic collisions to occur, the gas molecules in the tube can be dissociated and this initiates the excitation and ionization of atoms and molecules.

If the process of collision between electrons and gas particles is regarded as some kind of resistance of a medium in the equation describing the motion of electrons, then the equation of motion for a single electron in an alternating electric field defined by $F = F_0 \cos \omega t$, takes the form

$$\frac{dv_e}{dt} + \nu v_e = \frac{e}{m_e} F_0 \cos \omega t \qquad (3.31)$$

The term $m_e v_e \nu$ arising in eq. (3.31) from the resistance of the medium corresponds to the mean charge in electron impulses due to collisions per unit time.

The electron velocity is obtained from the equation of motion as

$$v_e = \frac{eF_0}{m(\nu^2 + \omega^2)^{1/2}} \sin(\omega t + \varphi) - Ce^{-\nu t} \qquad (3.32)$$

where $\varphi = \arctan(\nu/\omega)$ is the phase shift and C is the integrating constant. With the formula for current density per unit volume, which reads

$$j = n_e v_e e$$

and by eq. (3.32), we get for the power consumption per unit discharge volume

$$P = \frac{n_e e^2 F_0^2}{m\nu} \frac{\nu^2}{\nu^2 + \omega^2} \qquad (3.33)$$

This expression shows that the power consumed by the high frequency discharge is proportional to the square of the electric field, and that it depends on the electron concentration, the frequency of collisions and on the field frequency. It follows from the formula that for $\nu = 0$, that is, when no collision between electrons occurs, $P = 0$, thus there is no energy transfer from the field.

The collisions between electrons and gas particles lead to ionization only if the energy $E > E_i$ needed for ionization is imparted to them by the field over the time corresponding to their mean free path λ_e. Thus, the electrons must be accelerated over the distance $x = \lambda_e$ to a velocity

$$v_e \geq \sqrt{\frac{2}{m} eU_i} = \sqrt{\frac{2}{m} E_i}$$

required for the onset of the quasi neutral plasma of the high frequency discharge.

The ignition voltage U_b of the discharge has been evaluated [902] by using the expressions for v_e and x obtained from the equation of motion under the two conditions mentioned above in the collisionless case ($\nu = 0$) of the form

$$\frac{d^2 x}{dt^2} = \frac{e}{m} F_0 \cos \omega t$$

The values of the function for $U_b = U_b(f)$ were calculated with known values of $F_0 = F_0(f)$ and of electrode distances so that in given cases the ignition voltage of the high frequency gas discharge can be determined from the function $U_b(f)$.

The predicted values show a good qualitative agreement with the experimental data and the theoretical values can be utilized for the quantitative estimation of the value of U_b.

Figure 3.19 shows the curves for $U_b = U_b(f)$, as calculated for different gases at a pressure $p = 30~\mu$Hg.

In Fig. 3.20 the curves obtained for argon gas at different gas pressures can be seen. It is apparent that the minimum value of U_b and that of the associated frequency increase with increasing pressures.

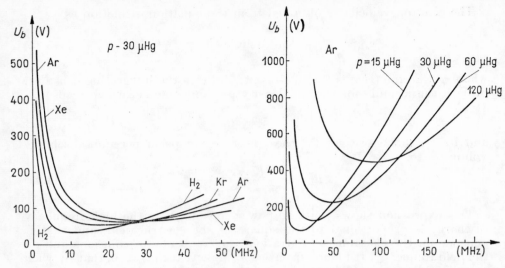

Fig. 3.19 Dependence of ignition voltage U_b on frequency of the discharge generating high frequency field for different gas media

Fig. 3.20 Dependence of ignition voltage U_b on frequency of the discharge generating high frequency field in argon gas at different values of pressure p

Qualitatively, the curves can be interpreted as follows. At low frequencies collisions occur within a small fraction of a period. Consequently, the amplitude of the high frequency field must be large in order to impart to the electrons in such a short time the energy needed for ionization. At high frequencies, on the other hand, the amplitude of the field must again be large because the electrons reach their maximum energy before collision and this decreases until the collision takes place. The lower energy which can be thus transferred by the electron in the collision must be higher than the ionization potential.

Experiments concerning the ignition voltage [914–916] have shown that at pressures $p < 10^{-2}$ mmHg the ignition voltage is independent of the nature and pressure of the gas in the discharge tube and that its value is primarily determined by the secondary electron emission coefficient of the discharge tube wall. This fact can be explained by the acceleration of the primary electrons present in the discharge tube in the high frequency field to velocities at which more than one secondary electron can be ejected by the electrons striking the wall ($\delta > 1$). The change in the polarity of the field causes the secondary electrons to fly against the opposite wall of the tube and to eject further secondary electrons. Since $\delta > 1$, the number of oscillating electrons continues to increase and upon reaching a given concentration a considerable ionization takes place. The increase in electron concentration is limited by the space charge in the discharge field. However, an electron concentration needed for the onset of an intense high frequency discharge can be brought about only by the resonance of secondary electrons. This means that the ejection of secondary electrons, i.e. the process of electron multiplication, is repeated in every half-period. In the case of resonance all the ejected secondaries strike the opposite wall and produce thereby more than just one secondary electron.

If the half-period of the high frequency field is of a longer duration than the time taken by the electrons to cross the discharge field and to eject secondary electrons, the multiplying process is interrupted and the discharge breaks down.

The frequency value below which no discharge can be initiated by the high frequency field is called the threshold frequency. The value of this threshold, as has been seen from the foregoing, depends on the dimensions of the discharge tube.

3.4.4.2 *Alternating magnetic field or ring high frequency discharge*
[508, 911–916]

If a discharge tube is placed inside the solenoid of a high frequency oscillator, a ring high frequency discharge is induced in the tube exposed to the alternating magnetic field. The investigations of this type of high frequency discharge [508, 911–913] have shown that circular currents are induced in the discharge, that is, the electrons which oscillate at the same frequency as that of the field are moving at the same time on a circular orbit. The condition of ignition for the ring high frequency discharge was found to be that the maximum value of $H = H_0$ of the applied alternating magnetic field $H = H_0 \sin \omega t$ must satisfy the equation [508]:

$$ r H_0 \frac{e}{m_e} = \frac{\dfrac{2eU_i}{m_e} + (\omega \lambda_e)^2}{\omega \lambda_e} \tag{3.34} $$

that is, the discharge is initiated if the electrons moving in the electric field induced by the alternating magnetic field are imparted the energy needed for ionization. It is apparent from expression (3.34) that the value of H_0 is at a minimum, if

$$ \frac{2eU_i}{m_e} = (\omega \lambda_e)^2 $$

and that the value of H_0 varies with the ionization potential, the field frequency and with the mean free path of electrons, that is with the gas pressure p, since $\lambda_e = \dfrac{1}{p}$.

Apart from the induced electric field strength, the initiation of the high frequency ring discharge is also affected by the electric field intensity of axial direction generated between the two ends of the solenoid [920]. The contributions from these two fields to the initiation and maintenance of the ring discharge depend always on the given experimental conditions. The discharge is often initiated by the axial field and transforms only later to a ring discharge affected mainly by the transversal electric field induced by the alternating magnetic field.

3.4.5 EFFECT OF STATIC MAGNETIC FIELDS ON GAS DISCHARGES

In the positive column of gas discharge, in the quasi-neutral plasma, the oppositely charged particles are present at nearly the same concentrations. The electron and ion currents leaving the plasma of the discharge are determined by diffusion (Paragraph 3.4.3). Now, if the gas discharge is exposed to a static magnetic field, the diffusion of charged particles will change.

If a d.c. gas discharge is exposed to a magnetic field with a direction parallel to that of the electric field, the charged particle current will decrease towards the tube wall in the direction normal to the magnetic field. This decrease is caused partly by the change in the movement of electrons normal to the axis under the action of the magnetic field. As the magnetic field increases, an ever smaller number of electrons are able to reach the tube wall as they move on a helical path with decreasing radius. The other cause of the decrease is understood from the diffusion theory since in the equation describing the electron current reaching the wall by diffusion as

$$j_e = -D_e \, \text{grad} \, n_e$$

the diffusion coefficient D_e decreases in a magnetic field in the direction normal to that of the field according to the ratio $D_H \sim \dfrac{D_{H=0}}{H^2}$ as given by formula (1.110b). At the same time, the value of the diffusion coefficient remains unchanged in the direction parallel to that of the magnetic field.

Because of the greater mass of the ions, their motion is only slightly affected by the magnetic field. However, for the conservation of quasi-neutrality, the upwards motion of the positive ions decreases in accordance with the decrease in the loss of electrons.

If the discharge is exposed to a magnetic field with a direction normal to that of the electric field, again two effects will assert themselves. One of these effects causes the positive column of the discharge to be deflected by the magnetic field towards one of the tube walls. On the side in the direction of the inclination the loss in charged particles is much more enhanced than the decrease in particle loss caused by the inclination on the opposite side. The other effect is the decrease in the diffusion of charged particles in the direction normal to the magnetic field, which in this case coincides with the directions of the axes of the electric field and the discharge, since by formula (1.110b) the diffusion coefficient decreases as $D_H \sim \dfrac{D_{H=0}}{H^2}$ in an increasing magnetic field.

It follows from the above considerations that in a d.c. discharge field the electrons cover a longer path under the action of a magnetic field which forces the electrons to circulate on a helical orbit at a Larmor frequency $f_H = \dfrac{eH}{2\pi mc}$. Consequently, the electrons have more chance to collide with and to ionize the gas particles. The magnetic field thus has the same effect as the increase in gas pressure in the discharge volume. In the presence of a magnetic field it is therefore possible to reduce the gas pressure in the

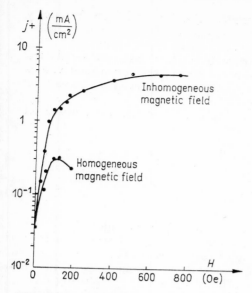

Fig. 3.21 Ion current density as a function of the homogeneous or inhomogeneous magnetic field applied to the discharge

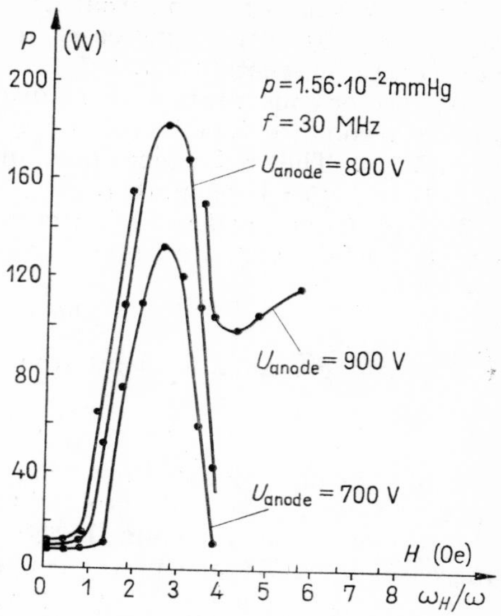

Fig. 3.22 Dependence of power consumption of discharge on the values of applied magnetic field and on ω_H/ω at different values of anode potential

discharge tube without causing thereby a change in the intensity of the discharge.

In the high frequency discharge with alternating electric field the charged particle current from the quasi-neutral plasma is determined by the diffusion process. If the linear high frequency discharge is exposed to a static magnetic field with a direction parallel to that of the electric field, the diffusion of the charged particles in the direction normal to that of the magnetic field decreases as given by formula (1.110b).

On the exposure of the linear high frequency discharge to a magnetic field in the transversal direction, it was observed [921, 922] that for $\omega_H = \dfrac{eH}{mc} = \omega$ the electron energies increase similarly to the case of resonance and that this involves a resonance-type increase in the power consumption of the discharge. The magnetic field parallel to the applied electric field of the discharge can be either homogeneous or inhomogeneous. It has been shown [712] that an inhomogeneous magnetic field applied in the direction of the axis is preferable to a homogeneous magnetic field as the ion current density in the linear high frequency discharge is substantially higher in the former case (Fig. 3.21). This difference in current density can be explained by the fact that the effect of plasma contraction, which contributes in a magnetic field to the ion current density in addition to the contribution from the decreased diffusion, is more pronounced in an inhomogeneous than in a homogeneous magnetic field.

The intensity of an alternating magnetic field induced high frequency discharge substantially increases in the presence of a longitudinal or transverse static magnetic field.

The power consumption of the discharge and the luminosity of the plasma show a resonance-type increase in a given interval of the static field values (Fig. 3.22). The resonance phenomenon can be observed at pressures $p = 10^{-3}$ to $5 \cdot 10^{-2}$ mmHg.

The resonance in the high frequency ring discharge has been observed [925] in a transversal magnetic field at frequencies

$$\omega_H = 1.5\omega\text{--}3\omega$$

while in longitudinal fields at frequencies

$$\omega_H = 3\omega\text{--}6\omega$$

The exposure of an alternating magnetic field induced high frequency discharge to a static magnetic field is also responsible for effects other than the resonance phenomenon. If the value of the magnetic field continues to increase, it causes the power consumption and the plasma concentration to increase as well. This phenomenon can be explained by the already mentioned decreasing diffusion in the direction normal to the magnetic field.

Experimental investigations of the ion current density in the high frequency discharge [712] were carried out in the absence of a magnetic field and in the presence of a magnetic field with a value H_{res} at which the resonance was found to be at a maximum. The efficiency of the power consumption in the discharge proved to be much higher in the latter case

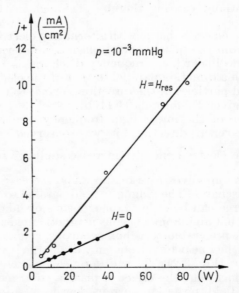

Fig. 3.23 Dependence of ion current density on the power consumption of discharge in the absence of magnetic field and in a magnetic field corresponding to the resonance value

since at the same value of power P the value of j_+ ion current density was considerably higher than in the absence of a magnetic field (Fig. 3.23).

The measurement of the ion current density distribution of a high frequency ring discharge exposed to a transverse static magnetic field [712] proved the ion current density distribution to be anisotropic, as was to be expected from the theoretical calculations (eq. (1.110b)). This means that the ion current density is higher towards the wall lying in the direction of the magnetic field than in the direction normal to the magnetic field.

3.5 HIGH VOLTAGE ARC-DISCHARGE ION SOURCES

An ion source utilizing the arc-discharge setting in at low gas pressure was first applied by Thomson for mass spectroscopic analysis [555, 556]. In this type of ion source (Fig. 3.24) the ions produced in the high voltage arc-discharge are driven by the applied high voltage towards the cathode and extracted through a channel bored into the cathode. This long channel makes it possible to obtain from the ion source a beam adequately collimated for the measurement and because of the transmission resistance of the channel, to have a higher pressure in the discharge volume than in the volume in which the ion beam is being analysed.

Many authors continued to develop and study the properties of high voltage arc-discharge ion sources [557–567, 1330, 1331]. An advanced version of the Thomson-type ion source is shown in Fig. 3.25 (Oliphant, Rutherford [558]). As is apparent from the figure, the interpenetrating discharge electrodes are of cylindrical shape.

The distance between the anode and cathode cylinders is chosen to be shorter than the mean free path for collision. In this way the discharge cannot set in within the space between the two cylinders. Ionization can occur in a considerable measure only close to the anode cylinder and this part of the volume appears in the dark space beside the anode as a thin luminous column. The concentration of the high voltage discharge plasma to the environment of the axis leads to higher ion currents through the channel in the cathode as compared with those obtained with the previous arrangement.

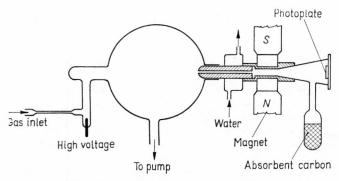

Fig. 3.24 Thomson's high voltage arc-discharge ion source

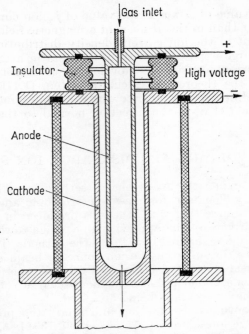

Fig. 3.25 High voltage arc-discharge ion source

The contributions from atomic ions to ion beams extracted from the high voltage arc-discharge ion sources are as high as 40 to 50 per cent. The energy spread of the extracted ions is comparable to the value of the high voltage maintaining the discharge, that is, it is very large. The gas and energy consumptions of this type of ion source are very high. The gas consumption varies from 30 to 60 cm³/h, while the power requirement from 1 to 2 kW. It seems from the above-mentioned facts that these arc-discharge ion sources have many unfavourable features. Thus, their still extensive use can be attributed to their simple construction, long lifetime and to the fact that owing to the concentration of the gas discharge, the solid angle of the extracted beam is very small and can therefore be well focused despite the large spread in ion velocities.

3.6 LOW VOLTAGE ARC-DISCHARGE ION SOURCES

It has been already mentioned in Paragraph 3.43 that the voltage needed for the ignition of arc-discharge can be considerably lowered by applying a heated cathode to increase the electron population of the discharge volume by thermionic electron emission. In this case the cathode fall, which can reach values of several times 10 kV in the case of cold electron emission, can be decreased to a few hundred volts.

However, plasma concentration is not induced by the low voltage arc-discharge, thus the low voltage operating ion sources do not yield ion currents above a few μA. To remedy this drawback, it is possible to enhance the ion concentration in a part of the plasma of the low voltage arc-discharge by a local narrowing of the discharge tube so that the plasma is mechanically concentrated over this short section. In the ion sources with locally narrowed discharge tubes the arc-discharge takes place between the hot cathode and the anode while the ions are extracted from the narrowed volume of the plasma at which the ion concentration is high.

Figure 3.26 shows a capillary arc-discharge ion source (Tuve, Dahl, Van Atta [613]; Fowler, Gibson [614]; Tuve, Dahl, Hafstadt [615]) in which the low voltage arc formed between the cathode and anode is led through an 8 mm long channel of 3.5 mm in diameter. The ions are extracted from the channel across a 5 mm diameter aperture on the side wall of the channel by use of an electrode at a negative potential relative to that of the plasma. The ions fly through a 1 mm diameter channel in the extracting electrode to the acceleration field. With an applied voltage $U_a = 100$ V, an arc current $I_a = 2.5$ A and an extracting voltage $U_{extr} = 7$ kV, the ion current obtained from this source was measured as $I_i = 1.6$ mA. The gas consumption of the source varies from 7 to 10 cm^3/h and the power consumption

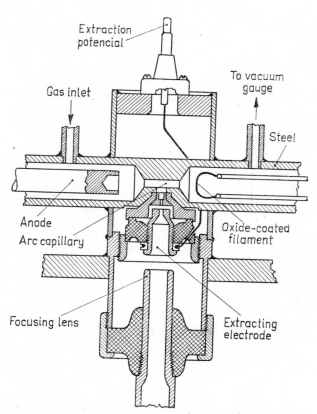

Fig. 3.26 Tuve's capillary arc-discharge ion source

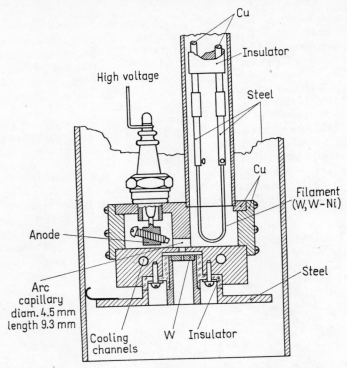

Fig. 3.27 Allison's capillary arc-discharge ion source

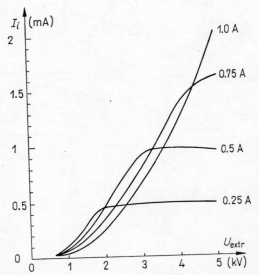

Fig. 3.28 Dependence of ion current from the capillary ion source shown in Fig. 3.27 on the extraction voltage U_{extr} at different values of arc-discharge current

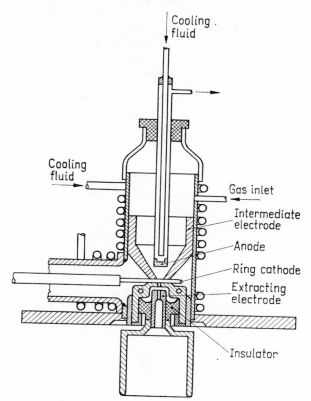

Fig. 3.29 Zinn's voltage arc-discharge ion source with ion extraction on the cathode side

is 250 W. The hot cathode of this source is a directly heated oxide cathode operated at a current of 16 A at voltages from 3 to 6 V.

The capillary arc-discharge ion source (Allison [625]), shown in Fig. 3.27, has similar operational parameters. The dependence of the ion current extracted through the channel (1 mm in diameter and 2 mm in length) on the voltage applied to the extracting electrode is represented graphically in Fig. 3.28 for different values of the arc-discharge current. It can be seen from the figure that the ion current leaving the electrode channel initially shows a rapid increase and eventually reaches a saturation value. The saturation value increases as the arc-discharge current increases. The low voltage arc-discharge ion sources with narrowed plasma have been modified and studied by many investigators [615–626]. As to the essential features there is almost no difference between the proposed versions.

Low voltage arc-discharge ion sources differing from the described types in the system and direction of ion extraction and in the formation of the narrowing in the plasma are to be seen in Figs 3.29 and 3.31. As is apparent from the drawings, the ions in this type of source are extracted from the discharge plasma not in the direction normal to that of the discharge but in the axial direction either from the cathode side [619, 626, 928] (Fig. 3.29) or from the anode side [921, 926] (Fig. 3.31).

In this geometry a plasma sphere confined by a double layer forms on the cathode side of the narrowing (see Paragraph 3.4.3) with a highly intense ionization in the interior [929, 930]. This region is more luminous than its environment in the discharge.

In the ion source of Fig. 3.29 (Zinn [619]), the narrowing in the arc-discharge between the ring-shaped cathode and the liquid-cooled anode is due to the thin slit between the electrodes and the 3 mm diameter hole of an intermediate electrode. The voltage fall in the arc-discharge $U_a = 65$ to 100 V, the arc-discharge current $I_b = 0.4$ to 2 A, while the gas pressure $p = (3 \text{ to } 12) \cdot 10^{-2}$ mmHg.

The ring-shaped cathode which emits the electrons is prepared from 1 mm diameter tungsten wire. It can be operated continuously for $\sim$50 hours. The cathode geometry is easy to set because of the sylphon coupling. The values of the ion current extracted through a 6 mm long channel, 1 mm in diameter, are plotted as a function of the extracting voltage in Fig. 3.30 for various values of the arc-current.

The low voltage arc-discharge ion source, called plasmatron, (Ardenne, Schiller [659]) is shown in Fig. 3.31. Here the helical cathode made of 1 mm diameter tungsten wire is mounted in the interior of the intermediate electrode. The 0.5 mm diameter tungsten anode is located at a distance of 3 mm from the 4 mm diameter aperture at the end of the intermediate electrode. In this ion source the plasma sphere with high ion concentration appears at the end of the intermediate electrode in front of the electron emitter helical cathode and the outlet aperture through which the ions are being extracted.

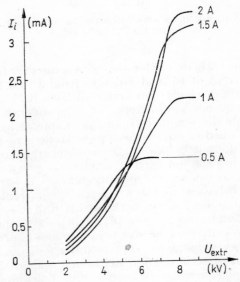

Fig. 3.30 Dependence of ion current from the source shown in Fig. 3.29 on the extracting voltage at different values of arc-discharge current

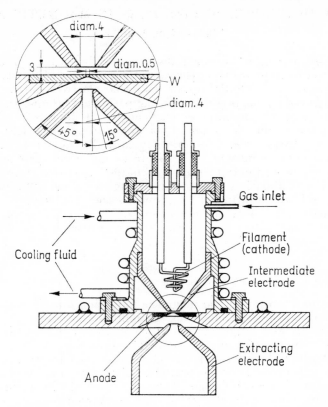

Fig. 3.31 Ardenne's low voltage arc-discharge ion source (plasmatron) with ion extraction on the anode side

Ion currents of the order of 10 mA can be obtained from this assembly at a gas pressure $p = (2-5) \cdot 10^{-2}$ mmHg, arc current $I_b = 0.3 - 2$ A, applied voltage $U_a = 30$–80 V and extracting voltages from 50–70 kV.

The spread in the energy of ions is not more than a few electron volts if the beam is extracted from an ion source with a narrowing of the plasma. These sources are therefore preferably used in experiments which require an intense ion beam of homogeneous energy. Another advantage of this type is the low power consumption since they can be operated at a few hundred watts. The reasons why this type of sources is not extensively employed despite the above advantages are the limited lifetime because of the heating filament and the poor atomic ion contribution ($\sim$20 per cent) to the extracted ion beam and if capillary arc-discharge with radial ion extraction is used, the arc-discharge is unstable and difficult to ignite. Thus, sources of this type are employed only if the ion sources operated in magnetic or high frequency fields are unsuitable for the given experimental work.

3.7 ARC-DISCHARGE ION SOURCES IN MAGNETIC FIELD
[627–681, 936–969]

The processes taking place in gas discharges induced by a magnetic field can be utilized to improve the operational parameters of arc-discharge ion sources. It has been seen in Paragraph 3.4.5 that electron and ion concentrations are enhanced by the presence of a magnetic field because the trajectory of the electrons in the discharge is lengthened and the loss in charged particles diminished under the action of the applied magnetic field.

The electron trajectories become substantially longer because of the almost helical path along which the electrons move between the two electrodes in a magnetic field. The loss in charged particles decreases partly because of the collimating effect of the magnetic field [931–933], partly because of the lowered diffusion rate towards the tube wall.

The collimation is due to the movement of the electrons with energy eU along a trajectory with a radius related to the intensity of the magnetic field as

$$r_e = \frac{1}{B} \sqrt{\frac{2m_e U}{e}} \qquad (3.35)$$

where B is the intensity of the magnetic induction and U is the potential difference accelerating the electrons. It follows that even in a relatively weak magnetic field a large fraction of the electrons cannot reach the wall of the discharge tube. Thus, the applied magnetic field can be increased to a value at which the intense discharge region of the arc with high electron and ion concentrations is concentrated to a narrow column of radius $\sim r_e$ around the axis of the discharge. With a sufficiently high magnetic field even the slow ions with charge n_e can be forced on a trajectory with radius

$$r_i = \frac{1}{B} \sqrt{\frac{2m_i U}{n_e}} \qquad (3.36)$$

which is smaller than the radius of the discharge tube.

The diffusion of charged particles in the direction normal to that of the magnetic field is due to the change of the diffusion coefficient according to formula (1.110b) as $D_H \sim \dfrac{D_{H=0}}{H^2}$.

The trajectory of the electrons in the discharge volume, that is the mean time spent by the electrons in this region, can be further increased if the electrons are somehow prevented from leaving the discharge volume. This can be done by applying an opposite electric field by which the electrons are driven back and thus forced to oscillate under the combined actions of the applied electric and magnetic fields.

The use of this oscillation of the electrons and the magnetic field increases the ionization probability per electron and leads therefore to higher ion concentrations for lower primary electron currents and lower gas pressures compared with sources utilizing applied magnetic field without forced electron oscillation.

3.7.1 ARC-DISCHARGE ION SOURCES IN MAGNETIC FIELD WITH HEATED CATHODE

The arc-discharge ion sources in magnetic field without forced electron oscillation are of two types, namely, sources with homogeneous and those with inhomogeneous applied magnetic field.

3.7.1.1 *The heated cathode arc-discharge ion sources with homogeneous magnetic field*

[637, 640, 649, 663, 964].

One version of the heated cathode arc-discharge ion sources with homogeneous magnetic field, is shown in Fig. 3.32 (Bailey, Drukey, Oppenheimer [640]). As is apparent from the figure, the directly heated cathode of the ion source is made of 1.5 mm diameter tungsten wire which has a lifetime of some hundreds of hours. The thermocathode is heated with an applied voltage of 2.5 V and a current of 60 A. The 5 mm aperture of the cylindrical anode chamber is drilled into the molybdenum disc ending the cylinder and facing the end of the narrow V-shaped cathode. In the molybdenum disc at the other end of the anode cylinder there is a slit of 1.5 mm in diameter for the extraction of the ions. The extraction slit faces the quasi-Pierce-type electric lens which focuses the extracted ions. The homogeneous magnetic field is generated by a solenoid surrounding the ion source.

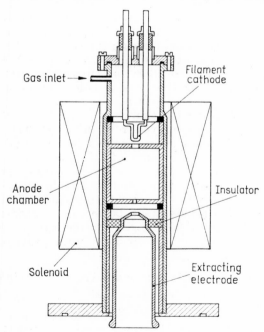

Fig. 3.32 Low intensity arc-discharge ion source with heated cathode in axially applied homogeneous magnetic field [640]

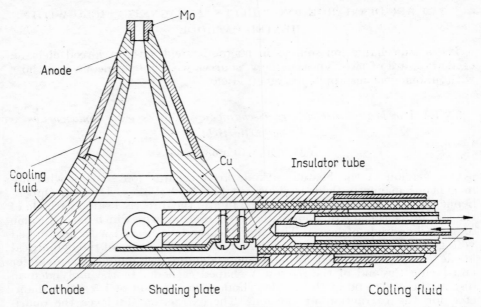

Fig. 3.33 High intensity arc-discharge ion source with heated cathode in axially applied homogeneous magnetic field [637]

The electrons collimated by the magnetic field produce the highest ion concentration in the axis of the anode cylinder, thus most of the ions formed in the chamber can be extracted through the slit located in the axis.

With a gas consumption of 8 cm³/h and with $B = 10^3$ gauss, a proton current of 500 μA and a molecular ion current of 100 μA, while with 32 cm³/h of gas and $B = 4 \cdot 10^2$ gauss, a molecular ion current of 200 μA and a proton current of 100 μA can be obtained.

Hot cathode arc-discharge ion sources without high intensity oscillating electrons can be utilized also in cyclotrons [637, 649, 964]. In this case the axially applied homogeneous magnetic field is generated by the poles of the magnet in the cyclotron. A source of this type can be seen in Fig. 3.33 (Atterling [637]). In this ion source the arc-discharge takes place between the hot tungsten cathode and the interior wall of the anode chamber. In the peaked end of the chamber the arc-discharge plasma is collimated to a thin column and thus a high intensity ion beam can be extracted through the molybdenum slit at the peaked end. Ion currents from 50 to 100 mA can be extracted from this source at a gas consumption rate of 60 cm³/h.

3.7.1.2 *Hot cathode arc-discharge ion sources in an inhomogeneous magnetic field*

Among the hot cathode arc-discharge ion sources in an inhomogeneous magnetic field [659, 662, 667–680, 926, 927, 936, 939–947] the so-called duoplasmatrons have the most favourable operational parameters (Ardenne [936]) [622, 667–680, 926, 927, 936, 939–947, 1292–1294]. In the duoplasmatron the highest ion concentration is obtainable near to the extraction slit and it can reach values as high as $5 \cdot 10^{14}$ ions/cm³. This high concentration is due to the double collimation of the arc-discharge plasma. The first narrowing is caused by the passage of the plasma through the narrow channel of the intermediate electrode, as described in the foregoing section on plasmatrons. This thin column is exposed to the strong inhomogeneous magnetic field formed in the channel of the intermediate electrode where the dimensions of the ion beam are reduced in the radial direction and the thus concentrated ion beam can be extracted through the small extraction slit on the anode. The configuration of the magnetic field in the channel of the intermediate electrode is such that it does not interfere with the formation of the plasma sphere on the cathode side of the channel and thus the electrons can be accelerated in the double layer at the plasma boundary (see Paragraph 3.4.3). The increase in electron energy and the double collimation of the plasma leads to a higher number of ionizations before the exit slit of the anode so that the ion concentration attains in this volume a high value. Under these conditions the ion beam density at the minimum cross section of the beam can be as high as 100 A/cm². In the duoplasmatron the gas is almost entirely ionized in the environment of the extraction slit and thus a gas utilization factor $\xi = \dfrac{Q_+}{Q_0}$ of 92 per cent is obtainable [680].

The duoplasmatron operates at gas pressures from $(0.5 \text{ to } 8) \cdot 10^{-2}$ mmHg. At pressures below or above this range the discharge becomes unstable and

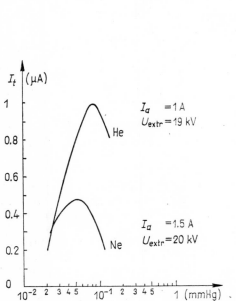

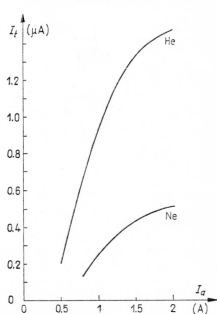

Fig. 3.34 Dependence of ion current on the gas pressure in duoplasmatron ion sources if helium or neon gas is used

Fig. 3.35 Dependence of ion current on arc-discharge current in low voltage duoplasmatron ion sources if helium or neon gas is used

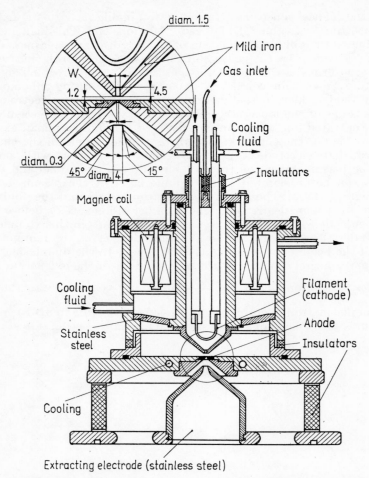

diam. 1.5

Mild iron

Gas inlet

W

1.2   4.5

Cooling fluid

Insulators

diam. 0.3

45° diam. 4   15°

Magnet coil

Cooling fluid

Filament (cathode)

Stainless steel

Anode

Insulators

Cooling

Extracting electrode (stainless steel)

Fig. 3.36 Mook's duoplasmatron-type ion source with electromagnet [662]

it is frequently quenched. The extracted ion current varies with the pressure of the ion source in a manner shown in Fig. 3.34. The maximum values of the ion current are associated with different values of gas pressure depending on the species of the gas used.

   The intensity and compositon of the ion beam extracted from the duoplasmatron depends to a great extent on the operational parameters and on the design of the source. The ion current extracted through a small slit can have a value of 100 mA, however, if the duoplasmatron operates with oscillating electrons, the ion current can attain values of 1 to 2 A (see Paragraph 3.8.2). In duoplasmatrons the value of the extracted ion current shows usually a dependence on the discharge current as plotted in Fig. 3.35 (Ciuti [942]). The small spread in extracted ion energy (~10 eV) is a function of the negative potential fall on the anode.

Duoplasmatrons are usually designed with an arrangement similar to that shown by the schematic diagram in Fig. 3.36 (Moak, Banta, Thurston *et al.* [662]). The arc-discharge forms between the hot oxide cathode and the anode. The plasma of the arc-discharge is contracted by its passage through the 1.5 mm diameter, 4.5 mm long channel of the intermediate electrode. The cathode is 8 to 10 mm wide, 0.1 mm thick and prepared from a barium oxide coated platinum or iridium band. The cathode is heated directly with a current of 30 to 60 A at a voltage of 2 to 3 V. The extraction slit is located in the tungsten disc serving as anode. This circular aperture is 0.2 to 0.8 mm in diameter. The magnetic field is generated by an electromagnet with a coil of ampere-turns from 2000 to 8000. The intermediate electrode, the metal housing surrounding the coil, which generates the magnetic field and the metal disc supporting the anode are prepared from mild iron of little carbon content, since these parts are the components of the magnetic circuit. The inhomogeneous magnetic field is generated between the peak of the intermediate electrode and the support of the anode. The value of the magnetic induction at the point of maximum field intensity can be as high as $(2-6) \cdot 10^4$ gauss. The magnetic coil, the intermediate electrode and the environment of the anode are cooled by a liquid coolant. The cathode, the intermediate electrode, the anode and the extracting electrode are electrically isolated from one another as each of them is at a different potential (Fig. 3.37). The lifetime of the duoplasmatron is determined by the lifetime of the oxide cathode and the tungsten anode with the extraction slit. With total ion currents of a few mA and an arc current of $\sim$2 A, the useful lifetime of the assembly can amount to 600–1000 hours.

In order to reduce the power requirement of the duoplasmatron the electronmagnet has been replaced by a permanent magnet in some designs (Fig. 3.38) (Ciuti [942]; Tawara, Suganmata, Suenmatsu [944]). The duoplasmatron in Fig. 3.38 also shows other modifications in addition to the application of a permanent magnet. The inner lining tube of the channel in the

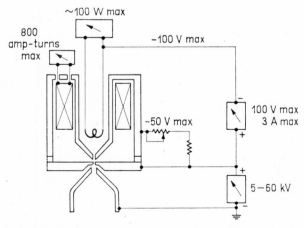

Fig. 3.37 Block diagram of power supply system for duoplasmatron shown in Fig. 3.36

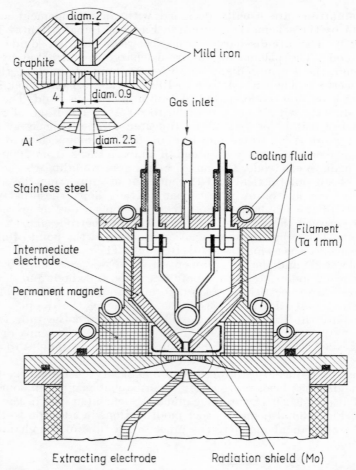

Fig. 3.38 Duoplasmatron-type ion source with permanent magnet [942, 944]

intermediate electrode and the disc with the 0.35 mm diameter extraction slit are prepared from graphite to decrease cathode sputtering during the discharge.

The divergence angle of the ion beam extracted from the usually employed duoplasmatron is generally large because of the already mentioned double plasma contraction. In order to diminish this angle, an expansion cup has been mounted by some designers after the extraction slit of the anode and kept at the same potential as that of the anode (Fig. 3.39) (Mineev, Kovpik [669]; Gabovits, Kutserenko [670, 948–950]; Gautherin [952]). This arrangement decreases the high ion concentration in the extraction slit of the anode.

This decrease, however, does not entail a similar reduction of the emitted ion current since the plasma which expands in the expansion cap is due to have a larger emitter surface area.

The particle density distribution and the shape of the emitter surface of the plasma expanded in the expansion chamber depend only on the geometry of the extracting system and on the electric and magnetic fields generated in this system [948–952]. Gases or vapours introduced into the plasma expanded in the cap can be also ionized. This phenomenon is called post-ionization (Gautherin [952]). The process of post-ionization can be exploited for highly efficient ionization of gases or vapours. If the partial pressure of the primary gas G_1 reaches a value of a few times 10^{-4} mmHg, the atoms of the gas G_2 introduced into the expansion cap are efficiently ionized [952]. Investigations have shown that a greater difference between the partial pressures of the gases G_1 and G_2 results in a greater fraction of G_2^+ ions in the beam leaving the expansion chamber. The ratio of G_1^+ to G_2^+ ions depends, of course, also on the physical properties of the ionized materials.

In the duoplasmatron of Fig. 3.39 the extraction slit of the anode is chosen to be small compared with the outlet aperture of the expansion cap in order to obtain the appropriate partial pressure ratio p_{G_1}/p_{G_2}. It is due to this arrangement that very little of gas G_2 will flow back through the anode slit and that thus the partial pressure (p_{G_2}) of the gas G_2 (or vapour) becomes sufficiently high without any appreciable penetration of G_2 atoms into the gas discharge volume through the anode slit. That is why the ion source equipped with an expansion cap as shown in Fig. 3.39, is suitable for the ionization of chemically active materials (oxygen, fluorine, and materials containing carbon) which cannot be introduced into the discharge tube of the ion source as they would cause the cathode to deteriorate or they would contaminate the tube walls.

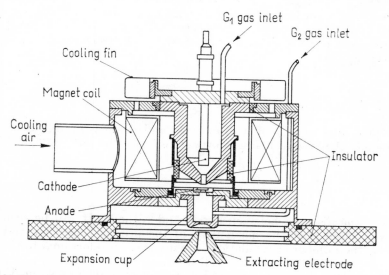

Fig. 3.39 Duoplasmatron-type ion source with expansion cap and electromagnet [952]

3.7.2 ARC-DISCHARGE ION SOURCES WITH HEATED CATHODE
AND OSCILLATING ELECTRONS IN MAGNETIC FIELD

The oscillation of electrons in hot cathode arc-discharge ion sources exposed to a magnetic field can be most conveniently achieved with the Penning-type [627, 628] electrode arrangement shown in Fig. 3.40a. The potential distribution pattern of this type of electrode assembly [643] can be seen in Fig. 3.40b. The plane electrodes K_1 and K_2 in the figure are at the same or nearly the same negative potential relative to the ring-shaped electrode A. The electric field generated between the electrodes drives the electrons in the direction of electrode A which functions as anode. If the electrode assembly of Fig. 3.40a is exposed to a magnetic field in a direction normal to the plane electrodes, the electrons will move along expanding helical trajectories. If the magnetic field is high enough, even the maximum radius of the electron trajectory will be smaller than that of the anode and thus the electrons are incapable of striking the anode but will continue to proceed towards the electrode on the other side of the anode. The electric field of the electrode at a negative potential prevents the electrons, which have lost part of their energy in collisions, from reaching the electrode surface and they are compelled to return. The process is repeated at the negative electrode on the other side of the anode and the cycle continues until the electrons have undergone a sufficient number of collisions to decrease their energy to the value at which they can strike the anode.

If a hot cathode is employed, one of the electrodes K_i functions as hot cathode, while the other is the electron reflector anti-cathode. In this case the primary electrons emitted from the hot cathode oscillate between the hot and the anti-cathode under the combined actions of the electric and magnetic fields according to the cycles described above.

The oscillation ν of the primary electrons can be estimated [953] from the expression for the mean path $\bar{x}$ covered by the electrons in the oscillation, given by

$$\bar{x} = \frac{1}{\omega_{ie} + \omega_e p} \tag{3.37}$$

If the probabilities ω_{ie} and ω_e of inelastic and elastic collisions, respectively, in the gas at pressure p are known, we can use the formula

$$\nu \approx \frac{\bar{x}}{2d} \tag{3.38}$$

where d is the distance between anode and cathode. It can be seen from these expressions that the mean path of the primary electrons and, consequently, the number of oscillations are inversely proportional to the cross sections for elastic and inelastic collisions.

It follows that the values of $\bar{x}$ and ν can substantially vary for different gas species. For example, for 100 eV energy electrons in N_2 gas $\bar{x}_{N_2} \simeq 140$ cm; $\nu_{N_2} \approx 6$, while in H_2 gas $\bar{x}_{H_2} \approx 500$ cm; $\nu_{N_2} \approx 20$, and in He gas $\bar{x}_{He} \approx \approx 800$ cm; $\nu_{He} \approx 32$.

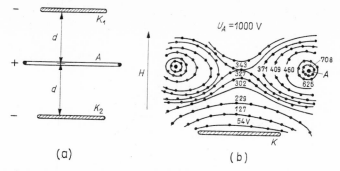

Fig. 3.40 Penning-type discharge with oscillating electrons; (a) electrode arrangement; (b) potential distribution between electrodes [643]

In the electrode arrangement shown in Fig. 3.40a the potential distribution in the case of discharge shows a markedly different pattern from that seen in Fig. 3.40b which has been obtained from measurements in the electrolyte tank [643]. Investigations of the discharge made with a thermoprobe have shown that, in addition to the small potential gradient along the discharge axis, there must be a potential change also in the radial direction because of the potential fall observed at the discharge axis [954, 955]. This phenomenon, as has already been mentioned, can be attributed to the decrease in electron diffusion in the direction normal to the magnetic field. The potential distribution decreasing in the radial direction towards the axis causes the positive ions to flow towards the axis and the cathode. As a result, the ion concentration is at a maximum in the axis of the discharge column and its value can be increased to some extent by increasing the magnetic field.

In the hot cathode arc-discharge ion sources with oscillating electrons, both homogeneous [649, 651, 956–962] and inhomogeneous [671, 963] magnetic fields can be applied. This type of source is useful in either linear or cyclical accelerators and in isotope separators. The gas consumption factor can be as high as $\xi = 60$ to 80 per cent. The intensities of the ion currents obtainable from these sources can be very high, varying from 10 to 1500 mA.

3.7.2.1 *Ion sources with oscillating electrons in a homogeneous magnetic field*

In the ion sources with homogeneous magnetic field with oscillating electrons the magnetic field is generated by solenoid [956] or one utilizes directly the magnetic field of the cyclotron in which the ion source operates [649, 651, 957–962, 1303, 1304].

One of the ion source types with a solenoid and oscillating electrons can be seen in Fig. 3.41 (Abell, Meckbach [956]). The electrons emitted from the hot cathode of the ion source reach the anode chamber across an aperture of 6 mm in diameter and are reflected from the anticathode at a 10 V negative bias relative to the cathode. The ions formed in the discharge are ex-

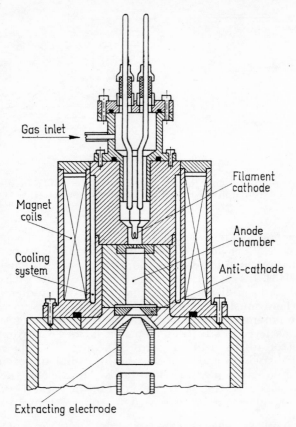

Gas inlet →

Magnet
coils

Cooling
system

Filament
cathode

Anode
chamber

Anti-cathode

Extracting electrode

Fig. 3.41 Low intensity ion source with heated cathode, oscillating electrons and
electromagnet [956]

tracted from the anode chamber through a 6 mm diameter slit in the direction parallel to that of the magnetic field and the axis of the discharge
column. The extracting system of the ion source is of quasi-Pierce type.
With a discharge current of 700 mA and at a hydrogen gas pressure of
$1.3 \cdot 10^{-3}$ mmHg an ion current of 42 mA can be extracted from this source.

In the arc-discharge ion sources with oscillating electrons which can be
used in cyclotrons and in isotope separators the ions are extracted in the
direction normal to that of the magnetic field and the axis of the discharge
[649, 651, 957–962]. A source of this type is shown in Figs 3.42 and 3.43
(Morozov, Makhov, Ioffe [651]; Pigarov, Morozov [959]). This ion source
yields a much more intense ion beam than the type described above. The
electrons are emitted here from a tungsten rod of $7 \times 7 \times 14$ mm in size
heated by electron bombardment. The bombarding electron beam has an
energy of 1000 eV and an intensity of 0.8 mA. The primary electrons are
reflected from the anti-cathode located at the opposite end of the anode
chamber and at the same potential as the emitter cathode. The dimensions

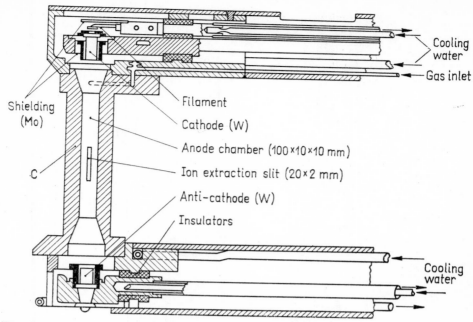

Fig. 3.42 Morozov's high intensity ion source with heated cathode and oscillating electrons [651, 959]; the axially applied magnetic field is generated by the magnet of the cyclotron

of the anode chamber are $10 \times 10 \times 100$ mm with a wall made of graphite or copper. The gas is introduced into the discharge tube through a perforated plate (gas distributor) facing the ion extraction slit. The perforated plate ensures the even distribution of the gas in the discharge volume.

The ions are extracted from the anode chamber through a slit of 2×20 mm in size with an extracting electrode at a potential of 30 kV mounted in front of the exit. The lifetime of the source is determined by the time over which the electron emitter tungsten filament of 1.2 mm in diameter is capable of operating (~ 24 hours). The ion source is also suitable for pulsed operation [561, 960, 961].

The strong magnetic field ($H \sim 7000$ Oe) needed for the operation of the ion source is the magnetic field of the cyclotron. With a gas consumption of 1 cm³/min, an arc-current $I_a = 4$ to 5 A and an arc-voltage $V_a = 750$ to 800 V, an ion current of ~ 30 mA is obtainable (Fig. 3.44). This value can be further increased by modifying the source parameters and the slit size with the result that current values of 200 to 700 mA obtained with this type of source have also been reported [959, 964]. The ion beam extracted from the source contains also multiply charged ions (see Paragraph 4.2).

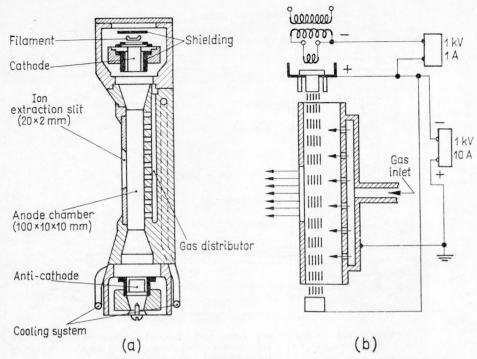

Fig. 3.43 Morozov's high intensity ion source with heated cathode and oscillating electrons: (a) cross sectional view of outlet slit of ion source; (b) block diagram of power supply system for ion source

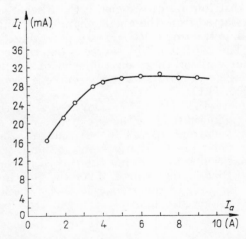

Fig. 3.44 Dependence of ion current I_i on discharge current I_a in Morozov's ion source (Figs 3.42 and 3.43)

3.7.2.2 *Ion source with hot cathode and oscillating electrons in an inhomogeneous magnetic field*

The ion source with hot cathode and oscillating electrons in an inhomogeneous magnetic field is similar in design to the duoplasmatron, as it can be seen also from Fig. 3.45 (Demirkhanov, Kursanov, Blagoveschenskii [671]; Antonov, Zinovev, Rasevskii [963]). The only difference is that an anti-cathode, at the same as, or at a more negatively biased potential than that of the intermediate electrode is mounted on the other side of the anode.

In this arrangement the electrons can oscillate between the intermediate electrode and the anti-cathode. This improves the ionization probability and thus the value of the extractable current as well.

The ion current varies with the decrease in the anti-cathode potential as shown in Fig. 3.46 where the ion current exhibits a considerable increase as the potential decreases [671]. The ion current also strongly depends on the arc-current and the magnetic field (Fig. 3.47). Experiments with arc-discharge ion sources with oscillating electrons in an inhomogeneous magnetic

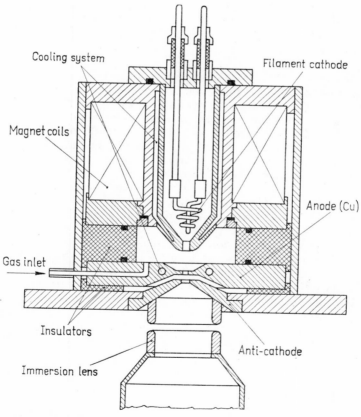

Fig. 3.45 Ion source with heated cathode, oscillating electrons in homogeneous magnetic field and with axial ion extraction [671]

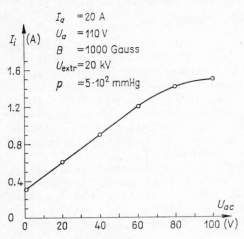

Fig. 3.46 Dependence of ion current I_i on the anti-cathode potential in the ion source shown in Fig. 3.45

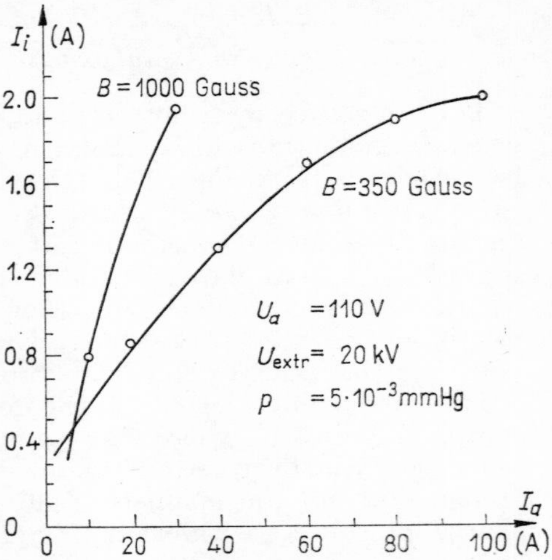

Fig. 3.47 Dependence of the ion current I_i on the arc-discharge current I_a for different sizes of magnetic field in the ion source shown in Fig. 3.45

field have shown that this type of source yields ion beams with particularly high intensities [671, 963, 1295, 1296].

The cathode of the ion source shown in Fig. 3.45 is a spiral made of 1.2 mm diameter tungsten wire which is heated to emission temperature with an a.c. of 60 to 70 A. The intermediate electrode is made of mild iron with a channel of 9 mm in diameter and 10 mm in length. The anode is of an anti-magnetic material (copper) with a transit slit of 7 mm in diameter. The distance from the intermediate electrode to the anode is 70 mm, and that from the latter to the anti-cathode is 8 mm. The anti-cathode is also made of mild iron and it also functions as one of the poles of the magnetic field. The exit slit of the anti-cathode is 6 mm in diameter. It is across this slit that the plasma expands into the volume at the same potential as that of the anti-cathode. This expansion of the plasma serves to increase the emitter surface of the discharge and to decrease the broadening of the ion beam as already explained in Paragraph 3.7.1.2.

The magnetic field applied to the ion source is generated by an electromagnet. The magnetic field in the axis of the discharge and in the environment of the anode can be as high as 1500 Oe.

With the parameter values $I_a = 20$ A, $U_a = 110$ V, $U_{extr} = 30$ kV, magnetic induction ~ 1000 gauss and $p = (5-7) \cdot 10^{-3}$ mmHg, the ion current was found to be 1.5 A [671].

3.7.3 COLD CATHODE ARC-DISCHARGE ION SOURCES WITH OSCILLATING ELECTRONS IN MAGNETIC FIELD

In the ion sources with heated cathode, the cathode lifetime limits that of the ion source and the construction of the assembly is also complicated by the use of heated cathode. For this reason, many investigators prefer cold cathode ion sources for the production of both low and high intensity ion beams. The latter have a longer lifetime, they are of simpler design and, in addition, their power requirement is lower. These types are usually known as cold cathode, Penning-type ion sources [635–639, 644–654, 657, 660, 661, 664–667, 674, 675, 965–970].

The electrode system of cold cathode ion sources with Penning-type discharge is essentially the same as that shown in Fig. 3.40a. In this case the electrodes K_1 and K_2 function as cold cathodes. If this electrode system is exposed to an inhomogeneous magnetic field with its axial direction parallel to that of the discharge axis, while a voltage higher than the discharge igniting voltage U_b is applied between the electrodes, self-maintained discharge will be generated in the discharge volume filled with low pressure gas.

As it has been seen in Paragraph 3.4, the condition of self-maintained discharge at low gas pressure is that each ion produced in the discharge volume by an electron emitted from the cathode strikes the cathode and ejects at least one new electron (formula (3.10)). The fulfilment of this condition is mainly influenced by the presence of the magnetic field, the number of electron oscillations v and by the secondary electron emission coefficient γ_i of the cathode surface.

The magnetic field is responsible for preventing the electrons emitted from the cathode and those formed in the discharge volume in the process of ionization from reaching the anode, and for decreasing the diffusion of charged particles in the direction normal to that of the magnetic field.

The electrons, which fly on a nearly helical trajectory under the action of the electric field generated between the electrodes, oscillate at the same time in the direction parallel to that of the magnetic field and to the discharge axis. Because of the complicated motion of the charged particles in the discharge volume, the above oscillation frequencies may vary over a wide range [643].

The trajectory of electrons substantially increases within the discharge volume under the combined action of electric and magnetic fields. This makes it possible for the electrons to produce a sufficient number of ionizations even if the distance of $2d$ between the electrodes K_1 and K_2 is less than the mean free path of electrons.

The secondary electron emission coefficient of the cathode surface depends, as it has been seen in Paragraphs 1.8 and 3.4, on the material of the cathode, the quality of the cathode surface and on the ion energy. Consequently the formation of a self-maintaining discharge may be influenced by the proper choice of the cathode material and the cathode surface.

Secondary electron emission induced by ions striking the cathode surface can be increased by coating the cathode surface with an oxide layer. In this case the ignition voltage U_b has a lower value. The metals suitable for use as cathode materials can be divided into two groups according to the required ignition voltage [645, 648, 674, 970, 972].

(a) *Materials requiring low ignition voltage*

	Ignition voltage U_b	Material lost to cathode sputtering in H_2, mg/ampere hour
Aluminium (Al) + O_2	350 V	< 29
Magnesium (Mg) + O_2	400 V	
Beryllium (Be) + O_2	300 V	
Iron (Fe)	400–500 V	68
Uranium (U)	500–800 V	< 30
Titanium (Ti)	800–1000 V	< 30

(b) *Materials requiring higher ignition voltage*

	Ignition voltage U_b	Material lost to cathode sputtering in H_2, mg/ampere hour
Nickel (Ni)	3600 V	65
Zinc (Zn)	3600 V	340
Aluminium (Al)	3500 V	29
Brass (Cu + Zn)	2800 V	
Monel	2800 V	
Copper (Cu)	2300 V	300
Carbon (C)	2300 V	262
Tungsten (W)	2100 V	57
Molybdenum (Mo)	1800 V	56
Tantalum (Ta)	1700 V	16
Magnesium (Mg)		9

Aluminium, magnesium and beryllium cathodes coated with oxides requiring low values of U_b gradually lose the oxide layer because of cathode sputtering. Consequently, ever higher voltages are needed for maintaining the discharge. This increase continues until the voltage value reaches that required for the ignition of the pure metal.

In the case of cathodes made of iron and uranium the discharge taking place at low applied voltage subsists for a longer operation time.

The aluminium, magnesium and beryllium cathodes coated with oxides can be operated for a longer time at a stable anode voltage if 2 per cent to 10 per cent oxygen gas is admixed with the ionized gas [635, 639, 674, 970].

The oxide layer deteriorated by cathode sputtering can be regenerated on the cathode surface if the discharge is operating for 10 to 30 minutes with oxygen gas, since it oxidizes the cathode surface again [645].

The value of U_b is not the only determinant parameter in the choice of cold cathode material for an ion source. The cathode lifetime, the anode voltage and discharge stability must also be considered. The cathode life-time is determined by the degree of cathode sputtering while the anode voltage and the stability of the discharge depend primarily on the main-tenance of the cathode surface properties.

The above requirements are best met by uranium (U) and tantalum (Ta). A cathode prepared from uranium ensures a stable discharge at low anode voltage and at a relatively low sputtering level even without the addition of oxygen [648, 657, 665].

In most cases tantalum cathodes are employed which operate at a rela-tively low applied voltage for a reasonably long time because of the low level of cathode sputtering [644, 646, 647, 652, 674, etc.]. Cathodes made of tantalum can be used in both high and low ion sources.

Apart from uranium and tantalum, magnesium (Mg) [635, 639, 660, 670, 674] aluminium (Al) [645, 674] and iron (Fe) [628, 638, 674] have been used as cold cathode material.

The ions are extracted from the cold cathode, Penning-type ion sources either axially, in the direction parallel to that of the discharge axis and of the magnetic field, or transversally, in the direction normal to the axial direction. The axial extraction is preferably applied in low intensity ion sources, while the transversal arrangement is chosen mostly for high inten-sity ion sources. However, also low intensity beams can be extracted with a transversal and high intensity beams with an axial arrangement. The contribution from atomic ions to the beams extracted from these types of ion source is rather small, not more than 10 to 50 per cent of the beam.

[3.7.3.1 Cold cathode Penning-type ion sources with axial ion extraction

One of the Penning-type ion sources with axial ion extraction [635–639, 644–648, 650, 652–654, 657, 660, 664–667, 674, 966–970] which produces low intensity beams utilizing a permanent magnet for the generation of the magnetic field is shown in Fig. 3.48 (Keller [638, 639, 970]).

The cathodes of the ion source seen in the figure are hollow to improve the stability of the discharge. Cathodes of this type are made of iron [638] or magnesium [639, 970]. In this source they are attached to the mild iron poles of the permanent magnet. The cathodes are hollowed out to an ex-perimentally determined optimum size of 8 mm in diameter and 6.5 mm in depth. The ions are extracted from the discharge volume at the anti-cathode aperture which is 1 to 2 mm in diameter. The ion current obtain-able from this source varies from 0.1 to 1 mA. The anode is a ring with 30 mm inside diameter made of iron [638] or of magnesium [639, 670]. The space between the cathodes is 25 mm. The ion source seen in Fig. 3.48 operates with iron cathodes in the anode voltage range $U_a = 470$ to 520 V and the discharge current is in the range $I_a = 10$ to 120 mA. If magnesium cathodes are used in hydrogen gas with 1–2 per cent oxygen content, a stable dis-charge can be obtained with $U_a = 250$ to 500 V [674]. If tantalum, molyb-denum, or titanium metals are used for cathode material, the discharge requires an anode voltage of the order $U_a = 2$–3 kV. The magnetic field applied to the discharge volume of this source is $\sim$1000 gauss.

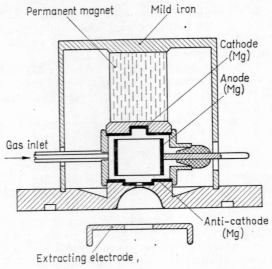

Fig. 3.48 Keller's version of a low intensity ion source with cold cathode, oscillating electrons and axial ion extraction

Figure 3.49 shows a cold cathode Penning-type ion source in which the magnetic field is generated by a solenoid (Prelee, Isalia [969]).

The cathodes of this source are made of 3 mm thick tantalum. The ions are extracted through a 0.7 mm slit in the anti-cathode. The anode is made of stainless steel and consists of two cylinders screwed together and attached to a metal disc soldered between the isolating rings. The metal disc at the same time serves for the input of the applied voltage. With this ion source

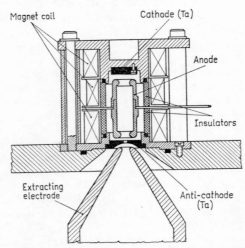

Fig. 3.49 Prelee–Isalia's version of low intensity ion source with cold cathode, oscillating electrons and axial ion extraction

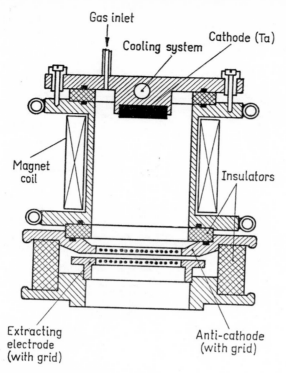

Fig. 3.50 Abdelaziz–Ghander's high intensity ion source with cold cathode, oscillating electrons and axial ion extraction. (A diaphragm with grid of large diameter functions as anti-cathode)

singly or multiply charged ions of carbon, nitrogen, neon, argon, krypton and xenon can be produced. The ion current extracted from the source varies from 2.5 to 5 mA with an extracting voltage of 16 kV, and the parameter values $U_a = $ 5–6 kV, $I_a = $ 0.5–1.5 A and H $\sim$ 270 Oe.

Two of the ion sources with similar axial ion extraction but yielding higher ion currents [648, 654, 657, 667, 669, 970, 973] are shown in Figs 3.50 (Flinta [657]; Abdelaziz, Ghander [973]) and 3.51 (Mineev, Kovpik [667]).

The source in Fig. 3.50 operates at low gas pressure ($\sim 10^{-3}$ mmHg) in a magnetic field of $\sim$ 250 to 300 Oe and yields ion currents from 200 to 600 mA through the large aperture of the anticáthode. The aperture is provided with a grid prepared from tungsten filament of a transparency of 88 per cent. The cathode is made of tantalum, the anode of copper and both are water-cooled. The source works also with a heated cathode [973]. In this case the hot cathode is mounted on insulating leads replacing the tantalum disc support. This ion source can be operated also as a pulsed source.

In the Penning-type ion source with axial ion extraction, shown in Fig. 3.51 [667, 669], the cathode is made of tantalum and the anode of copper. The tantalum cathodes are mounted here on the mild iron poles of the electromagnet. The discharge tube is 64 mm long and 6 mm in diameter. The

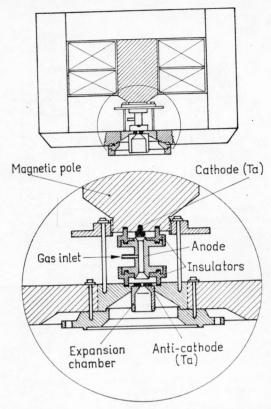

Fig. 3.51 Mineev–Kovpik's high intensity ion source with cold cathode, oscillating
electrons and axial ion extraction

gas is introduced through the side wall into the middle of the tube. Because
of the small diameter of the discharge tube, high magnetic fields have to be
applied. Consequently, the maximum value of the magnetic field between
the poles, spaced 70 mm apart, at 70 mm, can reach a value of 7 kgauss.
An expansion volume is provided after the outlet of the ion source (see
Paragraphs 3.7.1.2 and 3.7.2.2). This expansion cap is 30 mm long and 25 mm
in diameter. The discharge plasma swelling out through the 3 mm outlet on
the anti-cathode immediately expands and the ion emitter surface of the
plasma increases to a considerable extent. The ion current yield from this
source is about 100 mA. This arrangement lends itself to the production
of singly and multiply charged ions and to both continuous and pulsed
operation.

3.7.3.2 *Cold cathode Penning-type ion source with transversal ion extraction*

One of the low intensity Penning-type ion sources with transversal ion extraction [674, 940, 967, 968] can be seen in Fig. 3.52 (Heinicke, Bethge, Bauman [967]).

The source seen in the figure has 3 mm thick cathodes made of tantalum and attached to the poles of the electromagnet. The cathodes are electrically isolated from the magnetic poles. Their useful lifetime is $\sim$ 12 hours. The outlet aperture is on the side wall of the cylindrical molybdenum anode. The anode is attached to the electrode of the extracting system and it can be adjusted along with the latter. The magnetic field in the discharge volume varies between 0.5 and 1 kgauss. The gas pressure in the discharge chamber is $\sim 10^{-2}$ mmHg. The maximum ion current obtainable from this source is ~ 100 μA.

In the high intensity ion source the cold cathodes are exposed to intense ion bombardment. The temperature of the cathode surfaces can be raised by this bombardment which causes the electron emission to increase together with the intensity of the discharge. This effect can become more important if the heat conduction of the cathodes is poor. Then, the cathodes can be warmed by the ion bombardment to thermionic electron emission temperature and transform to thermocathodes. In this case the voltage drop in the discharge decreases to a few hundred volts [647].

Figure 3.53 shows one version (Anderson, Ehlers [652]) of the Penning-type, cold cathode ion sources with transversal ion extraction, suitable for the production of high intensity ion beams [647, 649, 652, 661, 965]. The

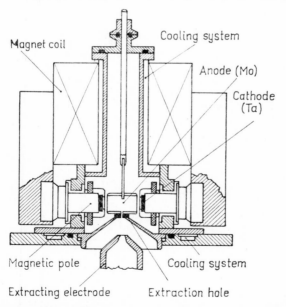

Fig. 3.52 Heiniche–Bethge–Bauman's low intensity ion source with cold cathode, oscillating electrons and transversal ion extraction

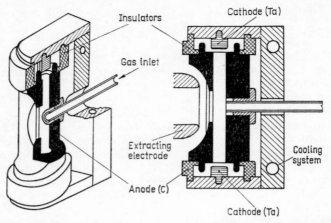

Fig. 3.53 Anderson–Ehlers-type high intensity ion source with cold cathode, oscillating electrons and transversal ion extraction

tantalum cathodes of the source shown in the figure are of cylindrical shape, 4.7 mm in height and 9.4 mm in diameter, and they can be screwed into the base plate of the cathode. The lifetime of the cathodes in a pulsed operation source ionizing nitrogen gas was found to be $\sim$ 80 hours. The inside dimensions of the anode chamber are 56 mm in length and 6.7 mm in diam-

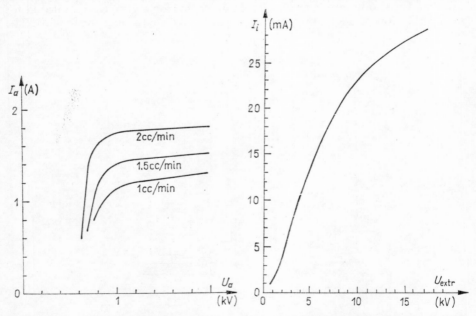

Fig. 3.54 Dependence of discharge current I_a of the ion source in Fig. 3.53 on discharge voltage U_a at different values of gas consumption

Fig. 3.55 Dependence of ion current I_i obtainable from the source in Fig. 3.53 on extraction voltage U_{extr}

eter. The anode is made of graphite and it has a 19 mm long and 1.5 mm wide outlet slit on its side, parallel to the discharge axis. The dependences of the discharge current I_a on the anode voltage are shown for different values of gas consumption in Fig. 3.54. The total values of the available ion current as a function of extracting voltage may be seen in Fig. 3.55. Singly and multiply charged ions of the elements hydrogen, helium, carbon, nitrogen, oxygen, neon, sulphur, argon and boron can be produced with this assembly.

The ion currents which can be extracted from the above described source (Fig. 3.53) with the use of a magnetic field of 4400 gauss and an extracting voltage $U_{extr} = 18$ kV, are listed below together with the specific parameter values.

Ionized gas	Discharge current I_a	Anode voltage U_a	Gas consumption		Ion current I_i	
Hydrogen	2.4 A	1750 V	H_2	9 cm³/min	43.5	mA
Helium	1.4 A	2000 V	He	6 cm³/min	29	mA
Carbon	1.3 A	1450 V	CO_2	0.5 cm³/min	11	mA
Nitrogen	1.3 A	2000 V	N_2	0.7 cm³/min	22.5	mA
Oxygen	1 A	2200 V	O_2	0.9 cm³/min	16.15	mA
Neon	1 A	2500 V	Ne	2 cm³/min	20	mA
Sulphur	0.8 A	2700 V	H_2S		6	mA
Argon	0.9 A	2900 V	Ar	0.7 cm³/min	10.1	mA

3.8 HIGH FREQUENCY GAS DISCHARGE ION SOURCES
[682–740, 974–976]

Gas discharge at high frequencies exhibits a number of features suitable for utilization in ion sources. Ion sources based on the exploitation of high frequency gas discharge can be operated at a low consumption of gas and energy, while yielding high intensity beams containing mainly atomic ions with a low spread of energy and remaining of a reliable stability for a reasonably long time.

Considering high frequency gas discharges (Paragraph 3.4.4), it has been seen that the values of the potential, the frequency of the applied high frequency field generating the discharge, the size of the discharge tube and the pressure in it are not independent. Consequently, the development of these ion-sources must be carried out with regard to this inter-connection when choosing the parameters. The frequency of the applied high frequency electric field has to be higher than the value of the limiting frequency corresponding to the size of the discharge tube (Paragraph 3.4.4.1). At a so rated frequency the high frequency power consumed by the gas discharge will be most efficient if the collision frequency is nearly equal to the field frequency of the electric field (Eq. (3.33)).

The collision frequency in the discharge is known to depend on the gas pressure, that is, higher gas pressure is associated with higher, lower gas pressure with lower collision frequencies. The high frequency voltage peak can be estimated from the value of the ignition voltage.

The power consumption efficiency of the high frequency gas discharge can be substantially improved by exposing the discharge to a static magnetic field. The increase in electron trajectories and the change of the diffusion caused by the presence of the magnetic field make it possible that the ion current density obtained for a given power P is much higher in a static magnetic field set to the value H_{res} than in the absence of a magnetic field (Paragraph 3.4.5, Fig. 3.33). The anisotropy of the ion current due to the change of diffusion can be exploited for an additional increase in the extracted ion current, if the directions of ion extraction and magnetic field are properly arranged.

The energies of the ions extracted from high frequency ion sources are almost independent of the voltage of the applied high frequency field since the ions of a relatively large mass are incapable of following the changes in the direction of the field, thus only a negligible energy transfer to the ions can take place. The energy of the ions is therefore determined by the voltage applied to the extraction system independently of the high frequency field.

In high frequency gas discharges the percentage of atomic ions is very high, possibly reaching 80 to 90 per cent. This can be attributed to the fact that the gas discharge plasma is not in contact with metal surfaces which have a high recombination coefficient. The exciting electrodes are mounted outside the discharge tube and the metal electrodes of the internal extracting system are covered with glass or quartz. This arrangement decreases strongly the surface recombination in the ion source because the glass and quartz surfaces in contact with the plasma have much lower recombination coefficients than metals (Paragraph 1.5, Appendix, Table 17).

The construction of high frequency ion sources is relatively simple. The most precise geometry is required by the ion extraction system. The discharge chamber is usually made of glass having a high melting point or of quartz. The exciting electrodes are mounted onto or around the discharge tube.

Two types of extraction system are applied in these sources. In one of them the extraction is done with a probe, in the other with the aid of a diaphragm which is a thin disc with a hole in the centre.

In the probe system the electrode at positive potential (anode) is mounted at one end of the discharge tube, while the extracting electrode (cathode) which provides the outlet channel for the ions is mounted at the other end of the tube. Both the anode and the extracting electrode are screened from the discharge plasma by a glass or quartz shielding to decrease the rate of recombination.

In the diaphragm system both electrodes are mounted at the same end of the discharge tube. The positive anode is here an electrode with a bore (diaphragm) located close to the extracting electrode and shielded from the discharge plasma by quartz or glass.

The ion beam components, the energy spread, the power and gas consumption are essentially similar in ion sources with either of the two extraction systems. The lifetimes of the two types are, however, different.

The lifetime of sources with probe extraction is considerably shorter compared with that of sources with diaphragm extraction. The reason for this is that the properties of interior walls in the extracting system and in the discharge tube are rather rapidly changed under operational conditions of probe extraction.

Owing to the sputtering of the electrode exposed to ion bombardment, the electrode material deposits on the quartz shielding of the extracting electrode and deteriorates the insulating properties. This leads to a modification of the electric field distribution at the plasma end of the extracting electrode. This modification in the electric field distribution affects the optimally formed emitter surface and the ion optics thereby causing an appreciable decrease in the ion current.

Owing to the sputtering of anode and extracting electrode, electrode material is also deposited on the wall of the discharge chamber. The metal coating of the interior wall increases the number of recombinations at the tube wall and thus decreases the ion concentration in the discharge volume.

The above effects of electrode sputtering which reduce the useful lifetime of ion sources influence much less the operating parameters in the case of diaphragm extraction.

Any change in the geometry of the extraction system involves a considerable modification of the ion current extracted from high frequency ion sources. This fact makes it possible to choose the parameters of the ion beam according to a given purpose.

3.8.1 HIGH FREQUENCY ION SOURCES WITH PROBE EXTRACTING SYSTEM

[682–685, 687, 690–693, 698, 700–705, 708–712, 715, 719, 720, 723–726, 974, 975, 1300]

The high frequency ion source with a probe extracting system proposed by Thonemann is shown in Fig. 3.56 [685]. The discharge chamber of this ion source is made of pyrex glass, 40 mm in diameter and 100 mm long. Between the two electrodes are mounted inside the discharge tube and in order to extract the ions there is a potential difference variable from 2 to 5 kV. The tungsten electrode at a positive potential (anode) protrudes at the upper end of the tube into a smaller space separated from the discharge volume by a narrowing. The ions are extracted through a 2.3 mm diameter and 19 mm long channel in the hard aluminium extracting electrode. The tubular extracting electrode which is shielded by a glass tube protrudes into the gas discharge plasma so that the ions can be extracted from the region of higher ion concentration. At the end of the extracting electrode at a negative potential a zone of positive space charge is formed between the shielding glass tube and the plasma. This zone on the plasma boundary acts as ion emitter surface and because of the electric field generated at the end of the extracting electrode, the configuration is such that the ions emerging from the surface are focused to the channel of the extracting electrode (for more details see Chapter VI).

Figure 3.56 shows an ion source in which the high frequency gas discharge is generated by an inductively coupled oscillator. The oscillator frequency is ~ 20 MHz, and with a gas consumption of 7.5 cm³/h, an ion current of 500 μA can be obtained. The contribution from atomic ions to the beam

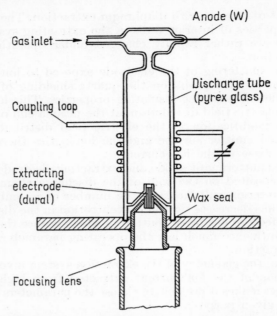

Fig. 3.56 Thoneman's low current, inductively coupled high frequency ion source with probe-type extraction

obtained from the ion source is relatively high and can reach 80 to 90 per cent. The energy spread of the ions varies from 40 to 50 eV.

Ion sources with the arrangement shown in Fig. 3.56 and its slightly modified versions have been studied by a number of authors [682, 683, 693, 698, 700, 708–711, 715, 720]. The most important modification has been the application of a permanent magnetic field to the discharge volume. The presence of the magnetic field diminishes the power and gas consumption without involving a decrease in the ion current because it promotes the introduction of electric power into the gas discharge and improves the power consumption efficiency (Paragraph 3.4.5).

Discovery of the resonance phenomenon in high frequency gas discharges exposed to a static magnetic field (Koch, Neuert [924]) suggested the application of static magnetic fields in both low and high intensity ion sources. The direction of the applied magnetic field in these sources is either normal (transversal) or parallel (axial) to the axis of the discharge tube.

A low current ion source arranged with a transversal static magnetic field [691, 692, 701–703] can be seen in Fig. 3.57 (Morozov [703]).

The discharge chamber of 50 mm in diameter and 90 mm long is made of quartz. The anode at positive potential is mounted in the narrowing at the upper end of the discharge tube. The anode and the support ring at its lower end are made of a metal (invar) with low thermal expansion coefficient. The metal is soldered to the quartz with tin. A 15 mm long channel of 1.1 mm in diameter is drilled into the aluminium extracting electrode which is shielded from the discharge chamber by a small quartz tube. The distance between the extracting electrode and the small quartz tube can be varied by means of the tubular diaphragm vacuum coupling mounted under the ion source. The high frequency gas discharge is generated by a coil surrounding the discharge tube and forming part of the oscillator circuit. The oscillator frequency is 50 MHz and its power output is 100 W. The transversal static magnetic field is generated by a permanent magnet the location of which has been experimentally determined. The reported value of the well focused ion beam current is 100 μA for a gas consumption of 1.5 cm³/h and $U_{extr} = 3$ kV. The ion beam is shaped in this source by use of a unipotential electrostatic lens.

The magnetic field applied in the axial direction is preferable to the transversal magnetic field because the anisotropy of the ion current induced by the resonance value of the former causes the ion concentration to increase in the environment of the extracting electrode. This higher ion concentration results from the concentration increase in the discharge plasma along the axis due to the decreased ion diffusion in the direction of the tube wall.

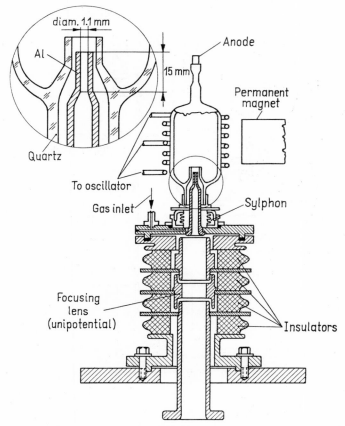

Fig. 3.57 Morozov's low current, inductively coupled high frequency ion source with probe type extraction and transversal magnetic field

In contrast, on applying a transversal magnetic field, the ion diffusion decreases in the direction normal to the magnetic field, that is, in the direction of the extracting electrode and thus the ions are not extracted from the volume of maximum ion concentration.

A specimen of low current ion sources with axial magnetic field [687, 690, 723, 975, 1300] can be seen in Fig. 3.58 (Moak, Roose, Good [690]).

The discharge tube of this ion source is 25 mm in diameter, 220 mm long and made of pyrex glass.

The aluminium anode with a cooling fin is mounted at the upper end of the discharge tube and cemented to the tube with polyvinylacetate. The cooling fin prevents the anode from excessive heating by secondary electrons accelerated in the direction of the anode. The aluminium extracting electrode has an ion extracting channel of 1.6 mm in diameter and 6.3 mm long. The extracting electrode is enveloped in a closely fitted quartz tube.

The high frequency gas discharge is generated through capacitive coupling by the two ring-shaped electrodes attached to the exterior wall of the

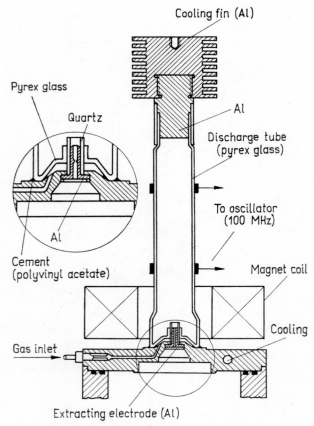

Fig. 3.58 Moak–Reese–Good's type low current, capacitively coupled high frequency ion source with probe type extraction and axial magnetic field

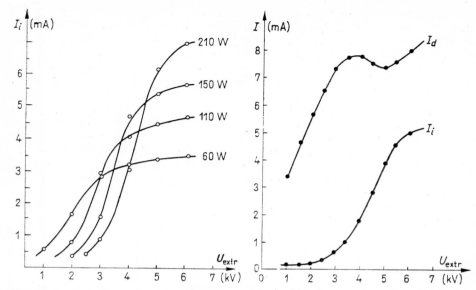

Fig. 3.59 Dependence of ion current I_i from high frequency ion sources on extracting voltage at different values of applied high frequency power output

Fig. 3.60 Discharge current I_d and ion current I_i curves plotted as a function of extracting voltage

discharge tube. The oscillator frequency is 100 MHz and its power output is 60 W. The axially directed static magnetic field is generated by the solenoid surrounding the lower part of the ion source. The thus generated magnetic field is inhomogeneous. The extracted ion current was measured as 1.2 mA for a gas consumption of 6 cm³/h with $U_{extr} = 5$ kV.

In high frequency ion sources the intensity of the ion beam can be increased by several methods. It can be achieved by increasing the diameter of the channel in the extracting electrode, by using simultaneously higher oscillator output power and higher extracting voltage or by optimizing the geometry of the extracting system.

In order to increase the ion current, the most simple method seems to be to increase the transparency of the ion extracting channel. However, this method is not always practicable as a high pumping speed vacuum system is required because of the increased gas consumption to keep the pressure of the ion source and the accelerating system at the proper level.

If the high frequency power output is increased simultaneously with the extracting voltage, the ion current can be substantially increased, as apparent also from Fig. 3.59. This method can be applied anywhere where the available high frequency output is not limited.

The geometry of the extracting system can be improved if the most appropriate configuration is determined and the geometrical precision of the setup carefully maintained, since even a small change in the geometry can cause an appreciable change in the value of the ion current. The most important functions of the ion optics which determine the performance of the extracting system, namely, the variations of $I_a(U_{extr})$ and $I_i(U_{extr})$ are shown in Fig. 3.60. It is apparent from the figure that the optimum ion optics of the extracting system are obtained if a rapid increase in the ion current is followed by a decrease in the anode current. This means that the ion current striking the wall of the extracting channel of the electrode is negligible, thus, the maximum available ion current is transmitted through the channel.

Figure 3.61 shows one of the high current intensity high frequency ion sources without applied magnetic field [698, 711, 720, 974]. The ion source in the figure (Eubank, Peck, Truell [698]) has a pyrex discharge tube and is divided by a narrowing into two parts. The larger part is the actual discharge chamber, ~ 50 mm in diameter and ~ 165 mm in length. The ~ 2.5 mm diameter tungsten electrode at positive potential is introduced into the smaller of the two parts. The ion extracting channel drilled into the aluminium electrode is 3.2 mm in diameter and 15.2 mm long, the high frequency discharge is generated by an inductively coupled oscillator, operated at a frequency of ~ 20 MHz with a power output of 300 W. The ion current extracted from this source was measured as 15 mA for a gas consumption of ~ 100 cm³/h and $U_{extr} = 5$ kV. This source combines all the three methods available for increasing the current intensity.

In this case reasonably high currents can be obtained even without the application of a magnetic field by the careful setting of the extracting system and using more gas and a higher high frequency output.

With a static magnetic field intense ion beams can be obtained for lower gas and power consumptions [705, 712, 719, 724–726] compared with those without magnetic field.

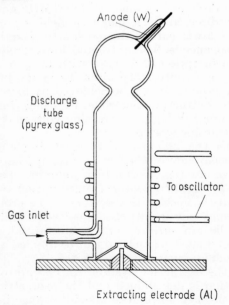

Fig. 3.61 Eubank–Pech–Truel's version of a high current, inductively coupled high frequency ion source with probe-type extraction

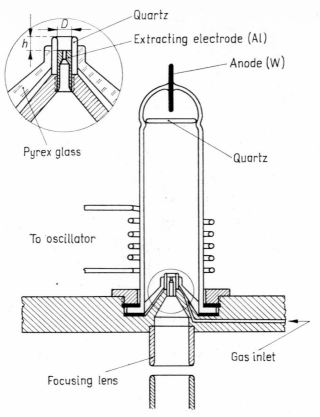

Fig. 3.62 Erő–Vályi's version of a high current, inductively coupled, high frequency
ion source with probe-type extraction

A high frequency ions source with a transversally applied static magnetic
field is to be seen in Fig. 3.62 (Erő, Vályi [705]). The discharge tube is 36 mm
in diameter and 150 mm long. The anode mounted on the upper end of the
tube is shielded from the discharge plasma by a quartz disc. The ion outlet
channel bored into the aluminium extracting electrode is 2 mm in diameter and 6 mm long. The extracting electrode is placed into a closely fitted
quartz tube. The high frequency gas discharge is generated by an inductively coupled oscillator operating at a frequency of $\sim$ 45 MHz and an output
power of 100 W.

The transversal static magnetic field is generated by a permanent magnet.
The ion source yields an ion current of 5 mA with a gas consumption of
$\sim$ 15 cm³/h and U_{extr} = 6 kV.

In the high frequency ion sources discussed up to now the ions are extracted in the direction parallel to that of the axes of the electrodes by which
the discharge is being generated. This arrangement is used for high frequency discharge ion sources with either inductively coupled alternating

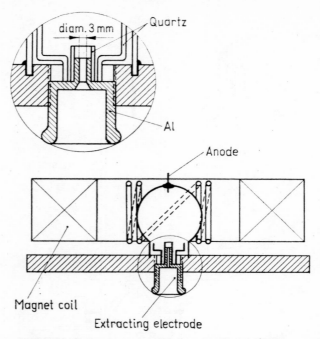

Fig. 3.63 Gabovits's version of a high current, inductively coupled high frequency
ion source with probe type extraction

magnetic field or capacitively coupled alternating electric field. Now, let
us consider the ion source shown in Fig. 3.63 where the ions are extracted
in a radial direction, normal to that of the axis of the exciting coil (Gabovits [712]). To increase the ion concentration, the ion source is exposed
here to a magnetic field in the same direction as that of the ion extraction.
With an assembly of this type of radial ion extraction, the ion current can
exceed by a factor of 3 that obtained with an ion extraction parallel to the
direction of the axis of the exciting coil. The gas discharge chamber is of
spherical shape with the anode mounted on its upper part. The ion channel
of the extracting electrode is 3 mm in diameter and 10 mm in length.

The high frequency discharge is generated by an inductively coupled
oscillator operating at a frequency of $\sim$40 MHz with a power consumption
of 300 W. For a gas consumption from 30 to 60 cm³/h and $U_{extr} = 5$ kV,
ion currents between 5 and 6 mA have been obtained from this source.

Figure 3.64 shows a high frequency ion source with a similar arrangement
of radial ion extraction which, however, yields much higher currents than
the above described assembly, if the high frequency output and the value
of U_{extr} are increased (Abdelaziz [974]). The discharge chamber of this ion
source is made of pyrex glass and the annular nickel anode is mounted into
the extension on the discharge tube facing the extracting electrode. The
extracting electrode is made of aluminium with a 3.5 mm diameter and 9 mm
long bored ion channel and it is shielded from the plasma by a pyrex cap.

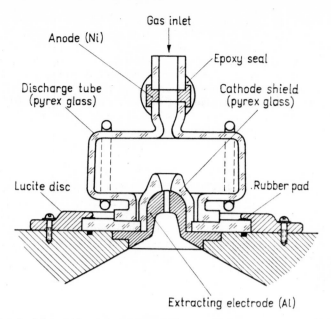

Fig. 3.64 Abdelaziz's version of a high current, inductively coupled high frequency ion source with probe-type extraction

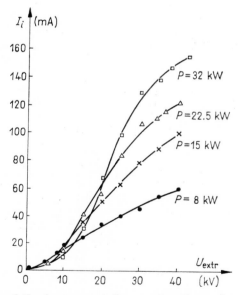

Fig. 3.65 Dependence of the ion current I_i on extraction voltage U_{extr} in the ion source of Fig. 3.64 at different values of high frequency power input

The end of the extracting system extends to the axis of the gas discharge to ensure that extraction takes place from a region with high ion concentration. The high frequency gas discharge is generated by a capacitively coupled oscillator operating at a frequency of 100 MHz energized from a pulsed generator with a power output from 30 to 50 kW. This pulsed ion source intensity has been reported to yield, in the peak of the pulse, an ion current of 150 mA for a gas consumption of 30 cm³/h and $U_{extr} = 40$ kV. The variation of the extracted current I_i with U_{extr} is shown in Fig. 3.65 for different high frequency power outputs.

3.8.2 HIGH FREQUENCY ION SOURCES WITH DIAPHRAGM-TYPE EXTRACTING SYSTEM

[686, 707, 721, 727, 730, 732, 733, 735, 738, 1298, 1299]

In the high frequency ion sources with diaphragm type extracting system, the anode and the extracting electrode mounted close to each other form a quasi Pierce-type system of ion optics. The plane, or spherical geometry of the anode and the extracting electrode can be implemented to high precision and the ion optics can be optimized in this way. In the design of a

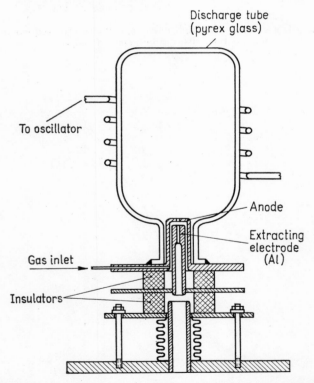

Fig. 3.66 Bayly–Ward's version of a low current, inductively coupled high frequency ion source with diaphragm type extraction

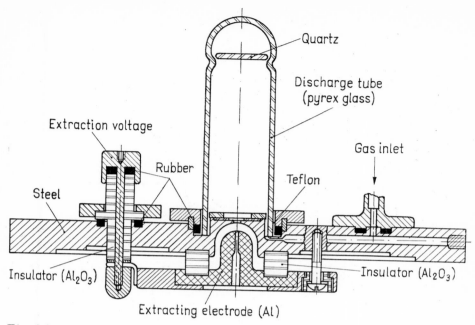

Fig. 3.67 Vályi's version of a low current, inductively coupled high frequency ion
source with diaphragm-type extraction

diaphragm type extraction system more care can be taken of the factors
increasing the lifetime of the ion source. Arrangements of this type have
been reported to work reliably for several hundred (300 to 1500) hours.

An extraction system with a diaphragm was first used by Bayly and
Ward [686] in the ion source shown in Fig. 3.66. The pyrex discharge chamber of the source is 75 mm in diameter and 11 mm long. The aluminium anode
and the duralumin extracting electrode are mounted into the narrowing
at the lower part of the tube. In the anode, which ends in a plane surface,
there is a 2 mm diameter and 1.5 mm long bore, while the bore in the extracting electrode is 2 mm in diameter and 12 mm long. The high frequency
discharge is generated by an inductively coupled oscillator of 15.5 MHz
frequency with a power consumption of 400 W. The ion current is 500μ A
for a gas consumption of 15 cm³/h and $U_{extr} = 1.2$ kV. These data show
that for a relatively high gas and power consumption the ion current obtained from this source is low.

Figure 3.67 shows an ion source (Vályi, Gombos, Roosz [730, 735]) which
yields about the same or somewhat higher ion currents but the values of
gas consumption are considerably lower and also the power requirement
of this source is less than that of the source in Fig. 3.66. The ion source is
made of pyrex glass with a discharge tube of 30 mm in diameter and 90 mm
long. The quartz disc mounted on the upper end of the tube protects the
upper wall from the secondary electrons. The extracting system has a spherical geometry (Fig. 3.68). The electrodes are made of high purity alumin-

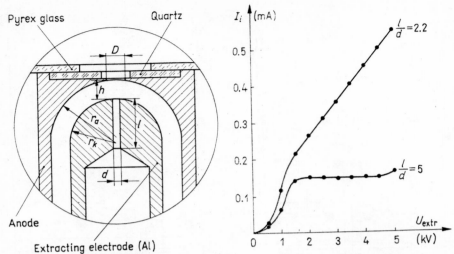

Fig. 3.68 Extraction system of spherical geometry in the ion source shown in Fig. 3.67

Fig. 3.69 Dependence of ion current I_i from the ion source in Fig. 3.67 on extracting voltage U_{extr} at different values of l/d [730, 735]

ium. The hemispherical anode with a radius $r_a = 9$ mm has an aperture of $D = 2$ mm in diameter and 0.2 mm long. The similarly hemispherical extracting electrode with a radius $r_k = 6$ mm is located at $h = 4.2$ mm distance from the anode and the bore forming the ion extracting channel is $d = 0.8$ mm in diameter and $l = 2$ to 7 mm long. The coaxiality of the anode and extracting electrode bores is kept within an accuracy of 0.02 to 0.03 mm. The anode surface facing the plasma is shielded by a quartz disc of 1 mm thickness which has a 4 mm diameter hole in the middle. The discharge plasma is at the anode potential. Ion current against extraction voltage curves, as plotted for two different channel lengths ($l = 4$ mm, $l = 2$ mm) can be seen in Fig. 3.69. The gas consumption is 2–3 cm³/h for $l = 4$ mm and 4–5 cm³/h for $l = 2$ mm. The high frequency gas discharge is generated in this source by an inductively coupled oscillator operating at a frequency of 45 MHz with a power output of 120 W.

Figure 3.70 shows one of the low current ion sources (Romanov, Serbinov [721]) in which a transversal magnetic field and an extracting system with diaphragm are applied [721, 732, 976].

The discharge tube of the ion source is made of pyrex glass. It is ~30 mm in diameter and ~80 mm long. The extracting system consists of a shielding disc pressed to a special shape from pyrex glass, an extracting electrode and an electrode at the plasma potential which supports the shielding pyrex disc. The hole in the pyrex disc is 1.2 mm in diameter, while the ion channel of the extracting electrode has a diameter of 0.6 mm and it is 3 mm long. The distance between the upper end of the extracting electrode and the plane of the pyrex disc facing the plasma is 0.8 to 1.0 mm. The high frequency discharge is generated by an inductively coupled oscillator operating at frequencies of 40 or 80 MHz, with an output power of 40 W and 100 W, respectively. The transversal magnetic field is generated by a permanent magnet. The ion current of this source was measured as 220 μA for a gas consumption from 1.2 to 1.6 cm^3/h at high frequency power of 100 W and $U_{extr} = 1.2$ kV.

Higher ion currents can be obtained from high frequency ion sources with a diaphragm type extraction system if an applied magnetic field [707, 717, 733, 735, 977] increases the ion concentration in the gas discharge.

One of the methods for producing high ion currents is the use of several ion outlet channels in the extracting system [717, 977]. A multichannel extraction system decreases the gas consumption of the ion source, while the minimum diameter of the ion beam is larger than that of the beam obtained with a single channel extraction. An ion source with several extracting channels is shown in Fig. 3.71 (Serbinov, Moroka [717]). The dis-

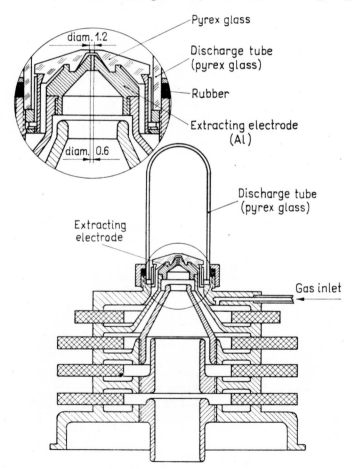

Fig. 3.70 Serbinov–Romanov's version of a low current, inductively coupled high frequency ion source with diaphragm-type extraction

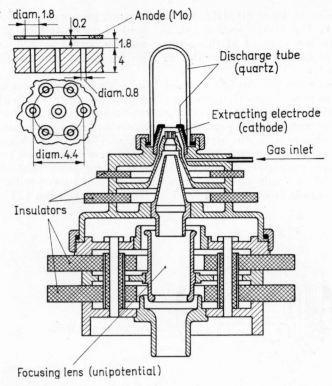

Fig. 3.71 Serbinov–Moroka's version of a high current, inductively coupled high frequency ion source with several ion extracting channels

charge chamber is made of pyrex glass with a diameter of 30 mm and a length of $\sim$80 mm. The extracting assembly is made of 0.2 mm thick molybdenum. The anode has 7 holes of 1.8 mm in diameter and it is mounted above the duralumin extracting electrode which contains 7 ion channels of 0.8 mm in diameter and 4 mm long. The distance between the two electrodes is 1.8 mm. The high frequency gas discharge is generated by an inductively coupled oscillator with a frequency of 80 MHz at a power of 300 W. The transversal magnetic field is generated by a permanent magnet. The ion current obtained from this source was measured as 16 mA for a gas consumption of 25 cm³/h, at $U_{extr} = 4.5$ kV.

Figure 3.72 shows a high frequency ion source with axial applied magnetic field, of high current yield (Vályi, Gombos, Roosz [733, 735]). The discharge tube of the ion source is made of quartz 30 mm in diameter and 90 mm long. The cooling fin placed at the upper end of the discharge tube dissipates the heat generated by the secondary electrons which are focused by the axial magnetic field to the upper end of the discharge tube. The extracting system is of spherical geometry (Fig. 3.68). The anode protrudes from the basic plate of the ion source to enable the ion extraction from a higher ion concentration region. The hemispherical anode is made of aluminium with a radius

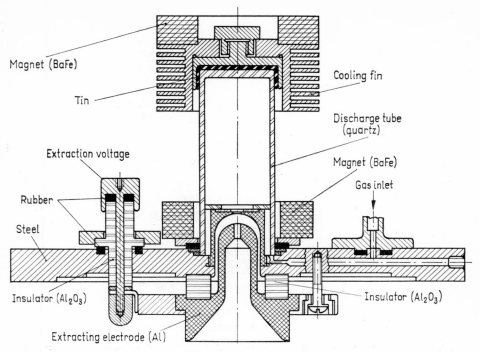

Magnet (BaFe)

Tin

Cooling fin

Discharge tube (quartz)

Extraction voltage

Magnet (BaFe)

Gas inlet

Rubber

Steel

Insulator (Al$_2$O$_3$)

Insulator (Al$_2$O$_3$)

Extracting electrode (Al)

Fig. 3.72 Vályi's version of a high current, inductively coupled high frequency ion source with diaphragm-type extraction

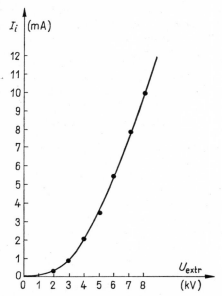

Fig. 3.73 Dependence of ion current I_i obtainable from the source in Fig. 3.73 on the extracting voltage U_{extr} [733]

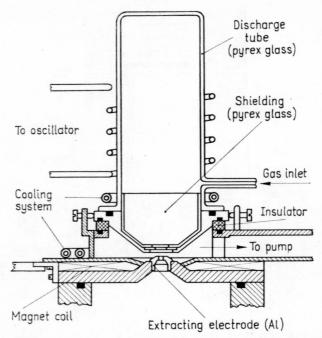

Discharge tube (pyrex glass)

Shielding (pyrex glass)

To oscillator

Gas inlet

Cooling system

Insulator

To pump

Magnet coil

Extracting electrode (Al)

Fig. 3.74 Harrison's version of a high current high frequency ion source with diaphragm type extraction

$r_a = 10$ mm and a concentrically located bore, 5 mm in diameter and 0.2 mm long. The channel of the extracting electrode is of a diameter $d = 2$ mm and a length $l = 5$ mm. The distance between the two electrodes $h = 3.5$ mm. The anode is isolated from the plasma with a quartz disc of 1 mm thickness and a bore of 7 mm in diameter. The high frequency discharge is generated by an inductively coupled oscillator operating at a frequency of 45 MHz with a power output of 140 W. The axial static magnetic field is generated by annular anisotropic barium-ferrite permanent magnets. The ion current from this source is 10 mA for a gas consumption of 30 cm³/h at $U_{extr} = 8$ kV. Higher ion currents can be obtained with higher extracting voltages since the $I_i(U_{extr})$ characteristic curve of the source (Fig. 3.73) shows that the value of 10 mA lies on the steeply rising section of the $I_i(U_{extr})$ curve.

If a more inhomogeneous static magnetic field is applied, not only resonance but also plasma contraction can be observed in the high frequency discharge (Paragraph 3.4.5). These two effects are exploited in the ion sources [707, 727, 728, 738] of the type shown in Fig. 3.74 (Thonemann, Harrison [707]; Harrison [728]). The pyrex discharge tube of this source is ∼45 mm in diameter and ∼110 mm long. The extracting assembly consists of a plane electrode with a bore of 10 mm in diameter, mounted above an extracting electrode at a distance of 6 mm. The extracting electrode is made of aluminium with an ion channel of 6 mm in both diameter and length. A specially shaped electromagnet, mounted close to the extracting electrode, generates a field of ∼7 · 10² gauss in the ion channel. The anode is separated from the discharge plasma by a pyrex glass shield. The high frequency discharge is generated by an inductively coupled oscillator operating at a frequency of 100 MHz and an output power of $\sim$400 W. The ion current obtained from this source was measured as 49 mA for a gas consumption of $\sim$100 cm³/h and $U_{extr} = 10$ kV.

3.9 ION SOURCES IONIZING SOLID STATE ELEMENTS

Ion beams obtained from solid state elements have lately become of increasingly extensive use. Initially, this type of ion source [978–986] was used mainly for mass spectroscopic studies and isotope separation; recently, they have become a tool in the research of interactions between solids and ions [993, 1012, 1028] and useful also in industrial technological procedures [1341].

One of the promising applications in the last group is the process of ion implantation. This technique enables one to introduce the accelerated ions into the surface layer of various materials. The most important use of this technique seems to be in microelectronics. Other hopeful applications are the production of conducting layers, high precision alloying of different materials or doping of a material with impurities of given concentrations [993, 1002, 1008, 1013, 1014].

In the development of ion sources suitable for the above applications, two important problems have to be coped with. One of them is the production of vapours of solid elements at an appropriate pressure and the other is the ionization of the vapour. Vaporization of solid elements involves not only a modified assembly by comparison with the ion sources considered up to now but it also requires utilization of physical properties of little importance in previously discussed arrangements

Ions of solid elements are most conveniently obtained from their gaseous compounds. If such compounds are available, the ionization of solid elements can be achieved in any type of the sources described in the foregoing. If gaseous compounds are used, first the molecules must be dissociated and then the atoms of the molecules have to be ionized. However, in the course of ionization each constituent of the molecule and then on-dissociated molecules are also ionized. Consequently, the ion beam obtained from these sources contains all these other ions, too. For this reason, the ion beam needed for any application other than for isotope separation or mass spectrometry must be isolated from the unnecessary components.

The ion sources of solid elements, in which the material is ionized in its pure elementary state, can be grouped according to the methods used for the production of the vapour and of the ions in question. The vapour pressure needed for the ionization of solid elements is obtained in most cases by heating the element in an evaporator oven. The elements can also be vaporized by electron or ion bombardment, and by arc discharge or spark discharge. The vaporized elements can be then ionized by any of the earlier described methods.

The evaporator oven can be mounted either inside or outside the ionization chamber. In the sources with an external evaporator oven there is a vapour outlet tube connected to the ionizer chamber. In this design the evaporator and ionizer can be set to the optimum parameter values independently of each other. Thus, the temperature of the evaporator and the ionizer, that is, the vapour pressure of the element they contain, can be chosen to be the most appropriate in either case.

If the evaporator is mounted within the ionizer, the free choice of the best parameter values is more restricted. On the other hand, the power consumption can be reduced and high temperatures can be obtained.

3.9.1 ION SOURCES WITH EVAPORATOR MOUNTED OUTSIDE OF THE IONIZER

In ion sources with an external evaporator an important condition of operational stability is that the transmission of the vaporized solid element to the ionizer should be independent of the quantity of material present in the evaporator — similarly to the case of the atom sources considered in Chapter II [992]. This condition requires that the length and cross section of the vapour outlet tube from the evaporator to the ionizer should be chosen such that the evaporation rate from the material exceeds the vapour transmission rate through the outlet tube, that is, the quantity $Q = C(p_{ov} - p_i)$ transmitted by the tube. This formula shows that the value of Q depends on the penetrability C of the inlet tube and on the difference between the pressures at its two ends. It follows that the pressure p_{ov}, that is, the temperature of the evaporator, must be set to a value depending on the pressure p_i required in the ionizer and on the value of C.

The temperature dependence of the pressure in the evaporator can be formulated [992] as

$$\log p_{ov} = -\frac{A}{T} + B \tag{3.39}$$

where A and B are constants [1033]. Temperature values required for obtaining given pressures in some materials of interest are listed in Appendix, Table 13. It follows from relation (3.39) that

$$\frac{dQ}{Q} = \frac{dp_{ov}}{p_{ov}} = \text{constant } \frac{dT}{T^2} \tag{3.40}$$

This expression shows that the vapour flow is of higher stability at higher than at lower temperatures. The stability and variability of the vapour flow rate depend on the heat capacity of the evaporator. The lower the heat capacity of the evaporator the easier it is to vary its temperature, thus, the vapour pressure of the element it contains.

To avoid gas condensation, not only the evaporator but also the vapour inlet tube and the ionizer have to be kept at a suitably high temperature.

In ion sources with low intensity ion currents and external evaporator, the ions are often produced by electron bombardment or by surface ionization.

One of the ion sources, ionizing by electron impact (Paragraph 2.2) is shown in Fig. 3.3. This arrangement is suitable for ionizing solid elements.

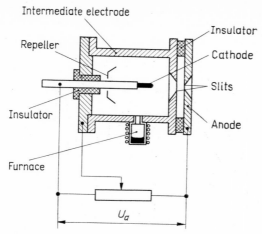

Fig. 3.75 Severac's version of a plasmatron-type ion source for production of ions from solid elements

It is preferably used in mass spectroscopic analysis which requires low ion currents.

One of the most advantageous arrangements of ion sources for utilizing surface ionization (Paragraph 3.3) is shown in Fig. 3.12. In this ion source a well-collimated atom beam leaves the evaporator and strikes the heated metal surface where the atoms are ionized.

More conventionally, the atoms of solid elements are ionized in a discharge process. The discharge is generated either in auxiliary noble gas or in the vapour of the ionized material at an appropriate pressure. In principle, all the gas discharge ion sources prepared from a suitably chosen material and equipped with an evaporator can be employed for the production of ion beams from solid elements.

A plasmatron without magnetic field (Paragraph 3.6), as modified for the production of ion beams from solid elements, is shown in Fig. 3.75 (Severac [1018]).

The ion outlet is a slit of 5×40 mm on the intermediate electrode of the ion source. The slit on the anode, which is at 1.5 mm from the intermediate electrode, has a size of 1×40 mm. The two electrodes are isolated from each other by an insulating distance rod. The evaporator of the solid material is directly coupled to the side wall of the intermediate electrode. Ionization takes place in the discharge plasma generated in the auxiliary argon gas. Thus the ion beam leaving the source also contains argon ions apart from the ions of the solid element. The total ion current at the collector is 1.5 mA. If the evaporated element is copper (Cu), the contribution from Cu^+ ions to the total current is 400 μA, if it is silver (Ag) the contribution from Ag^+ ions is 200 μA.

Figure 3.76 (Freeman [1009, 1036]) shows a version of the arc discharge ion sources with hot cathode and magnetic field modified for the ionization of solids. In this source the ions are extracted in the direction normal to

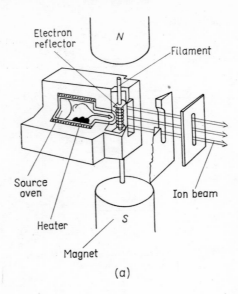

(a)

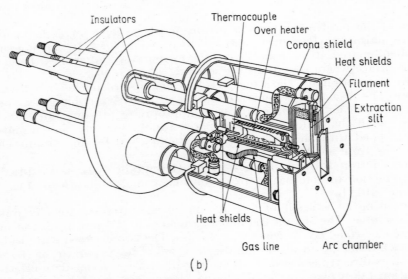

(b)

Fig. 3.76 (a) Schematic drawing of Freeman's version of an ion source with hot cathode producing ions from solid elements; (b) Freeman's version of an ion source with hot cathode producing ions from solid elements

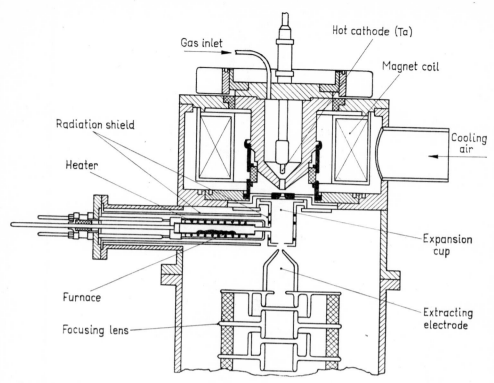

Fig. 3.77 Masic–Sautter–Warnecke's version of a duoplasmatron-type ion source with expansion cup for production of ions from solid elements

that of the magnetic field. The exit slit of 42 mm in length and from 1.5 to 2.5 mm in width is on the side wall of the tantalum discharge tube. The hot cathode is located at 3 mm from the exit slit, and parallel to the direction of the magnetic field. The cathode is electrically isolated from the discharge tube and kept at a potential which is more negative than that of the tube, differing from the latter by a value from 0 to 200 V. The cathode is made of a tantalum ribbon, sized $4.5 \times 1 \times 45$ mm; its useful lifetime is 100 hours. The electrons emitted from the cathode and released by ionization move on complicated trajectories in the discharge chamber due to the combined actions of the external magnetic field and the magnetic field generated by the current flowing through the cathode. Thus, their trajectories in the discharge plasma are considerably lengthened. The discharge plasma has the highest ion concentration in the region around the exit slit. The pressure required to maintain the discharge is obtained either with the mixture of the auxiliary noble gas and the vapour or with the vapour alone. The vaporized element is introduced from the evaporator into the ionizer by a vapour outlet tube. This ion source yields an ion current of 0.1 to 10 mA with the parameter values $I_a = 1$–6.5 A, $U_a = 40$–100 V, magnetic field 20–150 gauss, cathode current 150–200 A, $U_{\text{extr}} = 20$ kV.

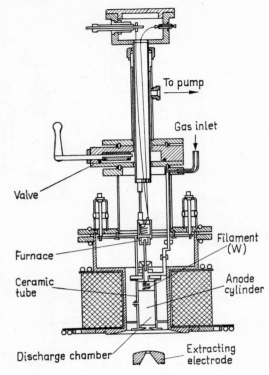

Fig. 3.78 Nielsen's version of an ion source of solid elements with oscillating electrons

The duoplasmatron with hot cathode, inhomogeneous magnetic field and expansion cup (Paragraph 3.7.1.2) also lends itself to the production of ion beams from solid elements [939, 987]. Figure 3.77 shows a duoplasmatron with expansion cup and an evaporator. The expansion cup is heated with a filament mounted around the cup. The vapour of the element to be ionized is introduced through a short, heated tube from the evaporator into the expansion cup. The vapour is then ionized in the helium or argon gas discharge plasma ballooning out from the duoplasmatron into the expansion cup (Paragraph 3.7.1.2). The ion current obtainable from this assembly is 0.4 to 1 mA.

The modification of a Penning-type gas discharge ion source with oscillating electrons and ion extraction in the direction of the magnetic field can be seen in Fig. 3.78 (Nielsen [992]).

The ion source is placed into a vacuum chamber of stainless steel which has a set of voltage input connectors to supply the discharge chamber, the magnet and the evaporator. A more detailed drawing of the discharge chamber is shown in Fig. 3.79. The base and top plates of the discharge chamber, which function as cathode, are made of stainless steel tubes separated by a cylindrical insulating ceramic spacer. The molybdenum anode is fixed to this ceramic spacer. The discharge chamber is kept in a fixed position with

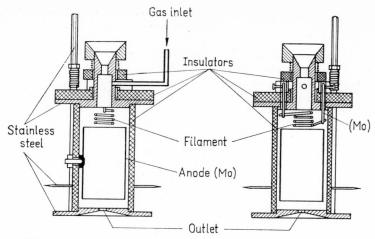

Fig. 3.79 Discharge chamber of the ion source of Fig. 3.78

stainless steel rods. The stainless steel ring mounted outside the ceramic insulator serves for the concentric alignment of the chamber. The tungsten heating filament is attached to the molybdenum rod support of 1.2 mm in diameter which is passed through a pyrophylite insulator disc and connected to a stainless steel block. The ion current intensity can be varied by varying the temperature of the filament. The centre of the filament is kept at a negative potential of ~5 V relative to that of the top and base plates of the discharge chamber. The auxiliary gas inlet is a thin-walled stainless steel tube. The maximum power consumption of the heating filaments was measured as being 600 W.

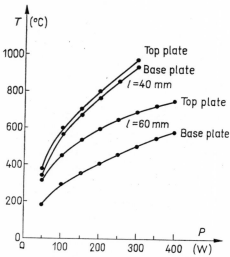

Fig. 3.80 Temperatures of base and top plates of the ion source in Fig. 3.78 as a function of power consumption of filament

The temperature curves for the top and base plates plotted against the power consumption of the filaments are shown for 40 mm and 60 mm tube lengths in Fig. 3.80.

A modified version of the discharge tube made of graphite in order to reach higher temperatures can be seen in Fig. 3.81.

The solid elements are evaporated in the evaporator composed of two parts, as is shown in Fig. 3.78. The lower part contains the material to be evaporated, the upper part houses the tungsten heating filament. The metal vapour is passed from the evaporator to the discharge tube through a thin tube. The flow rate of the metal is kept at a value which does not increase the pressure in the main vacuum chamber of the separator. If necessary, an auxiliary gas can be introduced into the discharge chamber. The magnetic field of 200 to 1000 gauss required for the Penning-type discharge is generated by a solenoid. The lifetime of this source is determined by that of the electron emitter tungsten filament and by the amount of material which can be placed in the evaporator. The ion current yield of this source varies between 0.03 and 1 mA.

An alternative type of the source shown in Fig. 3.78, which has been used for the magnetic separation of short-lived radioactive isotopes, has one discharge chamber and two evaporators (Uhler, Alvager [995]). One

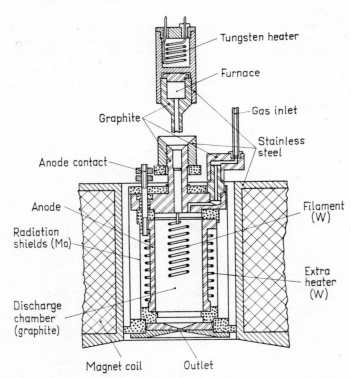

Fig. 3.81 Modified discharge chamber of the ion source shown in Fig. 3.78 to obtain higher temperatures

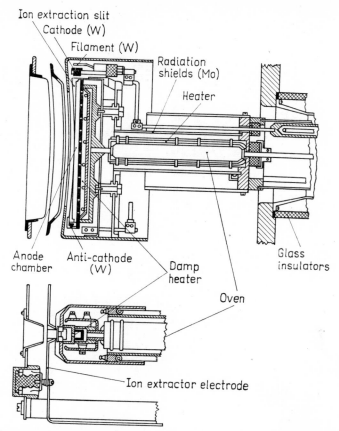

Fig. 3.82 Zolotarev–Ilin–Komar's version of an ion source with hot cathode and oscillating electrons for production of ions from solid elements

of the evaporators contains a relatively larger amount of the stable isotope used for the start and adjustment of the separator. The other evaporator contains the radioisotope. This evaporator is switched on only if the separator has already been brought to a steady operation.

Penning-type ion sources with ion extraction in the direction normal to that of the magnetic field are used preferably in high performance isotope separators and cyclotrons.

Figure 3.82 shows an ion source suitable for electromagnetic isotope separators (Zolotarev, Ilin, Komar [984]). This source operates with high utilization coefficient of material thus it has a high ionization efficiency, high current intensity and a number of other advantageous properties. The solid elements are evaporated in an evaporator oven and the vapour is introduced from there first into a vapour distributor through a short tube, then into the discharge tube. The vapour distributor ensures even vapour input distribution to the discharge chamber. The evaporator, the vapour tube and distributor as well as the discharge chamber are made of a material

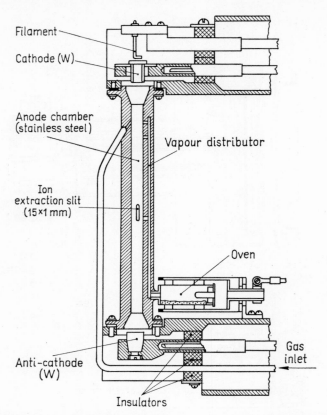

Filament

Cathode (W)

Anode chamber
(stainless steel)

Vapour distributor

Ion
extraction slit
(15×1 mm)

Oven

Anti-cathode
(W)

Gas
inlet

Insulators

Fig. 3.83 Pasyok–Vorobev–Ivanikov–Kuznyecov–Kytner–Tretakov's version of an
ion source for solid elements with hot cathode and oscillating electrons

which does not interact with the evaporated solids. The evaporator is heated
to the required temperature by a heater inserted into a double cylinder
made of molybdenum or stainless steel. In order to avoid vapour condensation, the vapour distributor, made of graphite, and the discharge chamber
are heated as well. The anode of the discharge is the wall of the discharge
chamber while a tungsten block functions as cathode. One of the cathodes
is heated by electron bombardment to the temperature required for electron
emission. The Penning-type discharge is generated in the strong magnetic
field of the isotope separator. The indirect heating of the cathode results
in a longer lifetime of the ion source. The atoms of the solid element are
ionized in the plasma formed in the discharge chamber. The ions are extracted from the source through a slit of variable width (2 to 8 mm) and
length (120 to 360 mm) according to the given purposes. The ions are accelerated to the required velocity by the voltage applied to the extracting and
accelerating electrodes.

Ion currents from 10 to 100 mA can be obtained with this setup from
various solid elements with a material utilization of 5 to 50 per cent [984].

With ion beams of such high intensity the broadening of the ion beam due to the space charge effect has to be coped with. One method to overcome this inconvenience is to neutralize the positive space charges with an electron beam [981, 986, 1024].

Figure 3.83 (Pasiok, Vorobev, Ivannikov *et al.* [1035]) shows an ion source with hot cathode Penning-type discharge and transversal ion extraction combined with an evaporator which can be applied in cyclotrons. The heat capacity of the evaporator is low and the required temperature is obtained by the variable direct current flowing through two coaxially mounted tubes connected in series. The low heat capacity of the evaporator permits the vapour pressure of the evaporated metals to be rapidly varied. The discharge chamber is made of stainless steel insulated thermally with titanium plates from the supports of the ion source. The vapour of the solids to be ionized is driven through a tube and a vapour distributor into the discharge chamber. The latter three components are kept at a higher temperature than that of the evaporator to prevent the vapour from condensation. The temperature is controlled with chromium-aluminium thermocouples mounted at the discharge chamber and at the evaporator.

The ions are extracted through a 1 mm wide and 15 mm long slit on the wall of the discharge chamber. Ion beams of a total intensity of 120 mA were obtained from calcium with the parameter values $I_a = 9.5$ A, $U_a = 700$ V, evaporator temperature 720°C, $U_{extr} = 15$–19 kV and ion beams of an intensity of ~110 mA with $I_a = 8.2$ A, $U_a = 470$ V, evaporator tem-

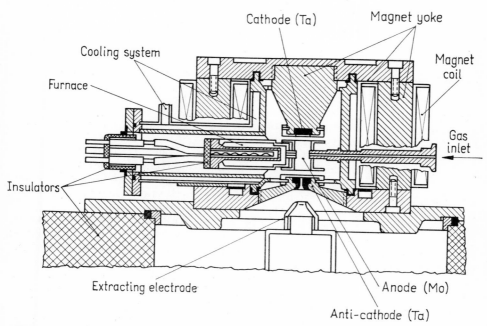

Fig. 3.84 Heinicke–Bauman's version of an ion source for solid elements with cold cathode and oscillating electrons

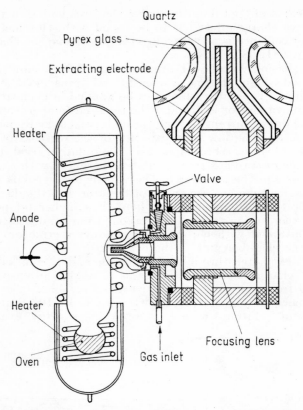

Quartz

Pyrex glass

Extracting electrode

Heater

Anode

Heater

Oven

Valve

Gas inlet

Focusing lens

Fig. 3.85 Kozlov–Marchenko–Fogel's high frequency ion source for solid elements

perature 390°C, $U_{extr} = 15\text{–}19$ kV using zinc. The major fraction of the ions was found to be multiply charged, thus Ca^{n+} ions, where $n = 1, \ldots 9$ and Z_n^{n+}, where $n = 1, \ldots, 10$.

Cold cathode Penning-type ion sources have been also combined with an evaporator (Heinicke, Bethge, Bauman [967]; Heinicke, Bauman [968]). In the ion source of Fig. 3.84, the ions are extracted in the direction parallel to those of the discharge axis and the magnetic field. The vapour is led through a short vapour driving tube into the discharge volume.

The auxiliary gas inlet is located in the side wall of the cylindrical anode so that it faces the vapour inlet. The magnetic field applied to the discharge is generated by an electromagnet. The thickness of the tantalum cathodes is ~ 3 mm. The discharge is generated in a mixture of hydrogen, nitrogen, helium, neon, argon or krypton auxiliary gas and the vapour of the ionized element at a pressure of 10^{-3} to 10^{-2} mmHg. Ion currents from 0.1 to 120 μA are obtainable from this source with the parameter values $I_a \sim 0.5$ A, $U_a \sim 0.5\text{–}1$ kV, magnetic field ~ 1000 gauss. The cathode lifetime varies from 10 to 30 hours. This type of ion source is used mainly for the ionization of elements with low melting point.

Ionization in high frequency discharge is utilized in the ion source of Fig. 3.85 (Kozlov, Marchenko, Fogel [999]). This type of source is suitable for the production of low intensity ion beams from low melting compounds of the required solid element. The atoms of the required element are ionized in the high frequency discharge generated in the vapour of various salts (HCl, NaCl, etc.). The material to be evaporated is placed into the spherical extension at the lower part of the discharge tube. Electrical heaters are placed around the evaporator and the discharge chamber to heat the evaporator and to prevent the condensation of vapour on the wall of the discharge chamber. As soon as the discharge tube is switched on, the salt starts to lose water. The steam is pumped out through the metal valve in the base plate. The metal valve is closed after the disappearance of hydrogen lines preceding the appearance of the metal spectrum. Initially, the pressure needed for the maintenance of the discharge is ensured by the addition of some suitable auxiliary gas. The introduction of auxiliary gas can be interrupted if the vapour pressure of the salt becomes high enough to maintain the discharge. The mass spectrum of the ion beam extracted from this ion source was found to contain mainly Na^+ ions with appreciable contributions from H_2^+, O_2^+, Cl^+ and $NaCl^+$ ions as well.

The mean lifetime of this ion source is $\sim$50 hours and the total ion current extractable from the source is $\sim$100 μA.

With the use of high frequency gas discharge ion sources Si^+ and Ge^+ ion beams have been also produced with intensities of $\sim$10 μA by evaporating $SiCl_4$ and GeJ_4, respectively [1000].

3.9.2 ION SOURCES WITH EVAPORATOR INSIDE THE IONIZER CHAMBER

The solids to be ionized in the ion source can also be evaporated in a crucible placed into the discharge chamber where the ionization actually takes place. Materials with higher melting point are usually ionized in this type of sources since in this way the required high temperature can be obtained with less power consumption.

A source of this type can be seen in Fig. 3.86 (Magnuson, Carlston, Mahadevan et al. [1017]). The 60 mm long and 25 mm diameter molybdenum chamber of the ion source contains the two tungsten heater spirals and the crucible. This chamber is also the anode of the discharge. The longer of the two tungsten spirals heats the chamber and it is kept at about the same potential as that of the anode, while the other heater spiral functions as hot cathode biased to a negative potential of $\sim$100 V relative to the anode potential. The crucible can be made of graphite, molybdenum, tungsten or tantalum. Since the latter three elements are soluble in liquid aluminium and because carbon forms with aluminium carbide crystals [477], as has been already mentioned in Paragraph 2.2.4.1, the crucible of this source is made of Al_2O_3 which has a high melting point. The cathode terminal outlet lead is insulated with Al_2O_3. The maximum operational temperature of the discharge chamber is 1600°C and the metal vapour pressure in the chamber is 10^{-3} mmHg. At this pressure and with an outlet slit of 3.2 mm in diameter, the material consumption of the ion source is $\sim$10 mg/h. The material

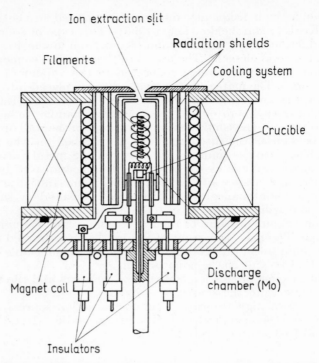

Fig. 3.86 Magnuson–Carlston–Mahadevan–Comeaux's version of an ion source for
high melting solid elements

which fills the crucible is sufficient for an operation of 70 hours. The evaporated material will be efficiently ionized if a magnetic field is applied. The
source has been used for the production of zinc, copper, silver, gold, iron
and aluminium ion beams. The beam intensity was found to be inversely
proportional to the square root of the quantity of evaporated material.
The ion current from zinc was measured as 60 μA. This ion source is of small
size, low power consumption, simple construction and it can be operated for
a relatively long time in a wide range of temperatures.

The most difficult problems proved to be the production of ions of elements belonging to the platinum and palladium groups, such as platinum,
iridium, osmium and palladium, rhodium, radon. The problem is partly due
to the high temperatures (2000° to 3100°K) needed to reach the metal
vapour pressure of $\sim 10^{-2}$ mmHg at which the ion source can operate.
For this reason only graphite crucibles can be used. Another difficulty is to
maintain a stable operation and the required temperature stability.

Figure 3.87 shows an ion source suitable for high operational temperatures
and thus primarily destined for the production of the platinum and palladium group ions (Gusev [983]). As is apparent from the figure, the discharge chamber of this source is made of graphite. It is 75 mm long with a
20 mm outside diameter and a 15 mm inside diameter at the cylindrical
part. The ion extracting slit in the upper part of the chamber is 2 × 40 mm

in size. The graphite discharge chamber is heated by electron bombardment to the required temperature.

The electron beam is collimated in a strong magnetic field and accelerated by 20 to 25 kV applied voltage. The power consumption can be reduced by surrounding the tube with a thermal reflector. Four graphite thermal reflector cylinders, spaced 5—6 mm apart, mounted around the chamber can reduce the power needed for heating by a factor of 5. The crucible with the evaporated material is placed inside the graphite chamber. The crucible is also made of graphite and it functions as the anode of the arc-discharge generated in the vapour of the material to be ionized. Evaporation is induced by the electrons striking the metal surface of the material in the crucible. The bombarding electrons are emitted from a prismatic tungsten block, heated by bombardment with an electron beam. The maximum operational temperature of the ion source was measured as 3173°K. The intensities of the palladium, osmium, iridium and platinum ion beams varied between 20 and 30 mA for a material utilization coefficient of ~5 per cent.

Another type of ion source suitable for similarly high operational temperatures has been also reported [1027]. In this ion source the evaporated material is introduced into the dent of the evaporator probe which can be moved along the axis of the chamber and thus taken to a position where it

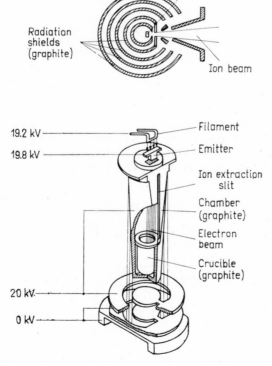

Fig. 3.87 Gusev's version of an ion source for high melting solid elements

can be heated to the optimum temperature by the thermal radiation of the filament.

Evaporation of the solid can be achieved for some metals even without the use of an extra evaporator.

In this case the tungsten or tantalum used as hot cathode material is alloyed with the metal to be ionized (Cu + W, Au + W or Ag + Ta). At the operational temperature of the hot cathode Cu, Au, or Ag leave the cathode at a diffusion rate which brings the vapour pressure to a value at which a stable discharge is maintained and also permits an ion current of a few mA to be extracted (Freeman [1026]).

3.9.3 ION SOURCES WITH METAL EVAPORATION BY ELECTRON BOMBARDMENT

The production of ions from solid elements is more simple if the temperature associated with the metal vapour pressure at which the source operates is lower than the melting point of the given material. In this case the electron beam which ionizes the metal vapour can be used also for the production of the required vapour pressure [990, 1007, 1011, 1029, 1332–1335].

Ion sources with Penning-type discharge, cold cathode and oscillating electrons are also suitable for the production of ions by the method considered above (Gabovits, Fedorus [990]). Molybdenum ions have been obtained from an ion source with molybdenum anode. It was found that while the molybdenum anode is cold, the discharge takes place only in the hydrogen gas used in the discharge tube of the cold cathode Penning-type source. Gradually, as the anode becomes sufficiently hot to evaporate as much molybdenum as needed for obtaining an appreciable partial molybdenum vapour pressure in the discharge chamber, the beam extracted from the ion source contains in addition to the hydrogen ions, also of $\sim$30 per cent Mo^+ ions.

Figure 3.88 shows an ion source in which the discharge could be generated without admixing any auxiliary gas with the vapour of the material to be ionized, but the voltage applied to the discharge was higher (Wolf [1007, 1011]). The ion source of Fig. 3.88 has a graphite rod mounted in its axis.

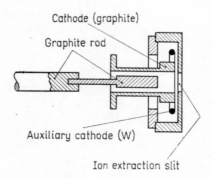

Fig. 3.88 Wolf's version of an ion source for high melting solid elements

At one end of the rod the surface is heated by electron bombardment to $\sim 3273\,^{\circ}$K to induce the emission of carbon atoms from the rod. The bombarding electrons are emitted from the thickened end of the graphite tube that encircles the end of the graphite rod. This tube is heated to electron emission temperature by electrons emitted from the tungsten filament surrounding the tube. The tungsten filament is kept at zero potential, the graphite tube at 1500 V and the graphite rod at 3000 V. The power consumption of the ion source varies from 500 to 600 W. The C^+ ion current extracted from this source was measured as 40 μA and the lifetime of the source was found to be ~ 2 hours.

Metal ions can be produced by electron bombardment also by exposure of the metal source to bombarding electron beam in the energy range from 30 to 60 keV at intensities from 10 to 20 mA. The metal surface is vaporized under the bombardment of the electron beam and the metal atoms in the vapour are ionized by collision with the bombarding electrons. This method was applied [1029] to produce currents of ~ 100 μA of Cu^+, Fe^+, Ag^+, Au^+ ions and somewhat lower currents of Si^{n+} and W^{n+} ions.

3.9.4 ION SOURCE WITH METAL VAPOUR SPUTTERED FROM THE CATHODE DUE TO ION BOMBARDMENT

Solid elements can also be ionized in the metal vapour produced by cathode sputtering.

In this case the electrode, made of the material to be ionized and kept at a negative potential, is introduced into the discharge generated in an additional gas medium. The positive ions generated in the discharge plasma by the negative potential applied to the electrode impinge on the electrode surface and induce cathode sputtering. As a result of this, a vapour with a given concentration of the atoms of the element to be ionized appears around the electrode surface and these atoms are then ionized in the discharge plasma. The ions bombarding the electrode surface can be obtained from the discharge generated in the additional gas [962, 965, 980, 1010, 1019, 1022, 1034, 1036, 1037, 1333–1335], or from an external ion source [1005].

An arc-discharge ion source with hot cathode and applied magnetic field, which is suitable for use in isotope separators and Cockroft–Walton type accelerators, can be seen in Fig. 3.89 (Hill, Nelson [1019]; Hill, Nelson, Francis [1037]). The vapour of the material to be ionized is obtained from the metal sputtered due to bombardment with ions generated in a gas discharge. The electrode made of the material to be ionized is introduced into a liquid-cooled holder made of copper or aluminium and kept at a negative potential of 1.5 kV. The anode is made of molybdenum with a tantalum-lined outlet slit on its side wall. The hot cathode is a tantalum or tungsten wire of 1 mm in diameter. The magnetic field, which is applied in order to contract the plasma for improving its concentration, is generated by an external electromagnet. The outlet slit of the ion source is 1 mm in diameter. The discharge is originally generated in an additional gas. The partial pressure of this gas can be gradually reduced as the metal vapour pressure due to cathode sputtering increases. Thus, it becomes possible after some time to maintain the discharge in pure metal vapour.

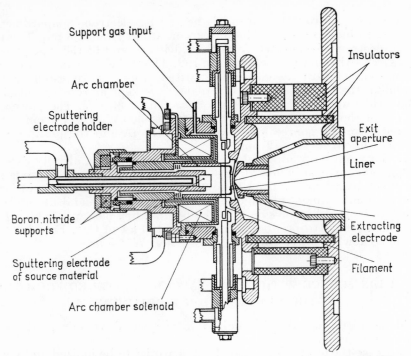

Fig. 3.89 Hill–Nelson's version of a sputtering ion source for high melting
solid elements

Under the experimental conditions $I_a = 1.5$ A, $U_a = 120$ V, magnetic
field $= 405$ gauss, $U_{extr} = 15$ kV, a total ion current of 500 μA can be obtained from this source with argon auxiliary gas. The current values measured on the target after acceleration were reported for the ionized elements
as 30 μA for Cu^+, 50 μA for Sb^+, 10 μA for P^+, 5 μA for B^+, 25 μA for Au^+,
5 μA for Ir^+, 5 μA for Pt^+, 50 μA for Ag^+, 10 μA for Al^+, 25 μA for Pb^+
and 10 μA for Cr^+.

A gas magnetron can be also used to produce ions from elements in the
solid state (Cobic, Tosic, Petrovic [1010]; Tosic [1022]; Nitschke [1038]).
An ion source of this type is shown in Fig. 3.90. The source assembly is
placed into a magnetic field directed parallel to the cathode. A discharge
current of a few amperes is generated in auxiliary argon gas between the
hot cathode and the graphite anode surrounding the former. The gas magnetron operates in a subcritical manner at a few hundred volts. The electrode
of $50 \times 20 \times 1$ mm size prepared from the metal to be ionized is introduced
into the discharge plasma. This electrode is at a negative potential of 800 V.
The metal atoms released from the cathode by the bombarding ion current
of $\sim$0.5 A are ionized in the discharge plasma. The metal ions are extracted
together with argon ions through the 2×50 mm extracting slit. If the
cathode of the ion source is prepared from a metal alloyed with the material
to be ionized, the extra electrode prepared from the latter becomes unnecessary. In this case ion beams with intensities from 50 to 200 μA can be

generated with the gas magnetron from iridium, nickel, cerium, lanthanum and neodymium.

Figure 3.91 shows a hot cathode, Penning-type ion source with transversal ion extraction, developed for use in cyclotrons (Tretyakov, Pasiok, Kulkina *et al.* [1034]). The ions inducing cathode sputtering are obtained in this ion source from discharge generated in neon, argon or xenon auxiliary gas. The electrode prepared from the metal to be ionized is mounted opposite to the 1×15 mm exit slit and kept at a negative potential of 400 to 1000 V. This high applied voltage leads to the maximum concentration of the metal ions in the environment of the exit. The ion current extracted from this source contains 60 to 100 per cent metal ions. With the parameter values $I_a = 7.5$–10 A, $U_a = 300$–600 V, $U_{\text{extr}} = 15$–18 kV and a cathode bombarding ion current of 1–3 A with an energy of 400–1000 eV, the ion current values have been reported as follows: magnesium, 140 mA; aluminium, 105 mA; calcium, 66 mA; titanium, 95 mA; copper, 150 mA; zinc, 160 mA; molybdenum, 100 mA; tantalum 54 mA and tungsten, 61 mA.

A hot cathode, Penning-type ion source with axial ion extraction can be seen in Fig. 3.92 (Rautenbach [997]). The discharge chamber of the ion source is heated by the heat radiation from the tungsten cathode and an extra heater. The electrode prepared from the material to be ionized is

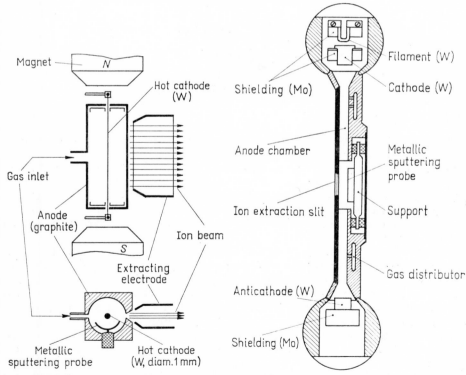

Magnet · N · Hot cathode (W) · Gas inlet · Anode (graphite) · S · Extracting electrode · Ion beam · Metallic sputtering probe · Hot cathode (W, diam.1 mm)

Shielding (Mo) · Filament (W) · Cathode (W) · Anode chamber · Metallic sputtering probe · Ion extraction slit · Support · Anticathode (W) · Gas distributor · Shielding (Mo)

Fig. 3.90 Tosic's version of a magnetron-type ion source for solid elements

Fig. 3.91 Tretakov–Pasyok–Kulkina–Kuznetsov's version of a sputtering ion source for solid elements

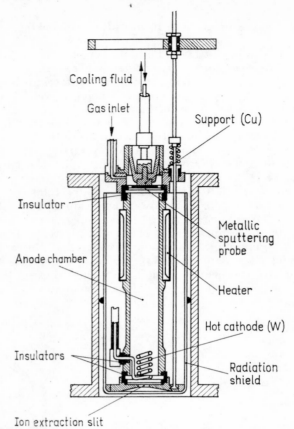

Cooling fluid

Gas inlet

Support (Cu)

Insulator

Metallic
sputtering
probe

Anode chamber

Heater

Hot cathode (W)

Insulators

Radiation
shield

Ion extraction slit

Fig. 3.92 Rautenbach's version of a sputtering ion source for solid elements

mounted on a copper block and kept at a negative potential of 500 V. The
discharge is generated in the mixture of the metal vapour and auxiliary
xenon gas. The ion current bombarding the cathode varies from 1 to 10 mA.
The metal ion current measured on the collector of the mass spectrometer
measured at these values of the bombarding ion current varied from 1 to
100 μA.

Figure 3.93 shows the dependence of the xenon and cadmium ion current
values on the partial pressure of xenon auxiliary gas for different values of
the cadmium electrode potential. It is apparent from the figure that the
contribution from cadmium ions to the extracted ion current increases with
increasing negative potential applied to the cadmium electrode or with decreasing partial xenon gas pressure in the discharge tube.

A cold cathode Penning-type source with transversal ion extraction can
be seen in Fig. 3.94 (Gavin 965]). The cathodes of this ion source are titanium rods placed in a copper holder. The anode chamber is assembled from
two parts made of stainless steel. The material to be ionized is placed in a
holder near and opposite to the outlet slit and kept at a negative potential
of 1.5 kV. With the parameter values $I_a = 1.3$ A, $U_a = 1500$ V, magnetic

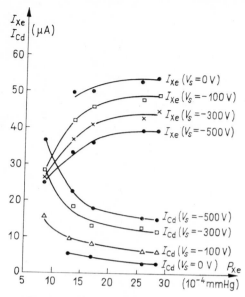

Fig. 3.93 Dependence of xenon and cadmium ions obtained from the source in Fig. 3.92 on pressure of xenon auxiliary gas at different values of sputtering target voltage V_s

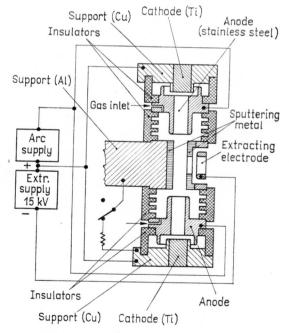

Fig. 3.94 Gavin's version of a sputtering ion source for solid elements

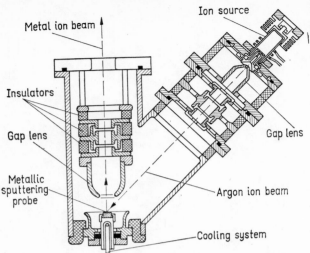

Fig. 3.95 Ion source of solid elements with separate gas ion source

field 4000 gauss and $U_{extr} = 15$ kV the following ion currents have been measured: beryllium, 525 μA; aluminium, 875 μA; calcium, 510 μA; titanium, 230 μA; niobium, 74 μA and gold, 510 μA.

It has been already mentioned that the ions bombarding the cathode can be obtained also from an external source. A setup with this arrangement is shown in Fig. 3.95. The argon ion beam of $\sim$10 mA is taken here from a high frequency gas discharge ion source (Fig. 3.67). The ion beam extracted from the source is led through an accelerating and focusing system consisting of a gap lens and an unipotential lens which accelerates the beam to the required energy from 10 to 20 keV and focuses it to the target prepared from the material to be ionized. The atoms emitted from the sputtering metal bombarded by the argon ions are ionized by collision with the ions. The thus formed metal ions are extracted by the negative potential applied to the extracting electrode mounted close to the electrode made of the metal to be ionized and are subsequently passed through a focusing lens to the mass separator.

Experiments performed with this source [1005] have shown that it can yield metal ion currents of $\sim$10^{-8} A if the metal surface is bombarded with argon ions having an intensity of 1 mA with energies from 10 to 12 keV.

It is of interest to note that the assembly shown in Fig. 3.95 lends itself to the analysis of metal surfaces.

In addition to the sources described above, ions of solid elements can be produced also with arc-discharge [988, 998, 1003, 1004, 1006, 1015, 1016, 1032] and spark-discharge [653, 664, 998, 999] type ion sources. In the arc-discharge type sources the element to be ionized is used as electrode and the ions are produced in the self-maintaining discharge generated between two metal or carbon electrodes. In the spark-discharge type sources the ions are produced in the short-time breakdown pulses of the discharge. The ion beams generated by either method are strongly contaminated and have a wide ion energy spread.

CHAPTER IV

SPECIAL ION SOURCES

4.1 NEGATIVE ION SOURCES

Ion sources yielding negative ion beams were initially used for the study of various atomic collision processes. Since the advent of electrostatic accelerators utilizing the charge transfer phenomenon [1074], the application field of negative ion sources has considerably expanded. This has led to the development of further types of sources with negative ion beams and to investigations concerning their physical properties and operational parameters.

Negative ions can be produced — as has been seen in Chapter I — by electron–molecule collisions, three-particle collisions, radiative electron capture (Paragraph 1.3) charge transfer processes (Paragraph 1.4) and by surface ionization (Paragraph 1.2.5).

In gas discharge the negative ions can be produced by any of the above three collision processes. Consequently, the discharge plasma contains, besides positive ions and electrons, negative ions as well. This makes it possible to obtain negative ions from the gas discharge plasma by direct extraction.

Negative ions are most frequently obtained from charge transfer processes. The principle of this method is to lead the positive ion beam, extracted from any of the various types of ion sources, across a gas, vapour or metal foil target in which a fraction of the positive ions is transformed to negative ions by capture of two electrons.

Surface ionization is also convenient for the production of negative ions. In this case the hot metal surface is bombarded by a neutral atom beam. The neutral atoms which capture the electrons released from the hot metal leave the surface as negative ions.

In view of the methods mentioned above, it is reasonable to divide the negative ion sources into three groups, according to the method used:

(a) the negative ions formed in gas discharge are directly extracted from the discharge plasma,

(b) the negative ions are produced by charge transfer,

(c) the negative ions are produced by surface ionization.

4.1.1 ION SOURCES WITH DIRECT EXTRACTION OF NEGATIVE IONS
[672, 1039, 1053–1055, 1058, 1060, 1067, 1069, 1323, 1342–1344]

Negative ions can be produced in the gas discharge plasma by the following types of collision between electrons and molecules:

$$AB + e \rightarrow A^- + B$$
$$AB + e \rightarrow A^- + B^+ + e$$

by three-particle collisions as

$$A + B + e \rightarrow A^- + B$$
$$A + e + e \rightarrow A^- + e$$

and by radiative electron capture of the form

$$A + e \rightarrow A^- + h\nu$$

The concentration of negative ions, depending on the plasma temperature, is higher if the plasma temperature is low. The reasons for this are the following. First, the maximum cross section for electron–molecule collision producing negative ions is at lower energies than that for processes yielding positive ions, second, the presence of negative ions is thermodynamically preferable to that of electrons if the plasma temperature is low. Consequently, gas discharges with low plasma temperature are the reasonable choice for negative ion sources with direct ion extraction.

For assemblies with direct extraction of negative ions most frequently a duoplasmatron with low voltage arc-discharge [1055, 1060, 1067, 1069, 1323, 1342–1344] or a Penning-type discharge ion source [672, 1053, 1054, 1058] is applied.

Figure 4.1 shows a duoplasmatron with direct ion extraction (Collins, Gobbet [1060]). The tantalum wire cathode is 1.5 mm in diameter. It has a useful lifetime of ∼200 hours. The channel in the intermediate electrode is 2.5 mm in diameter. The outlet slit of the tantalum or tungsten anode has a diameter $d = 1.5$ mm. The distance of the intermediate electrode from the

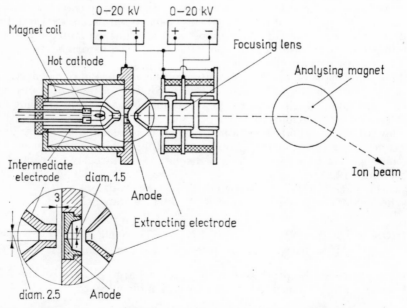

Fig. 4.1 Assembly for production of negative ions by direct ion extraction from duoplasmatron-type ion source

anode plate is $h = 3$ mm. The inhomogeneous magnetic field is generated by an electromagnet. The negative ions are focused with an unipotential lens and separated from the electrons by an applied magnetic field.

The ion source yields a negative ion current of 80 μA for the operational parameters $I_a = 10$ A, $U_{extr} = 16$ kV.

A higher intensity source of this type is to be seen in Fig. 4.2 (Ehlers, Gavin, Hubbard [672], Ehlers [1058]). The ion source is operated with a hot cathode Penning-type gas discharge. The hot cathode is made of a 3.5 mm thick tantalum plate and heated with a direct current of 380 A to the required temperature. A cylindrical tantalum electrode functions as anti-cathode.

The copper discharge chamber has an extraction slit of 1.17×12.5 mm at its side wall. The magnetic field of 4000 gauss needed for the operation of the ion source is generated by the magnet of the cyclotron. A negative deuterium ion beam of an intensity $I_i = 2$ mA can be extracted from this source with the operational parameter values $I_a = 5$ A, $U_a = 300$ V, gas consumption 23 cm³/h, $U_{extr} = 12.5$ kV.

4.1.2 NEGATIVE ION SOURCES UTILIZING CHARGE TRANSFER

For negative ion sources of this type usually two-electron capture induced charge transfer is exploited (Paragraph 1.4).

The capture process can be brought about by the passage of positive ions through a gas, vapour or metal foil target where a fraction of the positive ions transforms to negative ions due to the two-electron capture process described by formula (1.63) as

$$A^+ + B \rightarrow A^- + B^{2+} + \Delta E_2$$

The charge transfer process $A^+ \rightarrow A^-$ has a maximum probability in alkali metal vapour or metal foil targets (Fig. 1.64). Gases of greater atomic

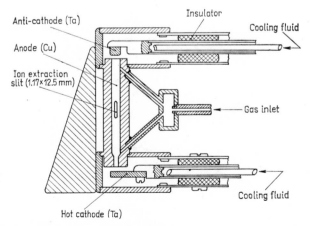

Fig. 4.2 Ehler's version of cold cathode ion source with oscillating electrons, producing negative ions

weight are more favourable targets since the maximum cross section for electron capture of gas atoms with more electrons is larger than that of gases of low atomic weight having a smaller number of electrons (Figs 1.56–1.61).

Negative ion sources with charge transfer can be divided into two groups according to the position of the charge transfer target.

In one group the long ion channel in the extracting electrode is utilized as the target for charge transfer, in the other the charge transfer target is a separate unit at some distance from the ion source.

4.1.2.1 *Negative ion sources with long extracting channel*
[1041, 1045, 1048, 1049, 1052, 1059, 1062]

Since in this type of ion sources the ion channel in the extracting electrode functions simultaneously as charge transfer target, its length must be chosen such that the gas flowing in the channel has the most appropriate 'thickness'. With this method, the charge transfer always occurs in the proper gas of the discharge which flows from the ion source into the extracting channel. This procedure has been most frequently applied in negative ion sources utilizing high frequency gas discharge [1041, 1045, 1048, 1059, 1062].

Figure 4.3 shows a negative ion source with linear high frequency discharge in an axially applied magnetic field (Khirnii [1045]). The quartz

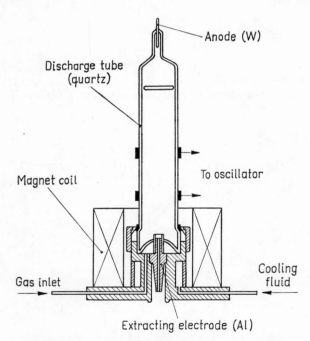

Fig. 4.3 Khirnii's version of a capacitively coupled high frequency ion source with charge transfer in long extraction channel, producing negative ions

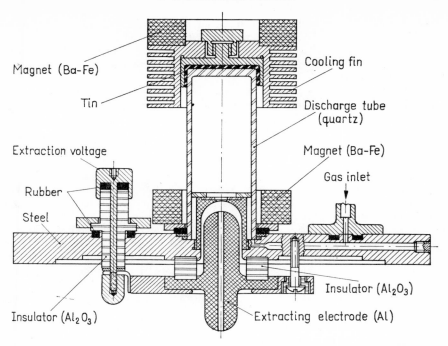

Fig. 4.4 Vályi's version of an inductively coupled high frequency ion source with charge transfer in long extraction channel, producing negative ions

discharge tube of the ion source is 26 mm in diameter and ∼200 mm long. The connection of the discharge tube is vacuum sealed with a thin indium ring and a probe type extraction system is applied. The aluminium extracting electrode has a 2.5 mm diameter, 45 mm long ion channel. The required axial magnetic field of 700 to 800 Oe is generated by a solenoid. The high frequency gas discharge is induced by a capacitively coupled oscillator. The lifetime of this ion source is ∼500 hours. H^- ion currents from 12 to 20 μA and D^- ion currents from 10 to 15 μA can be extracted from this ion source with a gas consumption of 28 cm³/h and extraction voltages from 6.5 to 7 kV.

The negative ion source in Fig. 4.4 is operated with a high frequency gas discharge of alternating magnetic field in an axially applied magnetic field (Gombos, Roosz, Vályi [1062]). The quartz discharge tube is 30 mm in diameter and ∼90 mm long. A cooling fin mounted on the top of the discharge tube prevents the upper part of the tube from inconvenient warming because of secondary electron bombardment. The diaphragm type extracting system of the ion source is of spherical geometry (Fig. 3.68). The aluminium anode has a bore of 4.5 mm diameter and a length of 0.2 mm. The anode is screened from the discharge plasma by a quartz disc of 1 mm thickness with a bore of 6 mm in diameter. The aluminium extracting electrode has an ion channel of diameter $d = 2.5$ mm and $l = 50$ mm. It is located at a distance of 7 mm from the anode. The high frequency gas discharge is generated by an inductively coupled oscillator with an output frequency of 45 MHz and an output power from 120 to 140 W. The axial magnetic field is generated by a barium–iron permanent magnetic ring. The negative ions extracted from the source are separated from the electrons in the applied magnetic field.

An H⁻ ion current of 10 μA can be extracted from this source with a gas consumption of 20 cm³/h at an extraction voltage of 10 kV. The lifetime of this source was found to be more than 500 hours.

4.1.2.2 *Negative ion sources with separate charge transfer target*

If separate charge transfer targets are used, the positive ions are **driven** through a gas, vapour or metal foil target and thus the negative ion sources of this type consist usually of an assembly containing a positive ion source beam forming an ion optical system and a charge transfer target. The most favourable operational conditions are obtained with the use of metal vapour or metal foil charge transfer targets. This is due to the higher charge transfer cross section values of the metal vapour or metal foil compared with other targets and to the fact that these charge transfer targets do not present an extra load on the vacuum system. In the other cases the pumping rate of the vacuum system must be substantially increased in order to continue the acceleration of the negative ion beam obtained from gas targets if one wants to avoid a loss of ions.

A type of negative ion sources with gas charge transfer target [1040, 1043, 1050, 1066, 1324, 1326] is to be seen in Fig. 4.5 (Weinman, Cameron [1043]). The positive ions are obtained from a hot cathode Penning-type discharge (Paragraph 3.7.2). The cathode is a tungsten spiral with three turns made of 1.5 mm diameter wire. The cylindrical anode is made of molybdenum. The applied magnetic field of 500 to 700 gauss is generated by a solenoid.

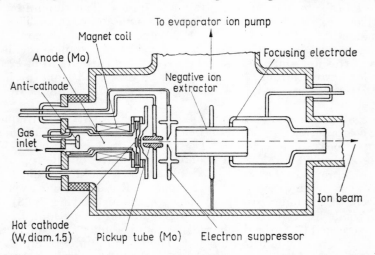

Fig. 4.5 Weinman–Cameron's version of a duoplasmatron-type ion source with separate gas target for producing negative ions

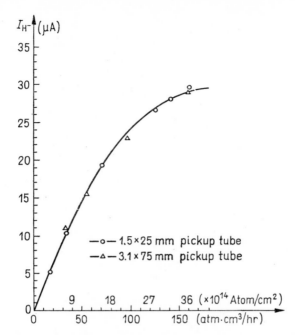

Fig. 4.6 Dependence of negative hydrogen ion current I_{H^-} on pressure in the gas target in ion sources of type in Fig. 4.5

The positive ion beam is extracted through an outlet slit of 3.5 mm in diameter. The ion beam is taken from the extracting slit to the input channel of the gas target at 3.9–4 mm from the ion exit. The charge transfer channel is 3.1 mm in diameter and 75 mm long. It is filled with H_2 target gas. The optimum gas pressure has been determined experimentally. Figure 4.6 shows the H^- ion current values plotted against gas target density, that is gas consumption, for two different sizes of charge transfer channel. It is apparent from the figure that the negative ion currents reach saturation values at gas consumptions between 160 and 200 cm³/h corresponding to gas densities of (3.5 to 4) · 10^{15} atoms/cm³.

The value of the extracted H^- ion current varies from 25 to 30 μA for charge transfer gas consumptions from 150 to 180 cm³/h at an extraction voltage of 17.5 kV.

For metal vapour charge transfer lithium [1066, 1327], potassium [1065, 1070], sodium [1328, 1329], caesium [774, 775, 1063, 1064, 1072] or mercury [362, 1046, 1047, 1051] vapours are used in the negative ion sources. It can be seen from Fig. 1.64 that the highest percentages of negative ions are obtained in the beams from caesium or potassium metal vapour targets.

Figure 4.7 (Rose, Tollefsrud, Richards [1064]) shows a negative helium ion source with caesium metal vapour charge transfer target. The positive helium ions are obtained from a high frequency ion source with probe extraction (Paragraph 3.8.1). The quartz discharge chamber is ∼30 mm in diameter and ∼160 mm in length. The aluminium extracting electrode has

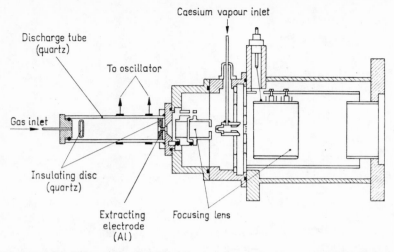

Fig. 4.7 Rose–Tollefsrud–Richard's version of a capacitively coupled high frequency ion source with separate caesium metal vapour charge transfer target for producing negative ions

an ion channel, ∼1.5 mm in diameter and ∼12.5 mm in length. A capacitively coupled oscillator generates the discharge. The positive ion beam is focused with a unipotential lens into the target chamber containing caesium vapour. The target chamber is 40 mm long, made of stainless steel and connected to the caesium evaporator furnace with a vapour conducting tube. A disc with a grid of ∼10 mm in diameter is mounted behind the charge

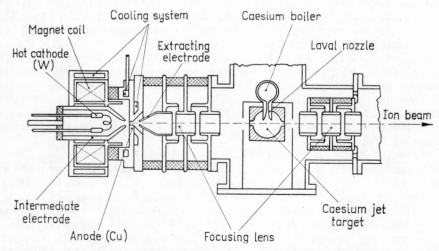

Fig. 4.8 Khirnii–Kotsemasova's version of a duoplasmatron type ion source with focusing lenses and separate caesium metal vapour charge transfer target for producing negative ions

transfer target. He$^-$ ion currents of $\sim$2 μA can be extracted from this type of source for a gas consumption of 5–10 cm^3/h and $U_{extr} = 4$–5 kV.

Higher negative ion currents can be obtained if greater attention is paid to focusing both the positive and the negative ion beams. Such a careful focusing system is provided in the ion sources shown in Figs 2.31 [522, 1065] and 4.8 [1068], and described in the reports referred to under [1063, 1066].

In the negative ion source in Fig. 2.31, the positive ions are obtained from a duoplasmatron and potassium vapour is used as charge transfer target (Rose [522]). The channel in which the charge transfer takes place is water-cooled and has a vapour inlet slit in the middle. The flow rate of the input potassium vapour is $\sim$30 mg/h. The positive ions from the duoplasmatron are accelerated from 15 to 20 keV energy by a system consisting of gap and unipotential lenses and focused to the charge transfer channel.

The output ion beam from the target channel is adapted with an immersion lens to the equipment utilizing the negative ions. For a positive helium ion current from 250 to 300 μA, the value of the negative helium current was measured as 1.2 to 1.5 μA.

By modifying the ion-optical system of the assembly shown in Fig. 2.31, the ratio He$^-$/He of the ion currents was able to be improved [1063] and this led to higher values of the He$^-$ ion current. In the modified version the focusing of the positive ions has been refined and the inlet and outlet slits of the target chamber have been decreased. Moreover a system of gap and unipotential lenses has also been applied for the forming of the beam extracted from the charge transfer chamber. With the modified ion optics the maximum obtainable He$^-$ ion current is of $\sim\mu$A order. The dependence of the He$^-$ current on the positive He$^+$ ion energy is to be seen in Fig. 4.9.

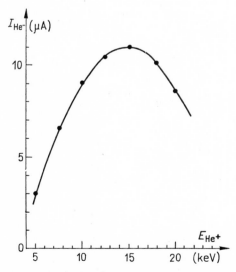

Fig. 4.9 Dependence of negative helium ion current I_{He^-} on the input energy of the He$^+$ ions flying out from duoplasmatron-type ion source to charge transfer target in the version shown in Fig. 2.32

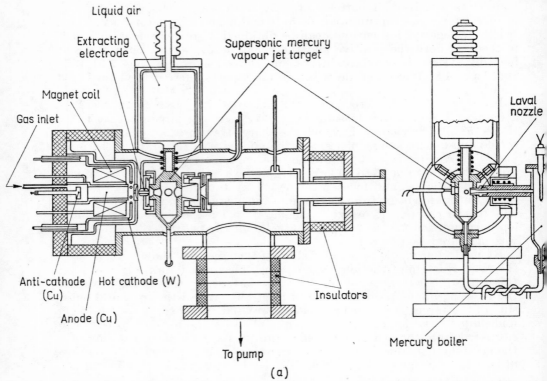

(a)

Fig. 4.10a Fogel–Koval–Timofeyev's version of a heated cathode ion source with oscillating electrons and separate mercury metal vapour charge transfer target for producing negative ions

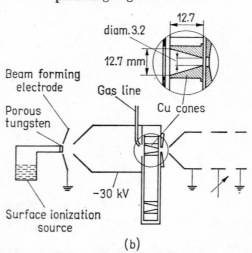

(b)

Fig. 4.10b Negative ion source operating with alkali metal vapour jet proposed by Middleton and Adams [1325].

In the charge transfer-type ion sources considered up to now the inflowing metal vapour reaches a thermodynamical equilibrium in the charge transfer target channel and it can escape from the target chamber through both the inlet and the outlet slits and thus contaminate the vacuum system and result in the formation of a metal layer on the insulating components.

This problem has been overcome in the assembly shown in Fig. 4.8 (Khirnii Kotsemasova [1068]) in which a caesium vapour jet of supersonic speed is used as charge transfer target. In this arrangement the metal vapour jet strikes the wall opposite the injector slit. The wall is kept at the melting temperature of caesium. The metal vapour condenses on the wall and trickles down to the funnel-shaped bottom of the chamber and is led from there back to the evaporator furnace.

In the ion source of Fig. 4.8 the positive He$^+$ ions are obtained from a duoplasmatron. The cathode of the duoplasmatron is an oxide coated nickel grid having a lifetime of 600 to 800 hours. The outlet slit of the intermediate electrode is 2.4 mm in diameter and 5 mm long. The outlet slit of the anode is 0.26 mm in diameter and 0.34 mm long. The required magnetic field is generated by a solenoid. The duoplasmatron yields a positive helium ion current of 350 μA for the operational parameter values: gas consumption from 10 to 15 cm^3/h, $I_a = 1.4$ A, $U_{extr} = 12$ kV. The supersonic caesium vapour jet is obtained from an evaporator similar to the equipment with a Laval nozzle in Fig. 4.10. Experiments have shown that the optimum charge transfer target thickness is obtained with an evaporator temperature from 250° to 270°C. The positive He$^+$ ion beam is accelerated and focused on the charge transfer target by a system of gap and unipotential lenses. The ion beam leaving the target is formed by a unipotential lens. The ion source shown in Fig. 4.8 yields He$^-$ ion beams of 8 μA for $I_a = 2$ A and $U_{extr} = 6$ kV.

A charge transfer target operated with a supersonic Hg metal vapour jet is applied in the negative ion source to be seen in Fig. 4.10 (Fogel, Koval, Timofeev [1047]). The positive ions for this negative ion source are taken from a Penning-type discharge. The anode is a copper cylinder, 50 mm in diameter and 100 mm long. The hot cathode is a flat, helical tungsten wire of 1 mm in diameter. The cathode is mounted at 7 mm from the exit slit of 5 mm in diameter. The anti-cathode is made of copper. The ion channel in the extracting electrode is 6 mm in diameter and 27 mm long. The positive ion source yields hydrogen ion currents of 4 mA for $I_a = 1.5$ A, $U_a = 220$ V, $U_{extr} = 25$ kV and a gas consumption rate of 80 cm^3/h. The positive ion beam is taken with the supersonic mercury vapour jet directly into the charge transfer target and the outgoing negative ions are focused by unipotential lenses. The supersonic mercury vapour jet is produced by the Laval nozzle type evaporator. It strikes the cooled wall of the target chamber where the vapour condenses and is then led back to the evaporator. The ion source of Fig. 4.10a is suitable for the production of ion currents up to 50 μA for H$^-$, 40 μA for O$^-$, 0.18 μA for He$^-$ and 0.1 μA for C$^-$.

The schematic drawing of a negative ion source of a novel type as compared with the previous variants is shown in Fig. 410b (Middleton, Adams [1325]). In this arrangement a positive alkali metal ion beam is generated with a surface ionization ion source of the type described in Paragraph 3.3 and shown in Fig. 3.11b. The ions are accelerated to energies from 8 to 20 keV

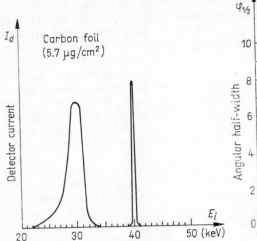

Fig. 4.11 Energy spread of 40 keV energy ion beam before and after the passage through a carbon foil of thickness 5.7 μ g/cm²

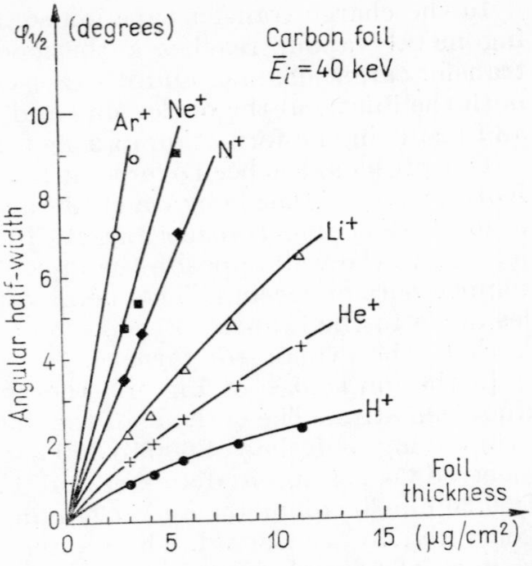

Fig. 4.12 Angle of divergence as a function of carbon foil thickness on charge transfer target, as measured on H⁺, He⁺, Li⁺, N⁺, Ne⁺ and Ar⁺ ion beams

before striking the surface of the cooled conical target. The alkali metal ions impinging on the conical target surface cause a sputtering of the target material and, owing to their scattering on the inside surface of the cone, form at the same time an alkali metal charge transfer target. This arrangement permits negative ions of different elements to be produced by varying the jetted gas and the conical target material. Thus, with cones prepared from graphite, copper, nickel, calcium or titanium ion currents to 15 μA for C⁻, 1 μA for Cu⁻, 1 μA for Ni⁻, 0.5 μA for Ca⁻ and 0.5 μA for T⁻ respectively, have been obtained. The value of the ion current could be increased to 7 μA for nickel and copper by jetting oxygen gas onto the surface of a nickel or copper cone through the gas inlet tube. On the jetting of H_2, D_2 or O_2 gases to titanium cone, ion currents of 10 μA for H⁻, 10 μA for D⁻ and 50 μA for O⁻ are obtainable, respectively. This source arrangement is also suitable for the generation of Li⁻, B⁻, F⁻, S⁻, Ca⁻, NH⁻ ion currents.

Negative ion currents can also be produced by charge transfer in metal or carbon foil targets [355, 357, 358, 386, 1042, 1051, 1071, 1073, 1075, 1077–79, 1316–1320].

This method is first of all useful for the production of negative ions by positive ions of energies above 10 keV. The minimum necessary positive ion energy depends on the thickness of the foil target. The generally used thickness varies from 3.5 to 10 μg/cm², as reported in a number of papers, e.g. [1042] and [1104, 1321] describing their preparation.

Applying foil charge transfer targets, especially with relatively low bombarding positive ion energies of a few times 10 keV, the energy loss and the increased energy spread of ions during the passage of ions through the foil can lead to a considerable broadening of the beam [1042, 1075, 1078, 1079].

The change of these beam parameters caused by the passage through a carbon foil target are illustrated in Figs 4.11 and 4.12 (Högberg, Norden, Berry [1078]).

Figure 4.11 shows the energies and their spread for a beam with $E_i = 40$ keV before and after the passage through a 5.7 $\mu g/cm^2$ thick carbon foil. Figure 4.12 shows the solid angles to which this beam is broadened by the passage through different thicknesses of the carbon foil target [1078].

It is obvious from the above that in the case of foil targets the beam must be collimated before and after the passage through the charge transfer target (Fig. 4.13). If foils of a few $\mu g/cm^2$ thickness are employed, it is advisable to put several foils into the target holder, since the thin foils are destroyed in a few hours by the bombarding beam. A useful method for the arrangement of the foils is shown in this figure where the foils are placed into the holes of a disc rotatable from the outside which permits them to be turned one after another into the path of the bombarding ions. The percentage of negative ions can be more than 10 per cent in the beam obtained from eiter a metal or a carbon foil bombarded with positive ions of energies $E_i < 100$ keV [1042, 1079].

It should be noted that negative ions can also be obtained from water steam [297, 1061] and oil vapour [1044] charge transfer targets; however, because of the numerous problems arising in their application, their use in practice has not become extensive.

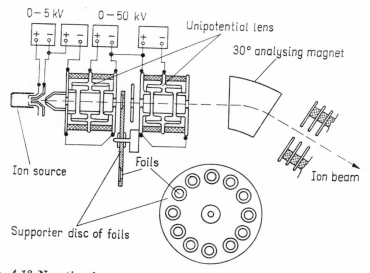

Fig. 4.13 Negative ion source with exchangeable charge transfer target

4.1.3 NEGATIVE ION SOURCES UTILIZING SURFACE IONIZATION

It has been seen from the considerations of the process of surface ionization (Paragraph 1.2.5) that the ionized particles leaving a hot metal surface struck by neutral atoms or molecules include negative ions besides the positive ions and the neutral particles. This fact is exploited in one of the methods for negative ion production. All the surface ionization sources described in Paragraph 3.3 are suitable for use as a negative ion source provided that the extraction voltage can be changed to the opposite polarity. Because of the inherently low ion currents, negative ion sources with surface ionization are useful in practice only for electron affinity measurements and for the study of negative surface ionization processes.

4.2 PRODUCTION OF MULTIPLY CHARGED IONS

Multiply charged ions have a particularly important role in the production of high energy heavy ions. Energetic heavy ions are needed in particle- and nuclear research to study the Coulomb collisions and interactions of heavy nuclei with various materials. High energy ions are most efficiently obtainable if multiply charged ions are accelerated in the available electrostatic particle accelerators. The maximum energy to which the multiply charged ions can be accelerated is the product of the number of electric charges per ion and the energy E_1 of singly charged ions, i.e.

$$E_n \leq nE_1$$

Multiply charged ions can be produced in ion sources or by stripping electrons from particles formed during the passage of singly charged ions of a few keV through a gas, or foil target. Both methods are useful for obtaining an ion beam composed of ions with a high degree of ionization.

4.2.1 PRODUCTION OF MULTIPLY CHARGED IONS IN ION SOURCES

Ionization processes in ion sources are generated mostly by electron–atom or electron–ion collisions. Accordingly, the processes in gas discharge which lead to multiple ionization can be divided into two groups, namely,

(a) single-step ionizations induced by a single collision,
(b) multi-step ionizations induced by several collisions.

In single collision processes several electrons are released from the atoms by a single impact. In multiple collision processes the electrons are released one by one as a result of successive impacts.

It has been seen that in ionization due to electron–atom or electron–ion collisions (Paragraph 1.2) the degree of ionization by single impact increases with increasing energies and concentrations of the striking electrons in the ionization volume as the ionization cross section σ_{0n} decreases strongly with increasing values of n (Figs 1.31, 1.39, 1.40 and Appendix, Table 5) and the maximum value of σ_{0n} is associated with ever higher electron energies E_e (Figs 1.31, 4.14 [1081] and Appendix, Table 4).

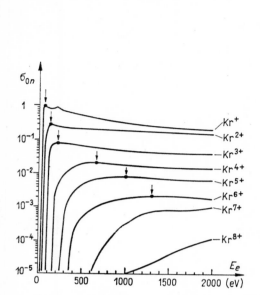

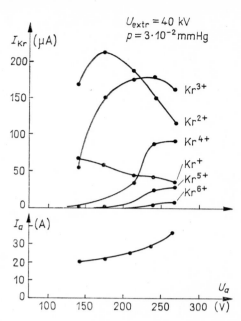

Fig. 4.14 Values of ionization cross section σ_{0n} of krypton atoms as a function of impact electron energy E_0 for different values of n

Fig. 4.15 Values of krypton ion-current I_{Kr} and of the discharge current I_a as a function of the ignition voltage U_a in the duoplasmatron-type ion source of Fig. 4.16

Multiple collisions present more favourable conditions of ionization to a high degree than single collision ionization processes, since

$$\sigma_{0n} \leq \sigma_{n-1,n}$$

and

$$E_{0n}^i \geq E_{n-1,n}^i$$

These two inequalities imply that ionizations in several steps at lower electron energies have a higher probability to produce multiply charged ions than ionizations by single impact (Fig. 1.32 and Appendix, Tables 4 and 5).

Multi-collision ionization processes can take place if the time interval between the inelastic collisions leading to ionization is shorter than the lifetime of the ions. This means that the degree of ionization attained in the preceding collisions must subsist when the impact occurs which leads to the degree of ionization one higher.

It is thus obvious that multiply charged ions can be most efficiently produced in ion sources with operational parameters which lead predominantly to multiple collision ionization processes.

Such operational parameters can be obtained in ion sources in which high electron and ion concentrations are available and the electron energies are higher than the value associated with the maximum value of the ionization cross section.

It has been seen in Paragraphs 3.7 and 3.8 that high electron and ion concentrations are obtainable in the discharge plasma of a duoplasmatron with arc discharge in a magnetic field, the hot and cold cathode Penning-type discharge and high frequency gas discharge ion sources.

Studies of multiply charged ions obtained from duoplasmatrons [662, 942, 945, 1083, 1084, 1297] have shown that highly ionized ions are produced in duoplasmatrons by both ionization processes.

In the case of low discharge currents the degree of ionization is of the order $n = 2$; 3 whereas for higher values of the discharge current, where multiple collisions have more chance to take place, ions ionized to degrees $n = 3$; 4; 5; 6 may also be obtained.

The most favourable conditions leading to the formation of multiply charged ions in a duoplasmatron are obtained if the arc discharge current I_a is increased simultaneously with the discharge voltage U_a within the permissible limits (Fig. 4.15; Krupp [1083]).

Ions ionized to a lower degree ($n = 2$; 3) can be produced also in the duoplasmatrons described in Paragraph 3.7.1.2. The higher currents needed for a higher degree of ionization require a more careful cooling of the intermediate electrode peak and the anode plate if one wants them to have a

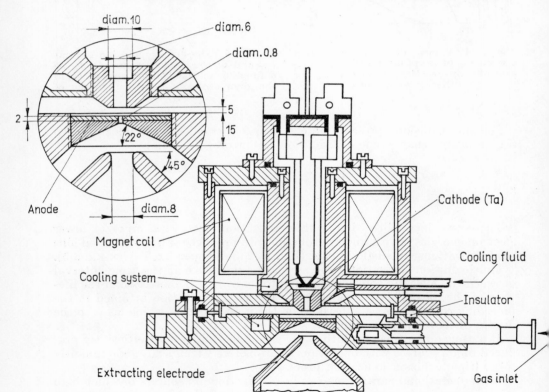

Fig. 4.16 Krupp's version of a duoplasmatron-type ion source for production of multiply charged ions

reasonably long lifetime. A duoplasmatron with efficient cooling is shown in Fig. 4.16 (Krupp [1083]). The cathode of the ion source is a tantalum wire of 1 mm in diameter. The exit slit of the intermediate electrode is 6 mm in diameter and 14 mm long. The anode slit is 0.8 mm in diameter and 2 mm long. The peak of the intermediate electrode as well as the anode plate and its support are prepared from Vacoflux which is a magnetic material of high Curie point (950 °C). An electromagnet generates the applied magnetic field required for the discharge. The peak of the intermediate electrode near the discharge arc and the anode are oil-cooled. With the operational parameters specified in Fig. 4.15 maximum 450 μA krypton ion current can be obtained in which the contributions from Kr^{6+} ions and Kr^{5+} ions are 4 μA and 87 μA, respectively. The ions of a higher degree of ionization ($n = 4; 5; 6; 7; 8; 9; 10$) with the highest intensity can be produced under the most favourable conditions in high intensity ion sources with oscillating electrons, Penning-type discharge using either hot [649, 651, 957, 959–962, 1034, 1035, 1305, 1306, 1312–1315] or cold [649, 652, 960, 961, 965, 1085, 1307–1310] cathode.

In these sources (Paragraph 3.7.2) the plasma of the arc discharge is concentrated into a small volume by the applied magnetic field of 4000 to 7000 gauss. Owing to electron oscillation between the cathodes and the discharge current of a few amperes, the frequency of electron impacts becomes so high that the multiple collision ionization process can yield ions charged up to $n = 10$.

The hot cathode versions of these ion sources can be seen in Figs 3.42, 3.53, 3.83, 3.91, 3.94. These types yield total ion currents from 1 to 100 mA. Because of their high power requirement, they are generally used in cyclotrons and linear accelerators, only.

The percentage of multiply charged ions increases in the output ion current with increasing values of I_a and U_a, as has also been confirmed by the experimental study of the source in Fig. 3.42 the result of which can be seen in Fig. 4.17.

The values of ion currents obtainable from other ion sources are shown for various elements and degrees of ionization in Figs 4.18 and 4.19.

Two Penning-type discharge ion sources with cold cathode of low power consumption [623, 669], suitable also for electrostatic accelerators, are shown in Figs 3.45 and 3.49. The current yields are of μA order with charges $n = = 2; 3; 4$.

High frequency ion sources, useful in electrostatic accelerators, can be utilized for the production of ion currents of $\sim\mu$A order with charges $n = = 2; 3$ [670, 738, 975, 1092–1097]. Sources of this type are shown in Figs 3.58 (Gabovits [670]) and 4.20 (Vályi [738]).

The gas discharge chamber of the high frequency ion source in Fig. 4.20 is made of quartz. It is 30 mm in diameter and 60 mm long. The extracting system of spherical geometry can be seen in Fig. 3.68. The semispherical anode has a radius of 9 mm, while its aperture is 5 mm in diameter and 0.2 mm long. The ion channel in the extracting electrode mounted at a distance of 4.7 mm from the anode is $d = 2$ mm in diameter and $l = 6$ mm long. The shielding quartz disc on the anode has a hole of 7 mm in diameter and 1 mm in length. The high frequency discharge is generated by a capacitively coupled oscillator. The cap on the top of the discharge chamber functions as one of the coupling electrodes. A coupling electrode of this type permits

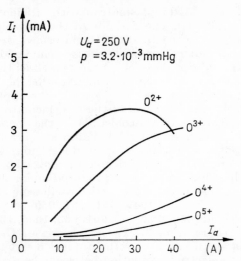

Fig. 4.17 Dependence of ion current output from the ion source shown in Fig. 3.42 on discharge current I_a

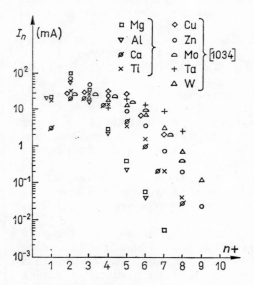

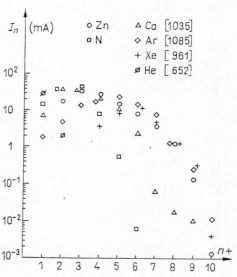

Fig. 4.18 Values of ion current I_n extracted from ion sources as a function of the ionic charge n + for magnesium, aluminium, calcium, titanium, copper, zinc, molybdenum, tantalum or tungsten

Fig. 4.19 Values of ion current I_n extracted from ion sources as a function of ionic charge n + for zinc, nitrogen, calcium, argon, xenon or helium

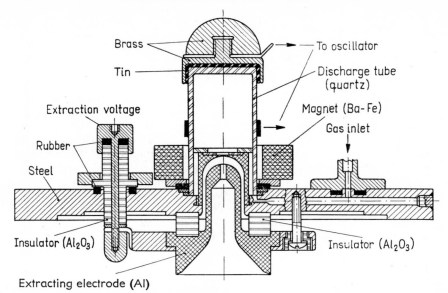

Fig. 4.20 Vályi's version of a capacitively coupled high frequency source of He²⁺ ions

a stable high frequency gas discharge to be maintained at a pressure $p \sim$ $\sim 10^{-3}$ mmHg. The oscillator has an output frequency of 120 MHz with an output power of 100 W.

The axially applied magnetic field is generated by an annular barium–iron permanent magnet. At an extraction voltage $U_{extr} = 5$ kV and a gas consumption rate of 4 to 5 cm³/h with proper focusing and separation a ³He²⁺ ion current of 1.5 μA can be obtained from this source. For focusing a system of gap and unipotential lenses, while for separation a system of transversal magnetic field and electric field are used (Fig. 4.21).

It has to be noted that the analysis of the He²⁺ ion beam involves a large uncertainty if ⁴He²⁺ ions are present since the ⁴He²⁺ ions of almost the same e/m and velocity as the always present H_2^+ ions cannot be separated from the latter with a magnetic or an electrostatic separator, thus, only the sum of the components ⁴He²⁺ and H_2^+ of the ion beam can be determined from the detected counts.

The unambiguous evaluation of the ⁴He²⁺ ion current is possible only if these ions are produced by charge transfer method. This method is to lead the ⁴He⁺ and H_2^+ ions emitted from the source and accelerated to a few hundred keV through a gas target, where a fraction of ⁴He⁺ ions is transformed because of electron loss to ⁴He⁺² ions. This fraction of ⁴He²⁺ ions leaving the gas target together with the H_2^+ ions can be separated from the latter due to their different velocities.

The above difficulties do not arise if ³He isotopes are used. For this reason, it is advisable to check the operation of He²⁺ ion sources with ³He isotope in order to see the reliability of the experimental results.

Multiply charged ions can also be produced by spark ion sources [653, 664]. In this type of ion sources, capacitors from 0.01 to 1 μF charged to voltages between 10 and 70 kV are discharged across the discharge volume.

The discharge current given by

$$I_a = \frac{U_a}{2\pi} \sqrt{\frac{C}{L}}$$

in the pulses of a duration from ~ 1 to 10 μsec reaches average values of 10^2 to 10^4 A. Multiply charged ions are produced in the discharge generated during the attenuated oscillation frequencies of 10^5 to 10^6 Hz.

A spark discharge ion source can be seen in Fig. 4.22 (Bolotin, Markin, Kuligin [664]). The discharge takes place in the channel of the aluminium nitride (AlN) tube containing the material to be ionized. The discharge plasma balloons out from the tube into the volume confined by the solenoid and the ions are extracted from this volume at a voltage from 15 to 20 kV applied to the extracting electrode. The magnetic field generated by the current flowing in the solenoid decreases the diffusion in the direction normal to the magnetic field and increases thereby the ion concentration in this

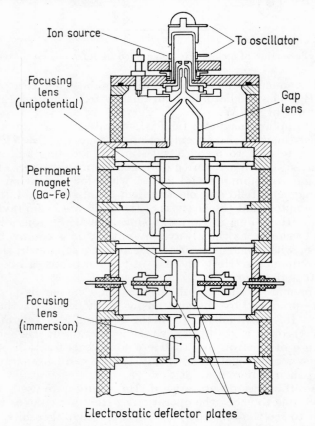

Fig. 4.21 Separator of ion beam from the He^{2+} ion source shown in Fig. 4.20. Separator consists of a focusing system of gap and unipotential lenses with transversal magnetic and electric fields

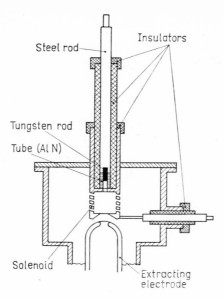

Fig. 4.22 Bolotin–Markon–Kuligin–Skoromij–Meleskov's version of a spark ion source producing multiply charged ions

volume. The total ion current extractable from this source at a pressure of 10^{-6} mmHg varies between 0.3 and 1.2 mA.

The contributions from multiply charged ions to currents extracted from the peak pulse vary between 100 and 400 μA with charges $n = 1$–5 for nickel; $n = 1$–4 for tungsten; $n = 1$–3 for oxygen or aluminium.

Multiply charged heavy ions can be produced also by use of toroidal [1350] and linear [1351] ion sources in which pulsed, high intensity electron beams of energies from 1 to 10 keV are applied, or with the help of ion sources with a hot plasma generated by pinch discharge [1352, 1353], magnetic mirror discharge [1357, 1358] or a laser beam [1354–1356].

4.2.2 PRODUCTION OF MULTIPLY CHARGED IONS BY ELECTRON STRIPPING

An alternative method for the production of multiply charged ions is the stripping of electrons in the interaction of singly charged positive ions of several hundred keV energies with the atoms of gas [355, 357, 385, 1086, 1098–1102, 1105, 1326] or foil [355–357, 1051, 1070–1079, 1103, 1106–1108, 1316–1320] targets. The electron stripping cross section values in this process depend on the ion energy. The electron stripping cross section values for He^+, Li^+ or B^+ ions interacting with the atoms of different gas targets can be seen in Figs 4.23 [385], 4.24 [1102] and 4.25 [1101]. It is apparent from the figures that the cross section values increase as the atomic weight of the target gas increases.

Carbon foils from 2.5 to 10 μg/cm² thickness are the most conventional

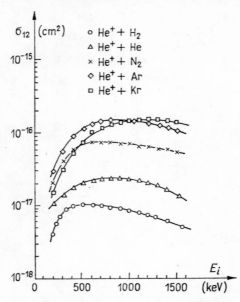

Fig. 4.23 Values of electron stripping cross section σ_{12} as a function of He^+ ion energy for hydrogen, helium, nitrogen, argon or krypton gas target

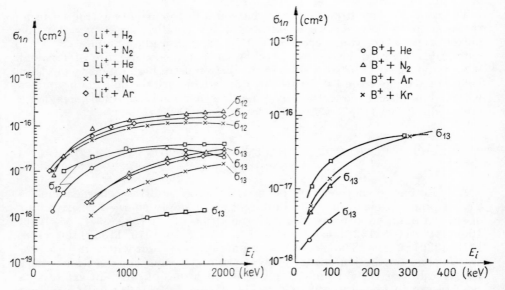

Fig. 4.24 Values of electron stripping cross section σ_{1n} as a function of Li^+ ion energy for hydrogen, nitrogen, helium, neon or argon gas target

Fig. 4.25 Values of electron stripping cross section σ_{1n} as a function of B^+ ion energy for helium, nitrogen, argon or krypton gas target

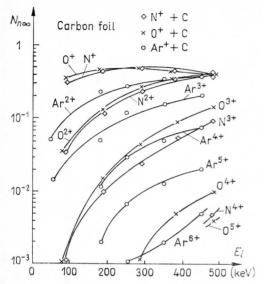

Fig. 4.26 Values of ion beam components charged to different degrees after the passage of N^+, O^+ or Ar^+ ions through carbon foil target

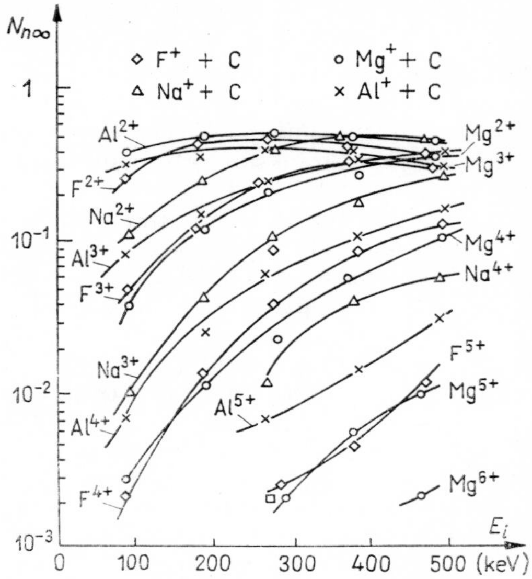

Fig. 4.27 Values of ion beam components charged to different degrees after passage of F^+, Na^+, Mg^+, or Al^+ ions through carbon foil target

electron stripping targets, however also Ni [1103], Al_2O_3 [1106] foils and Li [1327], Na [1328, 1329], Hg vapour jet [1107] targets have been used.

Gas targets — as has been already seen in the case of negative ion sources — require a high pumping rate vacuum system therefore the solid or condensing metal vapour targets seem to be preferable.

If the ions are passed through a foil target they can be multiply charged and ionized by electron stripping up to a degree of $n = 2–6$. The ions are charged during their passage to degrees of ionization in a percentage depending on the primary ion energies. Experiments on carbon foil targets (Hvelplund, Laegsgaard, Olsen et al. [1079]; Smith, Whaling [1110]) have led to the results shown in Fig. 4.26 for N^+, O^+, Ar^+ and for F^+, Na^+, Mg^+ and Al^+ ions in Fig. 4.27. It is apparent from the figures that the primary ion energies belonging to the maximum values of the percentual contribution from multiply charged ions show increasing values as the degree of ionization increases. It follows that the degree of ionization can be increased by using higher primary ion energies.

The degree of ionization can be even more increased if a positive ion beam consisting of heavy ions with sufficiently high energies is passed through several foils. With this method ions charged to $n \sim 40$ were obtained using a 10 MV tandem accelerator for protons with four foil targets for accelerating uranium ions to 222 MeV, while a 30 MV tandem accelerator for protons yielded under the same conditions uranium ions of energy 1263 MeV charged to $n \sim 70$ (Rose [1108]).

4.3 SOURCES OF PULSED ION BEAMS

Ion current pulses are utilized primarily by particle accelerators. Linear and cyclical accelerators cannot be operated but with ion sources yielding ion current pulses. Electrostatic accelerators operating with pulsed ion beams lend themselves to neutron and nuclear spectroscopic investigations, measurements of lifetimes of short-living reaction products and processes with high level background which cannot be detected if a continuous ion beam is employed.

The length Δt, the repetition frequency f_r and the energy spread ΔE of ion current pulses have to be chosen according to the requirement of the planned measurements.

The minimum length of ion current pulses is determined by the energy spread ΔE_i of ions obtained from the ion source and by the corrections arising from the pulsing system. They can be as short as $\Delta t \sim 10^{-9}$ sec. Such low values of Δt are highly desirable for measurements of short half-lives and flight times of fast neutrons as the accuracy of the measurement is strongly dependent on the time uncertainty of the start signal which depends on the length of the ion current pulse.

The choice of the repetition frequency also depends on several factors. Ion current pulses of lengths $(1-2) \cdot 10^{-9}$ sec used for measurements of very short half-lives or fast neutron flight times and the processes analysed in these cases allow higher (10^6 Hz) repetition frequencies to be used. If fission processes are studied with large sized scintillation detectors or flight times of neutrons with energies of keV order are measured, the repetition frequency of 10^5 Hz is allowed as the measurement of the effects takes a time of about 10^{-5} sec. The repetition frequency of ion current pulses is limited also by the signal-to-noise ratio, i.e. the counts of the measured effect as related to the background counts. It has to be noted that the signals due to the effect produced by the particles of the pulse are proportional to the peak current, while the background pulses are proportional to the average current, thus, the ratio of the former to the latter can be improved by decreasing the repetition frequency.

The number of particles in ion current pulses, that is, the value of the peak current, depends on the average current of the ion source and on the extent to which the pulses are bunched.

The energy spread of the particles in ion current pulses is determined by that of the ion source and by the pulsing or bunching method applied.

Several methods are available for the bunching of pulses. The most simple technique for the production of longer than μsec pulses is to pulse the gas discharge or the extracting voltage in the ion source, or simultaneously both of them.

Nanosecond long ion current pulses can be obtained by several methods which can be divided into two groups. In one of them the nanosecond pulse length is available at the target, in the other this pulse length is obtained in the duration of the nuclear reaction product output pulse from the target to the detector input. These methods are the following:

(a) Pulsing by oscilloscopic method;
(b) Methods for ion bunching:

1. velocity modulation by a.c. electric field applied to the modulator slit,
2. travelling wave bunching method with an electric field propagating at the final ion velocity,
3. bunching by magnet.

(c) Methods for the bunching of reaction products:

1. bunching with helical target method,
2. bunching with slanted target method.

4.3.1 ION SOURCES FOR THE PRODUCTION OF ION CURRENT PULSES OF LONGER THAN MICROSECOND DURATION

Ion current pulses are relatively easy to obtain if the pulses can be longer than a few microseconds. In this case ion current pulses can be generated by pulsing the gas discharge, the extracting voltage or the gas supply of the ion source. Any of these methods can be applied to the gas discharge ion sources discussed in the foregoing.

The gas discharge can be pulsed if either its igniting voltage or the oscillator generating the discharge are pulsed (Fig. 4.28) [651, 652, 669, 960, 962, 1051, 1153]. The minimum pulse length obtainable from a pulsed gas discharge is determined by the time needed for the discharge to become self-maintained and by the uncertainty of the time of ignition. Investigations of these limitations [1053] have shown that the uncertainty of the ignition time in a gas discharge is a function of the gas pressure in the discharge

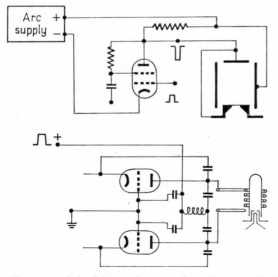

Fig. 4.28 Circuit diagram of igniting voltage pulser for gas discharge ion sources. Upper figure shows circuit diagram of a pulser for cold cathode ion source with oscillating electrons; lower figure shows that for a high frequency gas discharge ion source

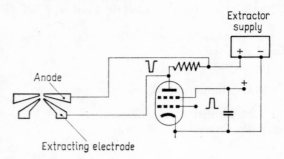

Fig. 4.29 Circuit diagram of an extracting voltage pulser

volume. At a gas pressure of $4.5 \cdot 10^{-2}$ mmHg and an oscillator frequency of 25 MHz the ignition time is ~ 5 μsec with an uncertainty of ~ 1 μsec, while at a gas pressure of $2 \cdot 10^{-2}$ mmHg the uncertainty of the ignition time can be as high as 10 μsec. Consequently, the method of pulsed gas discharge proves to be convenient for the production of ion current pulses with lengths not involving a significant error due to time uncertainty. For this reason, the method is generally applied for the production of ion current pulses longer than the range from 100 to 200 μsec.

The pulsing of the extracting voltage permits ion current pulses of ~ 1 μsec to be generated [1148, 1149, 1151]. In this method U_{extr} voltage pulses of the length $\varDelta t$ required for the experiment are applied to the extracting electrode at a repetition frequency f_r (Fig. 4.29).

If the extracting voltage is pulsed, ions also leave the ion sources in the intervals between pulses by diffusion from the discharge plasma through the extracting slit. No ion diffusion occurs if the gas discharge itself is being pulsed. Thus, if the measurement is affected by the background ion current flowing out in the intervals between pulses, it is best to combine the two types of pulsing (Fig. 4.30). In this case it is advisable to generate the igniting pulse 10 to 20 μsec earlier than the extracting voltage pulse and to choose it to be

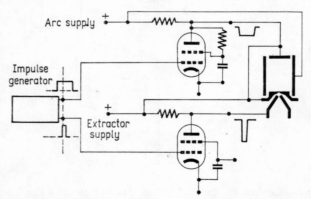

Fig. 4.30 Circuit diagram of an extracting voltage and gas discharge igniting voltage pulser

appropriately longer than the latter because in this way the time uncertainty will not introduce any appreciable error.

The pulse lengths obtainable from pulsed gas supply exceed those measured for the two other methods. This procedure is preferred if one wants to decrease the gas consumption. The gas supply can be pulsed either by mechanical [654] or by electrical [1154, 1155] devices. Ion sources pulsed with hydrogen gas-filled titanium disc are to be seen in Figs 4.31 (Ehlers, Gow, Ruby et al. [1154]), and 4.32 (Afanasev, Knazyatov, Fedotov [1155]). The gas pulses are delivered by the titanium disc heated by short high intensity current pulses which ignite simultaneously the gas discharge in the same pulsed manner. In the ion source of Fig. 4.32, 500 μsec long current pulses are applied, with a peak current varying between 200 and 300 A. In this source the value of the output ion current pulse in the peak is 10 mA for a gas consumption of 10^{-3} cm^3 per pulse.

Mass spectroscopical analysis of this type of source proved the output ion current to be composed predominantly of hydrogen ions with a negligible contribution from titanium ions.

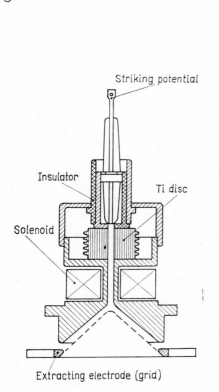

Fig. 4.31 Ehlers–Gow–Ruby's version of an ion source with titanium disc pulsing the gas supply

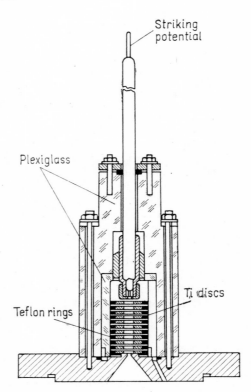

Fig. 4.32 Afanasev–Knazyatov–Fedotov's version of ion source with titanium disc pulsing the gas supply

4.3.2 METHODS FOR THE PRODUCTION OF NANOSECOND ION CURRENT PULSES

4.3.2.1 *Pulsing by oscilloscopic method*
[1111–1119, 1139, 1143, 1144, 1225, 1345]

One of the most simple and earliest techniques for the production of nanosecond ion current pulses is the oscilloscopic method. Theoretical and experimental studies of this method have been reported in numerous papers [1111–1119, 1139, 1143, 1144, 1225].

The most simple equipment required for this technique consists of a pair of deflector plates and a limiting slit (Fig. 4.33). The principle of operation of the equipment is the following.

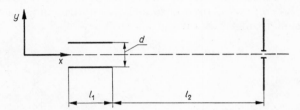

Fig. 4.33 Schematic diagram of ion beam pulse generator consisting of a pair of deflector plates and a limiting slit

The focused ion beam extracted from the ion source or the accelerator is led along the path between the deflector plates. The alternating voltage $U = U_0 \sin \omega t$ to the deflector plates forces the ions to periodical oscillation in the plane of the limiting slit. Accordingly, the length of the ion beam fractions, i.e. the length Δt of the ion current pulses escaping through the limiting slit depends on the beam velocity v_s at the slit and on the dimensions of the slit. The repetition frequency of the pulses depends on the frequency $f = \dfrac{\omega}{2\pi}$ of the alternating voltage applied to the deflector plates.

The motion of the ions driven in the direction of the axis x between the plates of length l_1 by the alternating voltage $U = U_0 \sin \omega t$ applied to the deflector plates is given if the effect of the stray field generated at the edges of the deflector plates, is neglected by the equation of motion

$$m\ddot{y} = \frac{e}{d} U_0 \sin \omega t \tag{4.1}$$

where e is the charge and m is the mass of ions, while d is the distance of the deflector plates. The motion of the ions excaping from between the deflector plates along the path of length l_2, where the electric field does not act any more can be described by the equation

$$m\ddot{y} = 0 \tag{4.2}$$

It can be readily shown that on ignoring the effect of the stray field at the edges of the deflector plates in the case of an axial ion beam and on assuming that the transit flight time τ_0 of the ions with energy E_0 between the deflector plates is small compared with the period time T, the voltage $U_0 \sin \omega t$ in eq. (4.1) can be replaced by the approximation to a linear function of the form $U_0(\omega t + \delta)$. In this case the displacement of the beam over the distance Δy, equal to the diameter of the limiting slit, can be described as

$$\Delta y = \frac{1}{4} l_1(l_1 + 2l_2) \frac{eU_0}{E_0 d} \omega \Delta t$$

This expression and the velocity v_s of the sweep of the beam at the slit give for the length Δt of the ion current pulse leaving the limiting slit the equation

$$\Delta t = \frac{\Delta y}{v_s} = \frac{4E_0 \Delta y d}{l_1(l_1 + 2l_2)eU_0\omega}$$

As is apparent from this expression, the value of Δt depends on a large number of parameters. However, for a given equipment, where the size and distance of the limiting slit as well as the dimensions of the deflector plates are given, the variable parameters are restricted to the ion energy and to the amplitude and frequency of the applied alternating electric field. The free choice of these variable parameters permits the value of Δt to be varied in a given interval.

Because of the finite diameter of the ion beam used in practice, its passage between the deflector plates causes a spread in the ion energies and a change in the position of the beam [1112, 1113].

The time spread arising from this effect can be neglected if one works with a narrow beam of low energy E_0 and of nearly parallel moving ions but at higher values of E_0, this effect can induce significant changes in the minimum value of Δt.

If the transit time τ_0 between the plates is comparable to the period time T, the calculation of Δt becomes more complicated because $U_0 \sin \omega t$ cannot be approximated by a linear function [1119].

Because of the unavoidable asymmetry of acceleration in particle accelerators care must be taken to transmit ion currents through the limiting slits only if the ion beams move in the same direction. This problem can be solved by use of an additional pair of deflector plates of d.c. voltage [1113, 1119] or by the generation of Lissajous curves [1120, 1121].

The d.c. voltage deflector plates are mounted between the plates of alternating voltage and the limiting slit (Fig. 4.34). The d.c. voltage applied to the former keeps the beam of positively charged ions always deflected in the direction of the plate at negative potential and the ion beam is let through the limiting slit only when its deflection in the direction of the positive plate is at a maximum in the alternating sweep. Thus, only a single ion current pulse per period is transmitted through the slit.

An alternative arrangement is shown in Fig. 4.34. In this assembly a second pair of plates is mounted at 90° relative to the first pair. An alternating voltage is applied to both pairs of plates with a given shift in phase and a given difference in frequency. With a proper choice of the relative fre-

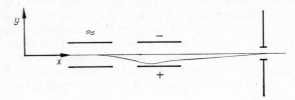

Fig. 4.34 Schematic diagram of ion beam pulse generator consisting of deflector
plates kept at a.c. and d.c. voltages and a limiting slit

quencies and the phase shift of the applied alternating voltages, the ion
beam shaped like a Lissajous curve in the plane of the limiting slit is transmitted only once in each period through the slit (Fig. 4.35).

The method of Lissajous curves enables one even to separate ions with
different masses by utilizing the fact that ions of the same energy but with
different masses move at different velocities and thus arrive at the second
pair of plates at times and in phases which generate other Lissajous curves
in the plane of the limiting slit than the ions with another mass (Fig. 4.35).

The minimum ion current pulse lengths obtainable by oscilloscopic method
vary from $\Delta t \sim (1–2) \cdot 10^{-9}$. The current of the pulse is equal to the average
ion current of the source.

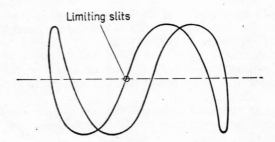

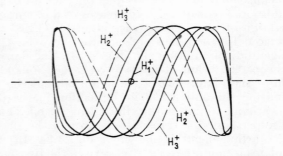

Fig. 4.35 Lissajous curves for single and three beam components obtainable for
frequency ratio 3 : 1 from pairs of deflector plates at alternating voltage placed normal
to and at a distance l from each other.

4.3.2.2 *Pulsation by bunching method*

(a) *Bunching of ions by velocity modulation in alternating electric field applied to the modulator slit*

[1117, 1120, 1128, 1131, 1139, 1141, 1142, 1145, 1146, 1150, 1152, 1225, 1288, 1289]

Velocity modulation is one of the methods by which continuous mono-energetic ion beams, evenly distributed in time, can be transformed into strongly bunched beam pulses. The principle of the method is the following. The velocities of the monoenergetic ions are modulated in a slit exposed to an applied alternating electric field. In this way the ion velocities change so that the distance S_0 from the modulator slit to the target is covered by the ions at different velocities and they arrive at the target in the same short time interval Δt. It can be readily shown that the modulating energy which has to be imparted to the ions in the modulator slit for bunching the total beam is given by

$$E_m = E_0\left[\left(1 - \frac{t}{\tau_0}\right)^{-2} - 1\right] \tag{4.3}$$

where E_m is the modulating energy, E_0 the ion energy before modulation, $\tau_0 = \dfrac{S_0}{v_0}$ is the time-of-flight and v_0 the velocity of ions with energy E_0, while t is the exit time from the modulation slit (Fig. 4.36).

If the total ion beam need not be utilized and if a utilization factor of less than 50 per cent is though to be sufficient, two modulator slits can be applied. In this case the modulator slits must be so placed that the transit flight time of ions between the two slits should be equal to the half period time of the alternating electric field.

The generation of the ideal modulation potential pulse is a highly complicated problem. That is why, in most cases, a sinusoidal modulating potential which is much simpler to generate, is applied to the modulator slit.

If a sinusoidal modulating voltage $U_m(t) = U_{m0} \sin \omega t$ is applied, the velocity of ions accelerated by the accelerating voltage U_0 is given at their arrival to the modulator slit by the expression

$$v_0 = \sqrt{\frac{2eU_0}{m}}$$

and the velocity at which the ions leave the modulator slit is

$$v(t) = \sqrt{\frac{2e}{m}}\sqrt{U_0 + U_{m0} \sin \omega t}$$

where e is the charge and m the mass of particles, while $f_m = \dfrac{\omega}{2\pi}$ is the frequency of the modulating alternating electric field. Consequently, the ions arrive to the target at distance S_0 (Fig. 4.36) at a time given by

$$\tau_T(t) = t + \frac{S_0}{v(t)} = t + \tau_0 \frac{1}{\sqrt{1 + \alpha \sin \omega t}} \tag{4.4}$$

where $\alpha = \dfrac{U_{m0}}{U_0}$ and the input ion current pulse to the target is

$$i_T(t) = i_0 \left[1 - \frac{\alpha \tau_0 \omega \cos \omega t}{2(1 + \alpha \sin \omega t)^{3/2}} \right]^{-1}$$

where i_0 is the average ion current. Introducing the notation $A = \dfrac{\alpha \tau_0 \omega}{2}$ into the ion current formula, we get

$$i_T(t) = i_0 \left[1 - \frac{A \cos \omega t}{(1 + \alpha \sin \omega t)^{3/2}} \right]^{-1} \tag{4.5}$$

It is apparent from expression (4.5) that the maximum bunching is obtained for $i_T(t) \to \infty$. This occurs if the term in brackets goes to zero, i.e.

$$A \cos \omega t = (1 + \alpha \sin \omega t)^{3/2} \tag{4.6}$$

The maximum of the current pulse goes to infinity if $A = 1$. This condition follows from the solution to eq. (4.6). If $A > 1$, the unique maximum transforms to two peaks and as the value of A increases the distance between the peaks also increases (Fig. 4.37).

Let us now define the phase angle of the ion beam packet arriving at the target as

$$\varphi_T(\omega t) = \omega(\tau_T - \tau_0)$$

By using equality (4.4), this formula can be written in the form

$$\varphi_T(\omega t) = \omega t + \omega \tau_0 [(1 + \alpha \sin \omega t)^{-1/2} - 1] \tag{4.7}$$

The bracketed term in (4.7) can be expanded into a power series if $\alpha \leq 0.1$ and thus, the phase angle at the arrival to the target can be expressed as

$$\varphi_T(\omega t) \approx \omega t - A \sin \omega t \tag{4.8}$$

The function (4.8) plotted for different values of A can be seen in Fig. 4.38.

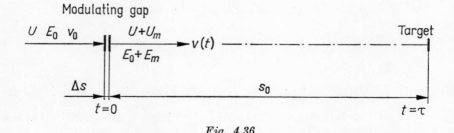

Fig. 4.36

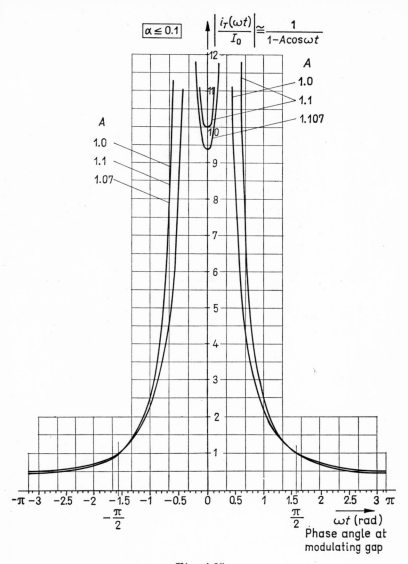

$$\boxed{\alpha \le 0.1} \qquad \left|\frac{i_T(\omega t)}{I_0}\right| \cong \frac{1}{1 - A\cos\omega t}$$

Fig. 4.37

In the general case, formula (4.7) can be used for the determination of the phase angle interval in which the ion beam fraction of length $(t_2 - t_1)$ lying in the phase angle interval from ωt_1 to ωt_2 is at the time of arrival at the target. This interval corresponds to the separation of the two current pulse maxima given by

$$\Delta\varphi_T = (\omega t_2 - \omega t_1) + \omega\tau_0[(1 + \alpha\sin\omega t_2)^{-1/2} - (1 + \alpha\sin\omega t_1)^{-1/2}]$$

At low modulating voltages when $U_m \ll U_0$, that is $\alpha \le 0.1$, formula (4.8)

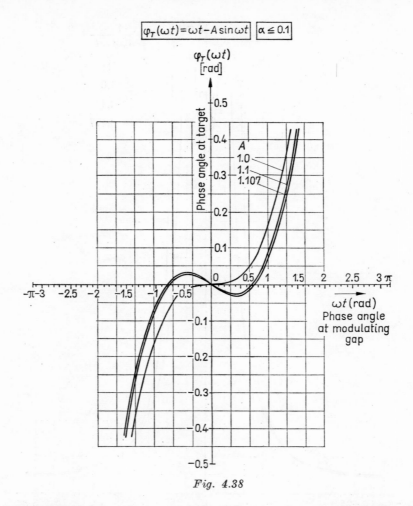

Fig. 4.38

can be used to express the separation of the ion current pulse in the simpler form

$$\Delta\varphi_T \approx (\omega t_2 - \omega t_1) - A(\sin \omega t_2 - \sin \omega t_1) \qquad (4.9)$$

The utilization factor of the continuous ion beam is lower in the case of a sinusoidal modulation compared with the application of the ideal modulating potential. This difference is due to the fact that only that part of the sinusoidal modulating pulse can be effectively utilized which resembles in shape the ideal modulating pulse. The utilization factor of the ion beam depends thus on the magnitude of the phase angle interval from $(\omega t_2 - \omega t_1)$ in which the ion beam fraction defined by $t_2 - t_1$ arrives at the target and it can be expressed as

$$\eta = \frac{100}{2\pi}(\omega t_2 - \omega t_1) = \frac{100}{2\pi}\Delta\omega t \qquad (4.10)$$

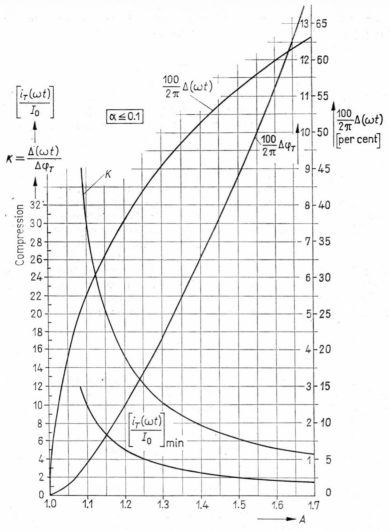

Fig. 4.39

This formula yields readily the utilization factor; e.g. if one uses the phase angle interval of the sinusoidal modulating voltage between $\omega t_1 = -0.89$ radian and $\omega t_2 = 0.89$ radian, determined by the axial crossing segment points of the curve for $A = 1.107$ in Fig. 4.38, the utilization factor of the ion beam will be $\eta \sim 28.4$ per cent.

The ion beam utilization factor combined with the beam pulse utilization factor

$$\xi = \frac{100}{2\pi}\, \Delta\varphi_T$$

gives for the time compression or bunching factor, characteristic of the bunching per beam fraction, the expression

$$K = \frac{\Delta \omega t}{\Delta \varphi_T} \tag{4.11}$$

The values in (4.11) plotted against A for $\alpha \leq 0.1$ can be seen in Fig. 4.39.

The applied interval of the modulating potential determines the energy spread ΔE_b of the ions in the current pulse striking the target, as

$$\Delta E_b = E_m(\sin \omega t_2 - \sin \omega t_1) \tag{4.12}$$

The energy spread is seen to be strongly dependent on the modulation energy E_m and it may attain even values of a few keV.

In bunching by velocity modulation, the energy spread is influenced also by the extent of bunching since it follows from the relation $\Delta E_i \Delta t_i \approx \Delta E_T \Delta t_T$, which holds in this case [1113, 1120], that the energy spread of the ions arriving to the target increases with increasing bunching.

The amplitude and length of the ion current pulse obtainable by ion bunching depend on the time spread of the input pulse to the target which is proportional to the energy spread ΔE_i of the ions extracted from the ion source. If the spread of ion energy is ΔE_i before the energy modulation $E_m \sin \omega t$, then the time spread of the ions at the target is given by

$$\Delta t_T = \frac{\Delta E_i}{\omega E_m} \tag{4.13}$$

The minimum ion current pulse length which can be achieved with bunching by velocity modulation varies as $\Delta t \sim (1\text{--}2) \cdot 10^{-9}$ sec at the target and the peak pulse current can be as high as 10 mA, depending on the extent of bunching.

(b) *Bunching by travelling wave with electric field propagating at the final ion velocity*
[1136, 1156]

It has been seen that the method discussed under (a) is suitable for the production of high intensity ion current pulses with lengths varying as $(1\text{--}2) \cdot 10^{-9}$ sec but with a large spread of ion energy at the target. Because of this drawback the velocity modulation technique can be applied only if the experiments allow a large spread of bombarding particle energies.

An alternative method permits the ions to be bunched so that their final velocity v_f is the same. This is the bunching by travelling wave. The principle of the method is the following (Kozinets, Sapiro, Stranik [1136]).

Evenly distributed ions with velocity v_0 are taken to the buncher input at $x = 0$ (Fig. 4.40). In order to bunch the ion beam of length l taken to the input in the time interval between $t = 0$ and $t = t_n$ (t_n is the start time of bunching) the velocity of each ion has to be changed by $v = v_f - v_0$ at the point $x = v_f(t - t_n)$ of the coordinate belonging to the corresponding time t, so that the bunched ion beam fraction of length l should be composed of ions having the same velocity v_f. This can be achieved by applying the electric field, which causes the velocities of the successive ions to change

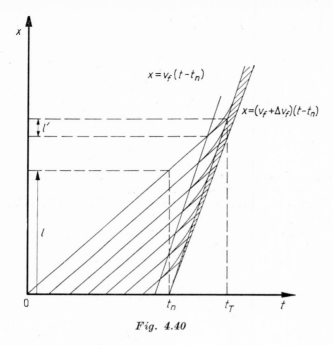

Fig. 4.40

by v, to the point $x = 0$ at time $t = t_n$, then run the field along the axis of the buncher at velocity v_f, the final velocity of the ions, up to the buncher output (Fig. 4.41).

By this technique the ion beam of length $l = v_0 t_0$ is bunched to length l'. The measure of bunching is given by the ratio l/l'.

The energy spread of ions in the current pulse at the target can be evaluated from the energy spread ΔE_i of the ions extracted from the source

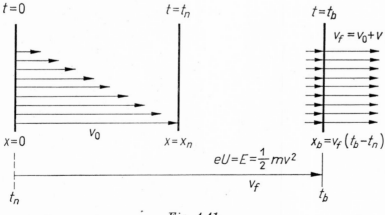

Fig. 4.41

and from the energy $E = \dfrac{1}{2} mv^2$ imparted to the ions as

$$\Delta E = \sqrt{\dfrac{E}{E_i}}\, \Delta E_i$$

hence the spread of the final ion velocity

$$\Delta v_f = \sqrt{\dfrac{2\Delta E}{m}}$$

The minimum length of the ion current pulse Δt at the target is determined by the value of the spread of ion velocity.

With this method and the parameter values $E_i = 2\,\text{keV}$, $\Delta E_i \approx 8\,\text{eV}$, pulse wave velocity $v_f = 2 \cdot 10^8$ cm/sec and amplitude $U = 10\,\text{kV}$, a pulse of length $l = 10$ cm can be bunched in a 20 cm long buncher to a pulse of $\Delta t \sim 2 \cdot 10^{-9}$ sec. In this case the ion beam utilization factor $\eta = 80$ per cent and the time compression factor $K \sim 80$.

(c) *Bunching by magnet*
[1119, 1123–1128, 1140, 1147]

The equipment needed for this method consists of a pair of deflector plates, a limiting slit and a Mobley-type compression magnet with homogeneous field (Fig. 4.42). It can be seen in this figure that the centre of the deflector plates and the target are placed each in one of the two focal points of the ion bunching magnet. The principle of the bunching method is the following.

The ion beam fraction moving between the pair of deflector plates and the limiting slit makes an angle relative to the direction of propagation, as to be seen in Fig. 4.42, due to the periodical deflection of the beam by the

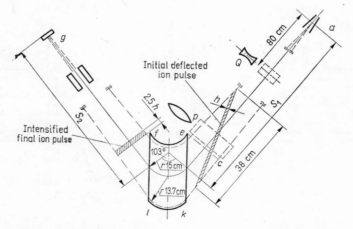

Fig. 4.42 Schematic diagram of ion beam buncher utilizing Mobley's method

alternating electric field applied to the deflector plates. The angle between the beam fraction and the optical axis depends on the velocity v_s of the ion beam motion. The geometry of the apparatus is chosen such that the lengths of ion trajectories depend on the position of the ions in the oblique ion beam fraction. Thus, the ions in the forepart of the beam which enter the magnet in the vicinity of the point K have longer trajectories than the subsequently entering ions which fly along the straight line $\overline{ke}$ in the homogeneous magnetic field normal to the plane of the figure applied to the area confined by the straight lines $\overline{ef}$ and $\overline{ik}$. Thus, the ions successively entering the magnetic field can be focused at one and the same time on the target at point g under the next approximate condition

$$ v_s \approx \left(\frac{\overline{ce}}{\overline{ck} + \overline{ld}} \right) v $$

The velocity of the ion beam motion can be readily evaluated from this formula for a given geometry and a given ion velocity v.

The length of the ion current pulse at the target is given by this method for an ion beam of diameter h, as

$$ \varDelta t \approx \frac{h}{v_s} = \frac{h}{v} \left(\frac{\overline{ck} + \overline{ld}}{\overline{ce}} \right) $$

and the time compression of the beam fraction can be approximated by the formula

$$ K \approx \frac{\overline{ce}}{h} $$

This formula clearly shows that the bunching increases as the diaphragm

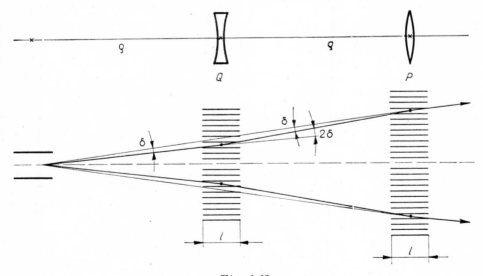

Fig. 4.43

size $\overline{ce}$ of the magnet increases or as the diameter of the beam decreases. The increase in the size of $\overline{ce}$ is limited by the nonlinear contribution from beam oscillation, while the decrease in ion beam diameter is restricted by effects arising from space charge and ion optical errors.

In order to increase the diaphragm size $\overline{ce}$ [1125] the ion beam fraction transmitted by the deflector plates is taken through the correcting electrostatic lenses Q and P shown in Figs 4.42 and 4.43. These electrostatic lenses, composed of numerous electrodes, generate a field which transforms the contribution from sinusoidal oscillation to a linear movement along the straight line $\overline{ce}$. In this way the diaphragm size as well as the bunching factor can be increased.

The energy spread of ions in the current pulse at the target arises in this method from the spread of energies due to the sweep of the beam and from the energy spread caused by the angular spread of the beam striking the target.

The minimum length of the ion current pulse achievable by this method is less than $\Delta t = 10^{-9}$ sec and the peak current of the pulse can exceed 4 mA. In this case the bunching factor is more than 40.

4.3.2.3 *Bunching of reaction products formed in the target*

(a) *Bunching with spiral-target method*
[1128–1130]

The detector counts in nuclear particle studies can be considerably increased if particles originating from nuclear reactions induced by the beam pulse reach the detector at the same time.

A method for increasing the particle intensity in this way is the use of a helically shaped target (Fig. 4.44) (Sapiro [1129]). The principle of the method is the following. The focused ion beam leaving the accelerator is led between two plates turned at 90° to each other. The alternating electric field applied to the pair of deflector plates with a phase shift corresponding to the ion velocities drives the ions along the helical target band.

The height h of the helical spiral in the target and the velocity of the beam movement at the target are so chosen that the neutrons produced by the reactions in the target with velocity v_n reach the detector at one and the same time independently of the point of their origin in the target.

The simultaneous arrival of neutrons of velocity v_n is made possible by choosing the target shape, the direction of the beam and the place of the detector such that the neutrons produced later have a shorter trajectory to the detector than the earlier released neutrons in the target. The neutrons with velocity v_n are bunched according to Fig. 4.44 in a cylindrical spiral target with a single thread of radius r in which the beam circles round at a frequency f while the beam position is defined by the angle $\Phi = 2\pi f t$, under the condition

$$\frac{\varrho_1}{v_i} + \frac{\varrho}{v_n} + \frac{\Phi}{2\pi f} = \frac{L_1}{v_i} + \frac{L}{v_n}$$

where v_i is the ion velocity.

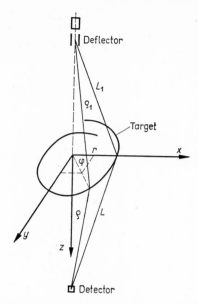

Fig. 4.44 Schematic diagram of ion beam buncher with spiral target

On the assumption that $v_n \ll v_i$, the equation for the spiral can be written as

$$z - a \frac{r^2 z^2}{2L^3} + \ldots = \frac{v_n \Phi}{2\pi f \left(\sqrt{1 - \frac{r^2}{L^2}} - \frac{v_n}{v_i} \sqrt{1 - \frac{r^2}{L_1^2}} \right)}$$

Since usually $r \ll L$ and $Z \ll L$, the product $r^2 z^2$ and the higher order terms can be neglected in the formula obtained from the expansion, thus, the target takes the form of a cylindrical spiral with a uniform upgrade h and the beam frequency can be expressed from the equation as

$$f = \frac{v_n}{h \left(\sqrt{1 - \frac{r^2}{L^2}} - \frac{v_n}{v_i} \sqrt{1 - \frac{r^2}{L^2}} \right)}$$

If $v_i \ll v_n$, that is, the ion energies are much lower than the neutron energies, the arrangement can be so designed that the ions incident on the target arrive simultaneously on the spiral with the appropriate thread. Then the neutrons of energy v_n produced in simultaneous reactions arrive simultaneously to the detector placed at about the same distance from every point of the target.

With the spiral target method the particle intensity can be increased by more than an order of magnitude compared with the simple oscilloscopic method. In this arrangement the pulse of bunched particles can have a shorter length than 10^{-9} sec.

(b) *Bunching with slanted target method*
[1130]

Similarly to the spiral target method, the slanted target technique increases the input intensity of reaction products to the detector (Grodzins, Rose, Van de Graaff [1130]). The principle of the method is the following.

The alternating electric field applied to the deflector plates sweeps the ion beam in a way that a fraction of the beam of given length is deflected by angle Φ to the original direction of the ion beam (Fig. 4.45). The geometry of the thin gas target is such that the plane of the target makes an angle Θ to the direction of the ion beam. The values of Φ and Θ are so chosen that the reaction products with velocity v_n originating from nuclear reactions induced by the ions of the beam pulse at different points of the target should appear simultaneously at the output window of the gas target.

The condition of a simultaneous output of particles with velocity v_n can be expressed as

$$\frac{v_n}{v_i} = \frac{\sin \Phi}{\sin (\Phi - \Theta)}$$

The time spread of the simultaneous output at the target window arises

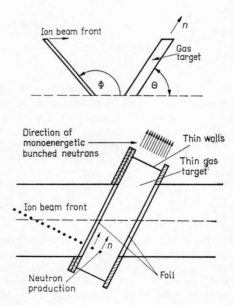

Fig. 4.45 Schematic diagram of ion beam buncher with slanted target

from the uncertainities due to the diameter d_B of the beam pulse, defined by

$$\Delta t(d_B) = \frac{d_B}{v_i \sin \Phi}$$

and to target thickness d_t, defined by

$$\Delta t(d_t) = \frac{d_t}{v_n \, \text{tg} \, (\Phi - \Theta)}$$

With the use of the slanted target method the minimum obtainable length of the bunched particle pulse is $\sim 2 \cdot 10^{-9}$ sec while the maximum bunching factor amounts to ~ 40.

4.3.3 SYSTEMS FOR THE PRODUCTION OF NANOSECOND ION CURRENT PULSES

All methods suitable for the production of nanosecond particle pulses have advantages as well as drawbacks. For this reason most of the apparatuses used in practice show a combination of several techniques. Preferably the oscilloscopic method is combined with other techniques as the intervals between the pulses generated by beam oscillation method are essentially free from background noise.

Nanosecond ion current pulses can be produced even by oscilloscopic technique alone but the low number of particles per pulse makes this method inadequate for the investigation of processes which have small cross sections.

In order to obtain particle pulses of higher intensity, the most convenient method seems to be to combine the oscilloscopic with the particle bunching technique. This combination permits the background contribution from the non-bunched beam to be eliminated since ion beam lengths corresponding to the beam utilization factor can be taken into the buncher by use of the oscilloscopic method. These longer fractions of the beam which are bunched to lengths of the order of $\sim 10^{-9}$ sec already contain a sufficient number of particles for experimental studies.

The bunching technique combined with the oscilloscopic method should always be chosen with regard to the physical measurements involved. In experiments where the spread of ion energy in the pulse does not affect the results of the measurement, it is advisable to combine the oscilloscopic method with velocity modulation. The equipment for the combination of these two techniques can be placed to operate either before acceleration or, if the energies are not too high, after acceleration. Of the two possibilities the pulsing and bunching before acceleration is preferable since in this case the changing of acceleration energy require merely the parameters of the buncher to be changed and the dimensions and power consumption of the devices to do this allow them to be mounted into the high voltage electrode of the accelerator apparatus.

It follows from the above that this combination is impracticable in the study of processes where the allowed energy spread is less than the minimum possible energy spread obtainable in the case of velocity modulation. In these cases the combination with travelling wave bunching or, if mono-energetic neutron pulses are needed, spiral or slanted target bunching can be applied along with the oscilloscopic method.

4.3.3.1 *Systems bunching by velocity modulation*
[1120, 1131–1134, 1137, 1139, 1141, 1142, 1145, 1146, 1150, 1152]

For the production of nanosecond ion current pulses, bunching by velocity modulation is seldom used alone. It is preferably combined with oscilloscopic pulsing technique. In systems where this combination is employed, the ion velocities in the beam fractions bunched by oscilloscopic technique have to be so modulated that the bunching takes place at the target placed at distance $S = \tau v_0$ from the modulator slit. Since this distance from the modulator slit to the target is covered by the ions at velocities other than v_0, the amplitude of the required modulating voltage U_m has to be evaluated when the equivalent lengths of the path fractions S_i covered with different veloc-

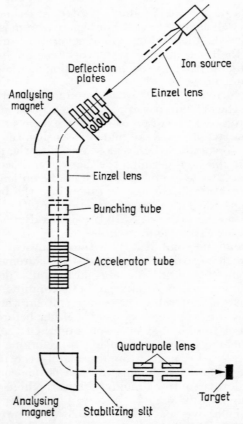

Fig. 4.46 Schematic diagram of assembly for producing nanosecond pulses by combination of oscilloscopic and velocity modulation techniques

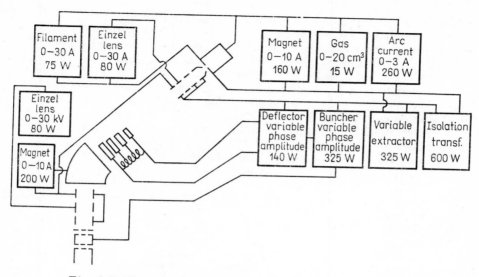

Fig. 4.47 Block diagram of the assembly shown in Fig. 4.46

ities have been already determined [1134]. The ions emitted from the source with energy E_0 fly with velocity v_0 from the modulator slit to the preaccelerating lens at distance S_0 from the modulator slit. Here energy E_1 is imparted to the ions and they fly over the distance S_1 to the accelerator tube input already with energy $E_0 + E_1$. The homogeneous field of the accelerator tube of length S_2 imparts to the ions energy E_2 and thus they fly over the distance S_3 from the accelerator tube output to the target with a velocity corresponding to energy $E_0 + E_1 + E_2$. In an assembly consisting of these elements, the equivalent length can be evaluated from the expression

$$S = S_0 + S_1' + S_2' + S_3' = S_0 + S_1\sqrt{\frac{E_0}{E_0 + E_1}} +$$

$$+ S_2\frac{2E_0}{E_2}\left(\sqrt{\frac{E_0}{E_0 + E_1}} - \sqrt{\frac{E_0}{E_0 + E_1 + E_2}}\right) + S_3\sqrt{\frac{E_0}{E_0 + E_1 + E_2}} \quad (4.14)$$

The value of the required velocity modulating pulse amplitude $U_{m0} = \alpha U_0$ can be calculated by use of the obtained equivalence and the formulae $\tau = \dfrac{S}{v_0}$, $A = \dfrac{1}{2}\alpha\tau\omega$ and $\alpha = \dfrac{U_{m0}}{U_0} = \dfrac{2A}{\tau\omega}$.

The equipment which combines the oscilloscopic and velocity modulation techniques for the production of nanosecond ion current pulses is placed, in most cases, between the ion source and the accelerator tube.

The schematic drawing of one of these arrangements and the block diagram of the associated electronics are shown in Figs 4.46 and 4.47, respectively (Anderson, Swann [1134]).

The ions for the pulsed ion beam are produced by capacitively coupled high frequency ion sources or by duoplasmatrons. The continuous ion beam emitted from the source is focused by a system of gap and unipotential lenses to the deflector plates of the pulser operating by oscilloscopic method. Here the ion beam is pulsed above the limiting slit by a deflector system consisting of 10 pairs instead of the conventional one or two pairs of plates. The ions are deflected by the square voltage signal of 200 V amplitude applied successively to the deflector plates with a delay chosen according to the ion velocities. In this way the difficulties of generating sufficiently fast rising pulses with amplitudes of a few kV can be avoided.

The beam fractions of $\sim$50 nsec length are led through a magnetic analyser which separates the different components of the beam from one another and transmits only the ions of given velocity across a unipotential lens to the buncher. The ion velocity is then modulated in the buncher so that the length of the ion current pulse striking the target is bunched to $\sim$2 nsec. The repetition frequency of the $\sim$50 nsec ion current pulses is $5 \cdot 10^6$ Hz/sec.

A magnetic field of maximum 2.5 kgauss is generated in the magnetic separator which has a radius of 100 cm and a sector field of 20°. The buncher unit has two modulating slits. The distance between the two slits can be varied from 15 to 26 cm and they are confined by a grid having a transparency of 65 per cent.

The length of proton current pulses obtainable at a repetition frequency of $5 \cdot 10^6$ Hz/sec were measured on the target as of $\sim$2 nsec length with current peaks as high as 1 mA. The energy spread of protons in a pulse was estimated as 5 keV for proton energies of 6 MeV. This energy spread is due to velocity modulation.

Another version of the apparatuses with pulsing and velocity modulation [1120, 1145, 1146] is illustrated by the schematic arrangement in Fig. 4.48 and by the block diagram of the electronics in Fig. 4.49 (Moak, Good, King et al. [1120]).

The ions are obtained from a capacitively coupled high frequency ion source or a duoplasmatron. The continuous ion beam is led through gap and unipotential lenses and a pulser and separator system which generates Lissajous curves. The ion beam is accelerated and formed by the gap lens placed after the ion source, it is then passed between the first pair of deflector plates to the unipotential lens which focuses the beam to the plane of the limiting slit.

Grids of h gh transparency are usually mounted into the apertures facing the deflector plates in the unipotential lens. These grids decrease the effect of the stray field generated on the deflector plates [1120] also the errors of the unipotential lens. The grid on the aperture considerably reduces the spherical aberrations. The minimum beam diameter in the plane of the limiting slit is smaller by a factor of almost 3 in the presence of a grid than that obtainable without it. This arrangement substantially improves the separation of the beam components. The ion beam fraction of 15 nsec length flies across the limiting slit into the velocity modulator where the ion velocities are modulated before their reaching the accelerator tube across the adapter immersion lens. The ions are then accelerated in the accelerator tube to the velocity required by the physical investigations. The accelerated ion beam is deflected to 90° and focused on the target with a quadrupole lens.

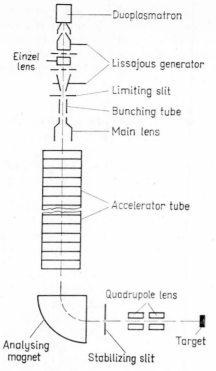

Fig. 4.48 Schematic diagram of assemb y for producing nanosecond ion beam pulses by combination of Lissajous curve generating and velocity modulation techniques

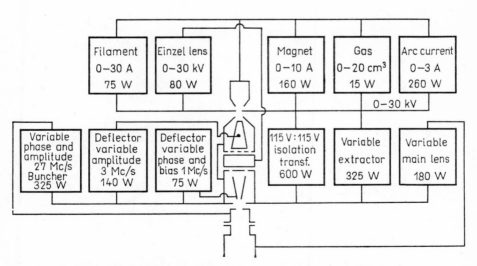

Fig. 4.49 Block diagram of the assembly shown in Fig. 4.48

Ion current pulse lengths on the target can be varied between 1 and 15 nsec by varying the amplitude of the modulator pulse U_m. The minimum pulse length measured on the target was found to be ~1 nsec with a current of ~10 mA in the peak. The spread of ion energies in a current pulse was estimated as ~10 keV, at 3 MeV proton energy and it could be attributed to the effect of velocity modulation.

4.3.3.2 *Systems bunching by magnetic compression*
[1126, 1127, 1147]

The schematic drawing and the block diagram of the electronics of an arrangement combining oscilloscopic and magnetic bunching techniques are shown in Figs 4.50 and 4.51.

The ion beam can be produced either from a high frequency ion source or from a duoplasmatron. The continuous ion beam of the source passes through a system consisting of a unipotential lens, pulser and a separator generating a Lissajous curve which produces ion current pulse lengths of 10 nsec with a given repetition frequency. These ion current pulses are passed across an adapter immersion lens to the accelerator tube where the ions are accelerated to the energy required for the physical experiments.

The ion beam sections are focused after the 90° deflection by means of a quadrupole lens on the input focal point of the Mobley-type magnetic buncher. Before reaching the deflector plates at the magnetic buncher input, the ions pass through a generator of synchronous input signals to the deflector plates of the buncher. The 10 nsec ion current pulse is then deflected

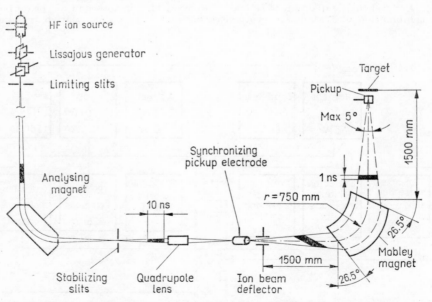

Fig. 4.50 Schematic diagram of assembly for producing nanosecond ion beam pulses by combination of Lissajous curve generating and Mobley's methods

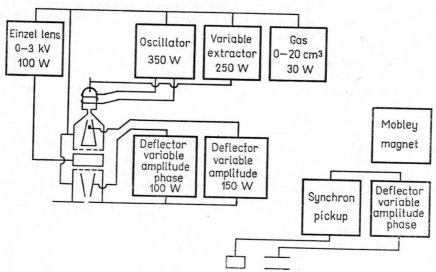

Fig. 4.51 Block diagram of the assembly shown in Fig. 4.50

by the synchronous input signal controlled plates in the way shown in the figure. This slant beam section is compressed by the magnetic buncher to a length of ∼1 nsec at the target.

In arrangements of this type the variation of the accelerating energy requires the change of many more operational parameters than in the previously described bunchers as the ions are bunched after their acceleration.

For this reason, the greatest problems in this arrangement are the design of the bunching magnet with a variable field over the large area of ∼0.5 m² between the homogeneous poles with the necessary high precision stabilizer and that of the high frequency power supply with an output power of 1 kW with amplitudes from 8 to 15 kV fed to the deflector plates.

With this arrangement ion current pulses of ∼1 nsec length with a current of 10 mA in the peak can be produced. For 3 MeV proton energy the spread of ion energies appearing at the deflector plates of the magnetic buncher input was measured as ∼6 keV on the target.

It has to be noted that when using this method the direction of ions striking the target is different and they penetrate the target to different depths. The thus different path lengths of the reaction products in the target can reduce the energy resolution of the equipment.

4.3.3.3 *Systems bunching by travelling wave*

The schematic drawing of an arrangement bunching by travelling wave technique can be seen in Fig. 4.52.

The continuous ion beam produced by a high frequency or a duoplasmatron type ion source passes through a focusing and pulser system composed of a gap lens, a unipotential lens and a pair of deflector plates. This system transforms the beam to pulses of appropriate input length to the buncher.

The bunching system is formed by a set of plane electrodes to which at time t_n a voltage pulse of amplitude U is applied which runs along the set at a velocity $v_f = v_0 + v$. The velocity of ions entering the buncher with velocity v_0, forming a pulse of length $(t_n - t_0)$ increases by v as they are overtaken by the travelling wave. This results in that the buncher output pulse at time t_b is generated simultaneously by the voltage pulse applied to generate the travelling field and by the bunched ions of velocity v_f. The symbols are the same as those used in Fig. 4.41.

The bunched ions enter the homogeneous field of the accelerator tube with the same velocity and are accelerated to the required experimental energy, then deflected by 90° and focused on the target with a quadrupole lens.

If this method is employed great care must be taken to stabilize the amplitude and velocity of the applied voltage pulse generating the bunching travelling wave and to minimize the energy spread of the ions flying out from the source through the pair of deflector plates.

The minimum pulse length obtainable with this arrangement is $2 \cdot 10^{-9}$ sec, while the maximum beam utilization and bunching factors are 80 per cent and $k \sim 80$, respectively.

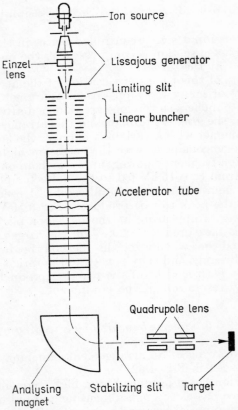

Fig. 4.52 Schematic diagram of assembly for producing nanosecond ion beam pulse by combination of Lissajous curve generating and linear bunching methods

4.4 NUCLEAR SPIN-POLARIZED ION SOURCES

4.4.1 INTRODUCTION

Interactions between nucleons and compound nuclei can be more thoroughly studied and better understood if nuclear interactions and their polarization phenomena are investigated using nuclear spin-polarized particle beams.

Conventional ion sources in particle accelerators yield unpolarized ion beams and new methods had to be developed for obtaining nuclear spin polarized ion beams.

In particle beams polarized with respect to nuclear spin, e.g. in the case of nucleons, the directions of the nuclear spins are partly or wholly the same. In the presence of an external magnetic field nuclear spins can have two directions according to the magnetic quantum numbers $m_I = \pm 1/2$, that is the spins can be either parallel or antiparallel to the direction of the magnetic field. In an external magnetic field the polarization of a nucleon beam can be defined as

$$ p = \frac{N\uparrow - N\downarrow}{N\uparrow + N\downarrow} $$

where $N\uparrow$ and $N\downarrow$ stand for the numbers of nucleons with parallel and antiparallel spins relative to the direction of the magnetic field, respectively.

Polarized nucleons can be obtained by scattering on certain nuclei and also from numerous nuclear reactions [782–785]. The idea of using polarized beams for nuclear studies has been suggested by the observation that polarized particle beams generated by various techniques can be accelerated without appreciable depolarization in electrostatic, linear and cyclical accelerators. The thus obtained polarized particle beams have several advantages over scattering on nucleus or reaction polarization products, namely, they are well-collimated, their energies can be varied and the intensity ratios are also more favourable.

Apart from the above mentioned nuclear physical methods, nuclear spin-polarized particles can be obtained by the following techniques:

(a) Separation of atoms in ground state with respect to nuclear spin state in an inhomogeneous magnetic field.

(b) Separation of excited atoms with respect to spin state by utilization of the metastable states and the Lamb shift.

(c) Separation of atoms with respect to spin state by "optical pumping" technique.

(d) Capture of polarized electrons by unpolarized ions.

Nuclear physical methods do not seem suitable for the production of collimated polarized particle beams with adequate intensity in the low energy range (0–10 MeV). Any other of the listed methods — which are in connection with atomic electron shell physics — can be utilized for the design of ion sources yielding polarized beams suitable for further acceleration.

Methods (a) and (b) have been most frequently used to date. Both lend themselves to the production of positively or negatively charged polarized ion beams. The equipment can be combined with electrostatic, linear and cyclical accelerators. Polarized positive ion currents in the range $(1–1.5) \cdot$ $\cdot 10^{-6}$ A, negative ion currents in the range $(0.2–5) \cdot 10^8$ A can be obtained from these sources. Polarization of protons to the degree $P \sim 0.45$ and if radiofrequency induced transitions are utilized, to $P \sim 0.75$, have been reported. For deuterons vector polarization to $P_3 \sim 0.29$ with induced transitions to $P_3 \sim 0.55$ and deuteron tensor polarization to $P_{33} \sim 0.75$ have been achieved.

4.4.2 THE HYPERFINE STRUCTURE OF ENERGY LEVELS IN HYDROGEN ATOMS
[1, 3, 426, 1164, 1183–1188]

To understand how nuclear spin polarization can be achieved, it seems worth considering briefly the fine structure of energy levels in hydrogen atoms with a comprehensive review of the spin states of hydrogen atoms.

Hydrogen atoms are composed of two particles, proton and electron, both with spin 1/2. The total angular momentum $\boldsymbol{F}$ of the atom is the sum of the total angular momentum $\boldsymbol{J}$ of the electron shell and the resultant momentum $\boldsymbol{I}$ of the proton, that is

$$\boldsymbol{F} = \boldsymbol{I} + \boldsymbol{J} \tag{4.15}$$

By the rules of space quantization, the possible values of the total angular momentum F are

$$F = I + J, \ I + J - 1, \ldots, I - J \tag{4.16}$$

In an external magnetic field the magnetic quantum number of the total angular momentum belonging to the projection on the direction of the external magnetic field can have the values

$$m_F = F, \ F - 1, \ldots, \ -F + 1, \ -F \tag{4.17}$$

For simplicity, the considerations will be restricted to the spin states associated with $J = 1/2$ and $I = 1/2$. In this case the total angular momentum F of the hydrogen atom has two possible values according to formula (4.17), namely

$$F = I + J = 1$$
$$F = I - J = 0 \tag{4.18}$$

that is, if the angular momentum of the nucleus is taken into account, the energy levels in the fine structure of hydrogen atoms split into two hyperfine energy sublevel groups. In the absence of an external magnetic field the distance between two sublevels belonging to the two hyperfine level structures is determined by the magnetic interaction between the atomic nucleus and the electron shells.

The value of this energy arising from the interaction of the magnetic moment μ_I of the nucleus, taken to be a point-like magnetic dipole, with the magnetic field $H(0)$, generated in the centre of the nucleus by the electron shell, can be expressed as

$$E_F = -\big(\mu_I H(0)\big) = -\mu_I \cos\big(\mu_I H(0)\big) \tag{4.19}$$

Since in the case of one electron, $H(0)$ and J are antiparallel, while μ_I and I are parallel, expression (4.19) takes the form

$$E_F = \mu_I H(0) \cos(IJ) = \mu_I H(0) \frac{F(F+1) - I(I+1) - J(J+1)}{2IJ} =$$

$$= \frac{\alpha}{2}\left[F(F+1) - I(I+1) - J(J+1)\right] \tag{4.20}$$

where

$$\alpha = \frac{\mu_I H(0)}{IJ}$$

On substituting the values $F = 1$ or 0, $I = 1/2$, $J = 1/2$ into (4.20), we have

$$E_F = \frac{\alpha}{4} \qquad \text{if} \quad F = 1$$

$$E_F = -\frac{3}{2}\frac{\alpha}{2} \quad \text{if} \quad F = 0 \tag{4.21}$$

The difference between the energy levels with $F = 1$ and $F = 0$ can be well approximated for hydrogen atoms, if $l = 0$ and $l > 0$, by using Bethe's equation [1]

$$\Delta E_F = \frac{4(2\mu_0)^2 z^3 g R_y m_e}{n^3 (2l+1)(j+1)m_p}\left(I + \frac{1}{2}\right) \quad \text{if} \quad j \leq I$$

It is obvious from this expression that ΔE_F rapidly decreases as the main quantum number n increases and that this decrease is slower with increasing values of the orbital and total angular momentum quantum numbers l and j. Thus we find for hydrogen ($Z = 1$; $g = 5.56$; $J = 1/2$).

	$1^2S_{1/2}$	$2^2S_{1/2}$	$2^2P_{1/2}$	
ΔE_F	$1420.4\,h$	$\dfrac{1420.4\,h}{8}$	$\dfrac{1420.4\,h}{24}$	(4.22)

In the presence of an external magnetic field the energy levels split into four components corresponding to the Zeeman-type hyperfine energy level structure given by formula (4.17) as a triplet state with $m_{F=1} = \pm 1$; 0 and a singlet state with $m_{F=0} = 0$ giving together four substates and four quantum numbers. The shift of the energy levels of the different states in the hyperfine structure relative to the energy levels of the fine structure can be obtained from the simultaneous solution to the eigenvalue equation $\mathcal{H}\psi = E\psi$ for the four states.

In the general case (if any external magnetic field H_0 is present) interactions between electron and nucleus (spin–orbit interaction), between electron and external field as well as interactions between nucleus and external field due to Zeeman effect have to be taken into account in the complementary term $\mathcal{H}_M$ of the Hamiltonian of the interaction. Considering the above interactions, the complementary term $\mathcal{H}_M$ of the Hamiltonian interaction of which takes account of the hyperfine contributions can be written down as [1, 3, 426, 1187]

$$\mathcal{H}_M = -\mu_I \boldsymbol{H}(0) - \mu_J \boldsymbol{H}_0 - \mu_I \boldsymbol{H}_0 \tag{4.23}$$

The general wave function, which holds in an arbitrary magnetic field, can be formulated by the superposition of the products of functions φ_{Im_I} and φ_{Jm_J} describing the nuclear and electron spins, in a strong magnetic field [1], respectively, that is

$$\psi_{IJ}F_m = \sum_i a_i\varphi_{Im_I}\varphi_{Jm_J} \tag{4.24}$$

where the coefficients a_i are the elements of the unitary matrix satisfying the normalizing condition $\sum_i |a_i|^2 = 1$.

The eigenvalue problem can be solved by the method usually applied to fourfold degeneration [1, 3, 1187, 1188], that is, we try to find the solution to the corresponding secular equation which yields the correction term in energy arising from spin-orbit and Zeeman-type interactions.

It can be shown that to obtain, e.g. the corrections for an external magnetic field arising in the energy levels $1^2S_{1/2}$, $2^2S_{1/2}$ and $2^2P_{1/2}$ in hydrogen and deuterium atoms, it is sufficient to solve the secular equation corresponding to the last submatrix of 2×2 configuration of the quasi-diagonal matrix with 4×4 configuration. The secular equation obtained from the matrix with a configuration of 2×2, has the form

$$\begin{vmatrix} \mathcal{H}_{M_{33}} - E_{F_m} & \mathcal{H}_{M_{34}} \\ \mathcal{H}_{M_{43}} & \mathcal{H}_{M_{44}} - E_{F_m} \end{vmatrix} = 0 \tag{4.25}$$

$\mathcal{H}_{M_{nk}}$ are here the matrix elements of $\mathcal{H}_M$ which give the corrections in the energies of the hyperfine levels E_{F_m}. The solution of the equation of second order obtained from the expansion of the determinant can be written in the form

$$E_{F_m} = -\frac{\Delta E_F}{2(2I+1)} - \frac{\mu_I}{I} H_0 m \pm \frac{\Delta E_F}{2} \sqrt{1 + \frac{4m}{2I+1}x + x^2} \tag{4.26}$$

where

$$x = \left(-\frac{\mu_J}{J} - \frac{\mu_I}{I}\right)\frac{H_0}{\Delta E_F}$$

and

$$\Delta E_F = E_{I+\frac{1}{2}}(H_0 = 0) - E_{I-\frac{1}{2}}(H_0 = 0)$$

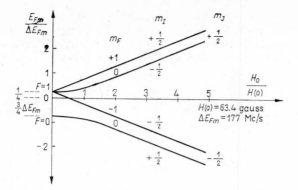

Fig. 4.53 Hyperfine structure of the $1S_{1/2}$ levels of hydrogen atoms in external magnetic field

is the energy difference between the levels belonging to $F = 1$ and $F = 0$. In formula (4.21) the sign is $+$, before the square root if $F = I + 1/2$, and it is $-$ if $F = I - 1/2$.

Formula (4.26) can describe all the four states of the Zeeman hyperfine structure as it becomes a linear equation in the case of $F = I + 1/2$ with the values of $m = \pm(I + 1/2)$ when the expression under the square root is the total square of $(x \pm 1)$. Consequently, formula (4.20) can be used to

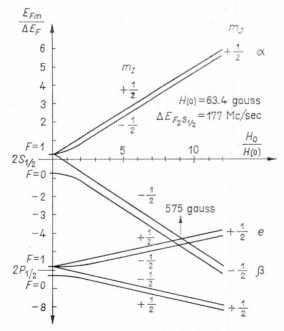

Fig. 4.54 Hyperfine structure of the $2S_{1/2}$ and $2P_{1/2}$ states of hydrogen atoms in external magnetic field

evaluate the energy corrections by taking into account both the internal and the external fields for the levels in the hyperfine structure of hydrogen atoms in states characterized by any of the possible values of m.

For the levels $1^2S_{1/2}$ and $2^2S_{1/2}$, $2^2P_{1/2}$ the following formulae can be used

$$E_{F_m1^2S_{1/2}} = \frac{-\Delta E_{F1^2S_{1/2}}}{2(2I+1)} - \frac{\mu_I}{I} H_0 m \pm \frac{\Delta E_{F1^2S_{1/2}}}{2} \sqrt{1 + \frac{4m}{2I+1} x + x^2}$$

$$(4.27)$$

$$E_{F_m2^2S_{1/2}} = \frac{-\Delta E_{F2^2S_{1/2}}}{2(2I+1)} - \frac{\mu_I}{I} H_0 m \pm \frac{\Delta E_{F2^2S_{1/2}}}{2} \sqrt{1 + \frac{4m}{2I+1} x + x^2}$$

$$(4.28)$$

$$E_{F_m2^2P_{1/2}} = \frac{-\Delta E_{F2^2P_{1/2}}}{2(2I+1)} - \frac{\mu_I}{I} H_0 m \pm \frac{\Delta E_{F2^2P_{1/2}}}{2} \sqrt{1 + \frac{4m}{2I+1} x + x^2}$$

$$(4.29)$$

On introducing into formulae (4.27), (4.28) and (4.29) the values of ΔE_F taken from formulae (4.22), the hyperfine structures of the energy levels belonging to the $1^2S_{1/2}$ ground state and to the $2^2S_{1/2}$ and $2^2P_{1/2}$ excited states of hydrogen atoms can be graphically represented as a function of the external magnetic field (Figs 4.53 and 4.54).

The variation caused by an external field in the magnetic moment μ_{eff} of the different atomic states relative to the direction of the magnetic field is given as

$$\mu_{\text{eff}} = -\frac{\partial E_{Fm}}{\partial H_0} = \frac{\mu_I}{I} m \mp \frac{x + \dfrac{2m}{2I+1}}{2\left[1 + \dfrac{4m}{2I+1} x + x^2\right]^{1/2}} \left(-\frac{\mu_J}{J} + \frac{\mu_I}{I}\right)$$

$$(4.30)$$

Since $\mu_I \ll \mu_J$, expression (4.30) can be well approximated by the form

$$\mu_{\text{eff}} = \mp \frac{x + \dfrac{2m}{2I+1}}{2\left[1 + \dfrac{4m}{2I+1} x + x^2\right]^{1/2}} \left(-\frac{\mu_J}{J}\right)$$

$$(4.31)$$

and here $x = \dfrac{\mu_J}{J} \dfrac{H_0}{\Delta E_F}$.

The graphical representation of the ratio μ_{eff}/μ_0 (where μ_0 stands for the Bohr magneton) for $m_{F=1} = \pm 1$; 0 and $m_{F=0} = 0$ can be seen in Fig. 4.55.

It is apparent from Fig. 4.53 that the shift of levels belonging to states with $m_{F=1} = \pm 1$ is a linear function of the external magnetic field, while

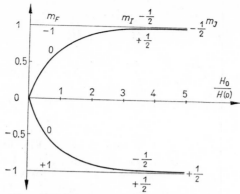

Fig. 4.55 Dependence of the ratio μ_{eff}/μ_0 on the external magnetic field H_0 in the case of hydrogen atoms

the components associated with the values $m_{F=1} = 0$ and $m_{F=0} = 0$ show a quadratic dependence in the case of weak external fields. The value of the effective magnetic moment is constant for $m_{F=1} = \pm 1$ and independent of the changes of the external field, whereas the values of the two other components vary as the external field increases approaching in a strong external field the value of the two components with total moment.

This variation is caused by the gradual decomposition of the coupling between the electron and proton spins. The coupling is eventually completely disrupted under the action of a strong external magnetic field. The coupling between the two spins persists as long as the interaction of the internal magnetic field $H(0)$, generated by the electron shell in the centre of the nucleus, with the nuclear magnetic moment remains negligible as compared with the interaction of the external magnetic field H_0 with the total mag-

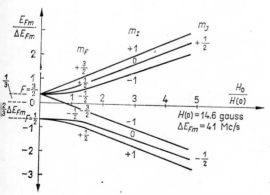

Fig. 4.56 Hyperfine structure of the levels in the $1S$ state of deuterium atoms in external magnetic field

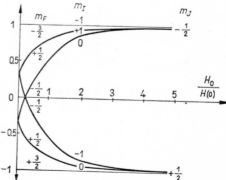

Fig. 4.57 Dependence of the ratio μ_{eff}/μ_0 on the external magnetic field H_0 in the case of deuterium atoms

netic moment of the atom, that is, as long as

$$\mu_I H(0) \ll \mu H_0$$

hence

$$H_0 \gg \frac{\mu_I H(0)}{\mu} \approx \frac{\Delta E_F}{2\mu_0} = H^* = 507 \text{ gauss} \qquad (4.32)$$

since for the $1^2S_{1/2}$ level in hydrogen atoms, $\Delta E_F = 2\mu H(0) = h. \ 1420.4$ Mc/sec.

In deuterium atoms the hyperfine Zeeman structure in a magnetic field can be composed of six states according to eq. (4.17). The energy levels of these states, similarly to those of the H_1^0 atoms, depend on the value of the external magnetic field. The dependence of the hyperfine energy levels on the external magnetic field in the $2S_{1/2}$ state of deuterium atoms is shown in Fig. 4.56 while the curve for the effective magnetic moment against the external magnetic field can be seen in Fig. 4.57.

4.4.3 SEPARATION OF ATOMIC COMPONENTS OF AN ATOMIC BEAM IN DIFFERENT SPIN STATES
[426, 471, 1157–1165]

It has been seen in the preceding section that the Zeeman hyperfine structure of energy levels is associated with four spin states of the $1^2S_{1/2}$ ground state and the $2^2S_{1/2}$ and $2^2P_{1/2}$ excited states in hydrogen and with six spin states in deuterium.

Two components having the same directions of nuclear spin (Figs 4.53 and 4.56) are found in hydrogen for the quantum number $m_{F=1} = \pm 1$ and in deuterium for $m_{F=3/2} = \pm 3/2$, while the states associated with other quantum numbers have components with different directions of nuclear spin.

The separation of atomic beam components in the different spin state of their ground state is made possible by Stern—Gerlach-type experiments. If an inhomogeneous magnetic field is applied to an atomic beam, the beam can be split into components with respect to different spin state and having a steric deflection relative to one another which permits the components to be separated by use of appropriate diaphragms.

On the assumption that E_{F_m} depends on the space coordinates only through the variation of the magnetic field, the force acting on the atomic beam in an inhomogeneous magnetic field is given by

$$F = -\text{ grad } E_{F_m} = -\frac{\partial E_{F_m}}{\partial H_0} \text{ grad } H_0 \qquad (4.33)$$

Using the equality $\mu_{\text{eff}} = -\frac{\partial E_{F_m}}{\partial H_0}$ given by expression (4.30), we get

$$F = \mu_{\text{eff}} \text{ grad } H_0$$

This formula shows that for an inhomogeneous magnetic field of given geometry the magnitude of the deflection of atomic beam components with specific spin states depends only on the value of μ_{eff} in the atomic state.

In states with unit moment the value of μ^i_{eff} is constant while it varies with the value of the magnetic field in the components with two mixed spin states (Fig. 4.55).

For the proper geometry of the inhomogeneous field it is the most convenient to apply multipole (4 or 6 poles) magnetic fields (Fig. 4.58).

Multipole fields of this type can be regarded as lenses of cylindrical symmetry by which the atoms with specific spin states are focused or defocused depending on the polarity of their magnetic moment (Fig. 4.59).

The magnetic field in the space between the poles of multipole magnets can be described by the equation

$$H_0 = - \text{grad } \varphi \qquad (4.34)$$

where $\varphi = a_n r^n \cos n\Theta$ is the magnetic potential with n being the number of pole pairs and r and Θ the cylindrical coordinates. Using formula (4.34), the magnetic field and the force acting on the atoms of the atomic beam are given for 4 and 6 poles as follows:

four poles: $n = 2$

$\varphi \sim r^2 \cos 2\Theta$

six poles: $n = 3$

$\varphi \sim r^3 \cos 3\Theta$

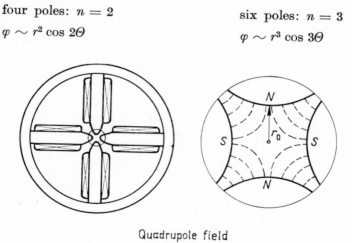

Quadrupole field

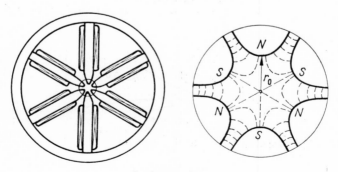

Sextupole field

Fig. 4.58 Arrangement of poles in quadrupole and sextupole magnets and configurations of magnetic fields generated at poles

$$|H_r| = \left(\frac{H_0}{r_0}\right) r \qquad\qquad\qquad |H_r| = \left(\frac{H_0}{r_0^2}\right) r^2$$

$$\text{grad } H = \frac{\partial |H_r|}{\partial r} = \frac{H_0}{r_0} \qquad\qquad \text{grad } H = \frac{\partial |H_r|}{\partial r} = \left(\frac{2H_0}{r_0^2}\right) r$$

$$F = \mu_{\text{eff}} \frac{\partial |H_r|}{\partial r} = \mu_{\text{eff}} \frac{H_0}{r_0} \qquad\qquad F = \mu_{\text{eff}} \left(\frac{2H_0}{r_0^2}\right) r$$

Atomic beams with the same initial parameters can be focused in inhomogeneous magnetic fields with either 4 or 6 poles under the condition that a force proportional to r acts on the atoms with negative magnetic moment. For the component with $m_{F=1} = +1$, which has a constant magnetic moment — when the linear Zeeman effect asserts itself ($E_{F_m} \sim H_0$), — this condition is satisfied in a field with 6 poles.

On the other hand, for the component with $m_{F=1} = 0$ and a weak magnetic field ($H_0 \leq H^*$) — when the quadratic Zeeman effect is dominant ($E_{F_m} \sim H_0^2$) — the condition of good focusing can be best satisfied in a quadrupole field.

However, it has to be noted that by use of appropriate discriminating diaphragms both the component with $m_{F=1} = +1$ and that with $m_{F=1} = 0$ can be separated with either of the above described inhomogeneous multipole magnetic fields.

In inhomogeneous quadrupole magnetic fields the position of the minimum diameter of the atomic beam depends on the initial velocity and angle of incidence of input atoms and on the distance r of the input from the axis of the magnetic field. The atoms entering the magnetic field under different initial conditions do not focus on a given point, but within a region of the axis which can be well determined if the input parameters of the beam are known. In a sextupole magnetic field the position of the minimum diameter of the atomic beam is determined only by the velocity of the atoms. In this case not all of the focusing atoms cross the symmetry axis of the magnetic field but focus on the focal point in the plane perpendicular to the axis at a distance $0 < r < r_0$ from the latter as determined by the r value of the input and by its angle to the plane crossing the axis.

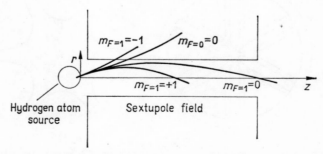

Fig. 4.59 Schematic diagram of separation process in inhomogeneous multipole magnetic fields

A considerable chromatic aberration is caused by the velocity distribution of the atomic beam emitted from the source. The image defects due to velocity distribution can be almost completely eliminated by decreasing the inhomogeneity of the magnetic field in the direction of beam propagation. The dependence of the focusing on the input geometry can be decreased by the use of an annular inlet diaphragm.

The separation with respect to spin states can be achieved with multipolar inhomogeneous magnets in either weak, $H_0 \lesssim H^*$, or strong, $H_0 \gg H^*$ magnetic fields.

In weak applied fields — as can be seen from the schematic drawing in Fig. 4.59, — each component can be singled out. Out of the states $m_{F=1} = +1$ and $m_{F=1} = -1$, which are of interest for the study of nuclear spin polarization, it is preferable to choose the focusable $m_{F=1} = +1$ component because the useful solid angle for good separation is rather small owing to the $H_0 \sim 0$ and grad $H \gg 0$ requirements. If the beam is polarized in a weak field the degree of polarization will be the higher the more the image defects due to initial conditions can be eliminated.

In the case of strong applied fields the force acting on the atoms of both the focusable component with $m_{F=1} = +1$ and that with $m_{F=1} = 0$ is the same since the value of μ_{eff} in the latter state becomes nearly equal to that of μ_0 if a strong field is present. Therefore, the above two components ($m_{F=1} = +1$ and $m_{F=1} = 0$) are focused together, while the component with $m_{F=1} = -1$ is defocused together with the component in the state $m_{F=1} = 0$.

However, these component pairs are polarized in a strong field only with respect to electron spin without being polarized in nuclear spin.

Nuclear polarization takes place during the slow (adiabatic) transition from the strong to the weak field at the end of the deflecting magnet since in the weak field the spins are not individually requantized but they are quantized according to the resultant spin values and thus the nuclear spins in the component with $m_{F=1} = 0$ show again a random distribution.

Figure 4.60 shows the magnetic field dependence of nuclear spin polarization. It can be seen from the figure that the maximum obtainable proton spin polarization in a strong field is 50 per cent if the adiabatic transition of the field can be secured.

The change of the field in the adiabatic transition has to be slow and con-

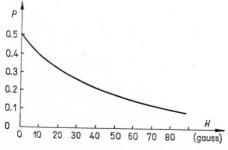

Fig. 4.60 Dependence of polarization of hydrogen atoms with respect to nuclear spin on the external magnetic field

tinuous so as to satisfy the condition

$$\frac{|\Delta H_0|}{\Delta t} < \frac{\Delta E \omega_L}{\mu_0} \tag{4.35}$$

where $\omega_L = \frac{\mu_0}{\hbar} H_0$ is the Larmor frequency, $\Delta E - \mu_0 H_0$ is the difference in energy between the states $F = 1$; $m_F = \pm 1$ and $F = 1$; $m_F = 0$. With these substitutions, we have

$$\frac{|\Delta H_0|}{\Delta t} < \frac{\mu_0 H_0^2}{\hbar} = (8.786 \cdot 10^6 \text{ gauss}^{-1} \text{ sec}^{-1}) H_0^2$$

If H_0 goes to zero over the distance l, then $(\Delta H) = H_0$ and $\Delta t = \frac{l}{v}$ (where v is the velocity of atoms in the beam) and thus, the inequality condition (4.35) can be estimated.

The transformation of this inequality expression to a more convenient expression gives

$$H_0 l > \frac{v}{8.786 \cdot 10^6} \text{ gauss sec} \approx 0.04 \text{ gauss cm}$$

In experiments this condition is generally easily obtainable if the region $H_0 \approx 0$ quite close to the axis is not utilized.

The 50 per cent polarization degree obtainable in strong fields can be increased if the induced transitions $F = 1$; $m_F = 0 \rightarrow F = 0$; $m_F = 0$ are applied (Fig. 4.61).

Such transitions can be induced by the radiofrequency method (Abragam, Winter [1167]). The principle of the method is the following. The atomic beam with components in states $m_{F=1} = +1$ and $m_{F=1} = 0$ is led through cavity resonator operating at a frequency ν_0 chosen with respect

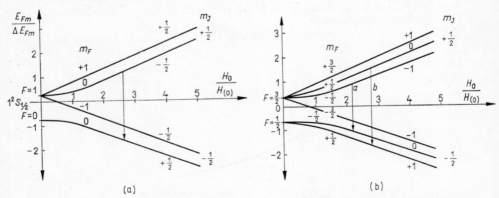

(a)   (b)

Fig. 4.61a Induced transition between states in the hyperfine structure of hydrogen atoms

Fig. 4.61b Induced transition between the hyperfine states of deuterium atoms

to the transition $F = 1$; $m_F = 0 \rightarrow F = 0$; $m_F = 0$; corresponding to the magnetic field H_0 in the centre of the resonator. The cavity is placed into a magnetic field so that the adiabatic change of the field is such that the atoms are exposed at the resonator input to a magnetic field $H_0 + \Delta H$ and at the output to a field $H_0 - \Delta H$, while the transition between these two values is nearly linear.

The magnetic field which can be described in the cavity resonator by the expression $H_1 e^{i\omega t}$, with $\omega = 2\pi\nu$, has to satisfy two conditions:

(1) adiabacity which requires that

$$\frac{2 \Delta H v}{l} < \frac{2\pi\nu_0}{4} \frac{H_1^2}{H^*} \frac{1}{x\sqrt{1 + x^2}}$$

where l is the length of the resonator, v is the velocity of atoms, ν_0 is the fine structure frequency and

$$x = \frac{H_0}{H^*}$$

(2) polarization, which requires that

$$\Delta H \gg H_1$$

In this case the polarization is given approximately as

$$P \approx 1 - \left(\frac{H_1}{\Delta H}\right)^2$$

It follows from this expression that a rather high degree of polarization can be obtained if the values of H_1 and ΔH are appropriately chosen and if the separation in a strong field is followed by an induced transition, since even for $\dfrac{H_1}{\Delta H} = \dfrac{1}{4}$ we already get $P \approx 0.93$.

The different spin states of deuterium atoms in the ground state can be separated from one another, similarly to hydrogen atoms, with multipole magnets having either 4 or 6 poles.

In deuterium the abundances of nuclei with spin states corresponding to quantum numbers $m_I = \pm 1$; 0 and $m_j = +1/2$ in the presence of a magnetic field can be expressed as

$$N_{+1} = \frac{1}{3} \left(1 + |a_{1,\frac{1}{2}}|^2\right)$$

$$N_0 = \frac{1}{3} \left(|a_{2,\frac{1}{2}}|^2 + |a_{1,-\frac{1}{2}}|^2\right)$$

$$N_{-1} = \frac{1}{3} |a_{2,-\frac{1}{2}}|^2$$

where a_{im_F} are the elements of the unitary matrix.

The abundances N_{mI} plotted as the function of against the external magnetic field yield the curve shown in Fig. 4.62.

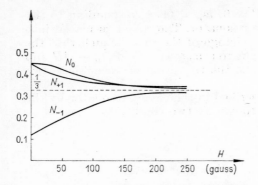

Fig. 4.62 Dependence of the population N_{+1}, N_0, N_{-1} on the external magnetic field for nuclei in spin states corresponding to the quantum numbers $m_I = +1/2$ and $m_I = \mp 1$; 0 in deuterium atoms

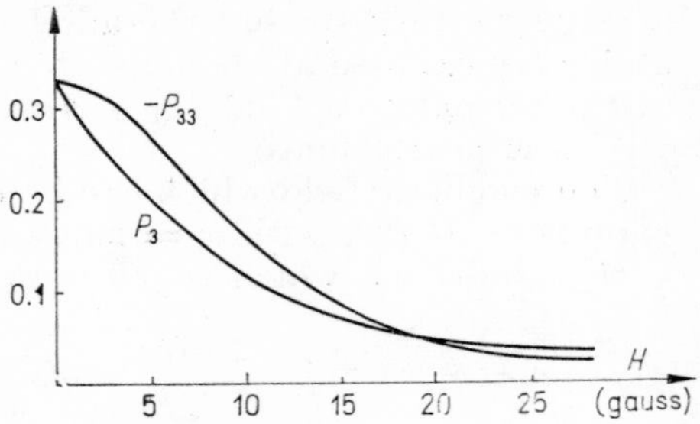

Fig. 4.63 Values of the P_3 vector and P_{33} tensor polarization of deuterium atoms as a function of external magnetic field

The spin state of the deuterium beam with spin 1 is uniquely described by the density matrix ϱ consisting of $(2I + 1)$ rows and columns [1191, 1192]. The density matrix ϱ is known to contain $(2I + 1)^2 - 1$ independent real terms, thus, 8 parameters are required to describe the spin state of the deuterium beam. These parameters can be expressed in terms of the polarization vector P and the polarization tensor P_{ij} which are defined by the spin operators S as

$$P = \frac{1}{\hbar}\langle S \rangle$$

$$P_{ij} = \frac{3}{2\hbar^2}(\langle S_i S_j \rangle + \langle S_j S_i \rangle) - 2\delta_{ij}$$

Here i and j can have the values 1, 2 and 3 corresponding to the axes x_1, x_2 and x_3 of the Cartesian coordinate system and δ_{ij} is the Kronecker symbol.

If the polarization vector has the same direction as the homogeneous magnetic field in the region of ionization, that is, the x_3 coordinate is parallel to the direction of the magnetic field H_i, we get for P_i and P_{ij} the values

$$P_1 = P_2 = 0 \qquad\qquad P_3 = N_{+1} - N_{-1}$$

$$P_{12} = P_{13} = P_{23} = 0 \qquad\qquad P_{11} = P_{22} = \frac{1}{2}(3N_0 - 1)$$

$$P_{33} = 3(N_{+1} + N_{-1})$$

Since the abundances N_{mI} depend on the value of the external magnetic field, the values of P_3, P_{11}, P_{22} and P_{33} are also dependent on the magnetic field. The curves for P_3 and P_{33} are shown in Fig. 4.63. It can be seen that the maximum polarizations of the vector P_3 and the tensor P_{33}, that is 33 per cent and -33 per cent respectively, can be achieved with $H_0 = 0$. By applying induced transitions, P_3 vector polarizations have been obtained up to 45–55 per cent [501, 512, 543, 549, 550, 1162].

4.4.4 NUCLEAR SPIN POLARIZATION OF HYDROGEN ATOM BEAMS EXCITED TO $2S_{1/2}$ STATE

Nuclear spin states of the Zeeman hyperfine structure of the $2S_{1/2}$ level of excited hydrogen atoms (Fig. 4.54) can be separated by utilizing the Lamb shift between the energies of the hyperfine levels $2S_{1/2}$ and $2P_{1/2}$ and the metastable behaviour of the level $2S_{1/2}$.

The lifetime of the $2S_{1/2}$ metastable state in a fieldless space ($\Delta 1 = 0$) can be estimated as $\tau_{2S} \sim 1/7$ sec [42] if the transition probabilities of 1- and 2-photon radiations are taken into account. This value is much higher than the meanlife of the $2P$ state, estimated as

$$\tau_{2P} = 1.6 \cdot 10^{-9} \text{ sec}$$

In an external electric field, which causes a shift in the energies of the atomic levels due to the Stark effect, the $2S_{1/2}$ and $2P_{1/2}$ states are mixed and this reduces the value of τ_{2S}.

If the electrostatic field F is so weak that the level shift due to Stark effect is small compared with the splitting of the fine structure levels but comparable with the Lamb shift of the levels, the lifetime of the $2S_{1/2}$ state depends on the electric field F as given by [1]

$$\tau_{2S}(F) \approx \left(\frac{475 \text{ V/cm}}{F} \right)^2 \tau_{2P} \tag{4.36}$$

This expression shows that at a field intensity of 475 V/cm, the Lamb shift and the Stark splitting between the levels $2S_{1/2}$ and $2P_{1/2}$ become equal and that in this field the lifetime of the $2S_{1/2}$ state approaches that of the $2P_{1/2}$ state.

The Zeeman hyperfine structure of the atomic levels $2S_{1/2}$ and $2P_{1/2}$ in an external homogeneous magnetic field is shown in Fig. 4.54. It can be seen that the pair of levels symbolized by β, with $m_j = -1/2$ of the $2^2S_{1/2}$ state and their pair symbolized by e, with $m_j = +1/2$ of the $2^2P_{1/2}$ state cross each other at a well calculable value of the applied magnetic field. In this interval of the magnetic field values the pair of levels symbolized by α, with $m_j = +1/2$ in the $2^2S_{1/2}$ state does not approach any component of the $2^2P_{1/2}$ state. This fact leads to a considerable difference between the lifetimes of the components α and β of the metastable $2^2S_{1/2}$ level. The atoms of the component α move in this interval of the magnetic field values like in an electric field $\boldsymbol{F} = \dfrac{1}{c} [\boldsymbol{v} \times \boldsymbol{H_0}]$ and their lifetimes decrease only because of the Stark effect, while the atoms of the component β are exposed not only to the Stark effect but also to the Zeeman effect in the external magnetic field H_0 which causes the admixture of e levels to the β levels. Consequently, the interaction between these levels with an energy difference

$h\nu \approx 0$ decreases the lifetime of the β component to a value comparable with the lifetime of the $2^2P_{1/2}$ state. In the case of pure Stark effect this phenomenon would occur only for an electrostatic field strength of $F = = 475\ V/\text{cm}$.

The magnetic field dependence of the lifetimes of the α and β states of the $2^2S_{1/2}$ level can be expressed by a formula [494] which takes account of the allowed transitions in the form

$$\tau_{2S} = \left(\frac{\alpha_F}{3}\right)^2 \left(\frac{c}{v}\right)^2 \frac{1}{\gamma x^2} \left(\sum_{i=1}^{3} \frac{A_i}{B_i^2 + D}\right)^{-1} \tag{4.37}$$

where $\alpha_F = \dfrac{e^2}{\hbar c} \approx \dfrac{1}{137}$ is the fine structure constant, $\gamma = 6.26 \cdot 10^8/\text{sec}$ is the decay rate of the $2P_{1/2}$ state, v is the atomic velocity,

$$x = 3\mu_0 \frac{H_0}{2\varDelta E}$$

A_i and B_i are the transitions allowed by the selection rules for the states α and β, respectively [494, 1193].

The values of lifetime calculated from eq. 4.37 and plotted as the function of the value of the external homogeneous magnetic field are to be seen in Fig. 4.64.

Considering the above, the principle of the method used for the separation of the hyperfine components of the metastable $2^2S_{1/2}$ level is the following.

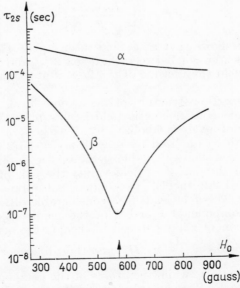

Fig. 4.64 Lifetimes of hyperfine states corresponding to the quantum numbers $m_J = +1/2$, denoted by α and $m_J = -1/2$, denoted by β, of the $2^2S_{1/2}$ level of hydrogen atoms (Fig. 4.54) in external magnetic field

The beam of hydrogen atoms excited to the level $n = 2$ is led into a homogeneous magnetic field of 575 gauss. The magnetic field induces in the atoms the onset of the spin states corresponding to the hyperfine level structure (Fig. 4.54). The atoms excited to the $2P$ level have a lifetime $\tau_{2P} = 1.6 \cdot 10^{-9}$ sec and decay to the $1S_{1/2}$ ground state together with the admixed atoms of the $2^2S_{1/2}$ level which are in the hyperfine states with $m_{F=1} = -1$ and $m_{F=0} = 0$. Thus, only the atoms in the hyperfine states with $m_{F=1} = +1$ and $m_{F=1} = 0$ remain in the excited metastable $2^2S_{1/2}$ state. Now, if this beam of atoms in the metastable $2^2S_{1/2}$ state and wholly polarized in electron spin is led by adiabatic transition into a weak magnetic field, the atoms can be polarized in nuclear spin up to 50 per cent.

The polarization with respect to the nuclear spin can be further increased by applying the induced transitions already considered in the preceding paragraph.

In the case of deuterium atom beams nuclear spin polarization is obtainable in a similar way to that of hydrogen atoms. In deuterium the three hyperfine levels with quantum numbers $m_I = 0; \pm 1$ of the $2S_{1/2}$ state with $m_J = -1/2$ will cross the hyperfine levels with quantum numbers $m_I = 0; \pm 1$ of the $2P_{1/2}$ state with $m_J = +1/2$ if a magnetic field of ≈ 575 gauss is applied to the deuterium beam. Consequently, only the spin states with $m_I = 0; \pm 1$ of the longer lifetime $2S_{1/2}$ metastable state with $m_J = +1/2$ remain at excited levels and the adiabatic transition of the excited atoms into a weak magnetic field can bring about a vector polarization up to $P_3 = 0.33$ and a tensor polarization up to $P_{33} = -0.33$ in the deuterium atoms. These degrees of polarization can be further increased by induced transitions.

4.4.5 NUCLEAR SPIN POLARIZATION OF ATOMS BY OPTICAL PUMPING
[1173–1182, 1226]

The method known as optical pumping induces transitions between atomic levels due to the absorption of circularly polarized light, thus it can be used to change the number of atoms at given energy levels. If the optical pumping is performed on atoms in a weak magnetic field with a light of appropriate frequency, it becomes possible to polarize the atoms with respect to nuclear spin.

This can be achieved as follows.

The atoms are exposed to a weak magnetic field which causes the formation of the hyperfine Zeeman-type level structure in the atoms. Transitions are now induced with the use of circularly polarized light at a given frequency, e.g. from the spin states with $F = 1$ to those with $F = 2$. If dextro-circularly polarized light is used, then, according to the selection rule $\Delta m_F = +1$ (Fig. 4.65), transitions can occur from the $m_F = 0; \pm 1$ states of the level $F = 1$ to the $m_F = 0; +1; +2$ states of the level $F = 2$. From the levels of the $F = 2$ state there are more available transitions to those of the $F = 1$ state, depending on the direction of rotation in the circular polarization of the light emitted by the atoms during their decay to a lower energy level. The result of induced transitions is that the atoms are "pumped" from the states with relatively lower quantum numbers of the

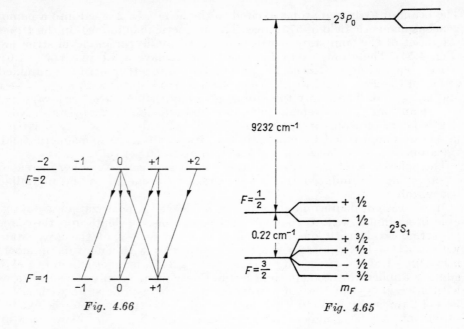

Fig. 4.66   Fig. 4.65

level with $F = 1$ to a state with higher quantum number and they thus become polarized also in nuclear spin.

The optical pumping technique has been applied to induce nuclear spin polarization of ^{3}He atoms [1178–1182]. In this case transition could be induced from the hyperfine levels of the 2^3S_1 metastable state to those of the 2^3P_0 state (Fig. 4.66). With dextro-circularly polarized light the possible transitions conform to the selection rule $\Delta m_F = +1$. The circularly polarized light used for optical pumping induces the transition of the spin states with $m_F = -1/2; -3/2$ of the 2^3S_1 metastable level to the spin states of the 2^3P_0 level. Subsequent to the transition, the atoms can return by spontaneous transition to any of the spin states of the 2^3S_1 level with equal probability. Thus, the states with quantum numbers $m_F = -1/2; -3/2$ of the 2^3S_1 level are "pumped" into the states with the higher quantum numbers $m_F = +1/2; +3/2$ by which the atoms in the metastable state 2^3S_1 become polarized with respect to nuclear spin to an extent depending on the value of the applied magnetic field.

4.4.6 NUCLEAR SPIN POLARIZATION BY POLARIZED ELECTRON CAPTURE

[1159, 1163]

Polarized electrons can be captured if an unpolarized ion beam is led through a ferromagnetic foil magnetized to saturation. The thus obtained neutral atom beam is polarized with respect to electron spin. If this atomic beam is adiabatically led into a weak magnetic field and ionized by passing through another foil placed into the magnetic field, we get an ion beam which is to some extent polarized with respect to nuclear spin.

The maximum spin polarization obtainable by this method is half the value of the electron spin polarization of the neutral atom beam produced from the ion beam by electron capture while the ions are passed through a ferromagnetic foil. The nuclear spin polarization of the ion beam is thus dependent on the type of ferromagnetic material used.

On assuming that the $3d$ and s electrons in the foil are captured by the ions with equal probabilities, a proton beam led through an iron foil magnetized to saturation will be polarized to ~ 15 per cent. The degree of polarization can also be increased in this case by induced transitions.

4.4.7 NUCLEAR SPIN POLARIZED ION BEAM SOURCES

Ion beams polarized with respect to the nuclear spin can be generated by the above described polarization methods. Of these methods the most frequently employed are separation techniques which utilize either the splitting of an atomic beam in an inhomogeneous magnetic field or the Lamb shift between the levels of the $2S_{1/2}$ metastable state. Either of these techniques is suitable for producing positively or negatively charged ion beams polarized in nuclear spin.

4.4.7.1 *Polarized ion sources using inhomogeneous magnetic field for separation*
[434, 441–444, 501–504, 509–513, 549–552, 1160–1169]

Ion sources of nuclear spin polarized beams require an assembly consisting of three main units:

(a) atom source
(b) separator
(c) ionizer

The production of atom beams is most easily achieved with a separation in an inhomogeneous magnetic field as in this case a beam of atoms in the ground state can be used.

Hydrogen or deuterium atoms are most conveniently obtained from the dissociators with multi-collimator or Laval nozzle-type outlet slits (described in Paragraph 2.2.5). Both methods of collimation can produce beams consisiting of atoms flying out from the dissociator mostly parallel or at a small angle to the axis of the magnetic separator.

Dissociators of these sources are built predominantly with high frequency gas discharge in either alternating magnetic [434, 441, 444, 503, 511, 513, 544, 550–552] or alternating electric field [443–445, 495, 502, 512, 549]. Both types of dissociators can generate an atomic current of $\sim 10^{16}$ atoms/sec.

Separators in which a well-collimated atomic beam can be separated with respect to spin states are magnets generating inhomogeneous fields with either 4 [441, 442, 444, 501, 509, 511, 513] or 6 poles [434, 443, 502, 503, 510, 512, 550]. In Paragraph 4.4.3, it has been shown that the atoms in their ground state separate in a weak inhomogeneous magnetic field into a number of components corresponding to the spin states in the Zeeman hyperfine structure (Fig. 4.59).

In a strong inhomogeneous field atoms split only according to the electron spins. In a weak magnetic field any of the beam components containing atoms of two focusable spin states can be singled out with appropriate magnetic geometry and a diaphragm [509, 517]. Nevertheless, weak fields are seldom applied in practice because very long magnets are required for a reasonable degree of polarization.

In the assemblies with an inhomogeneous magnetic field, strong fields are applied for the production of polarized ion beams. In this case the beam is first separated according to electron spins only, thus at this stage it is unpolarized with respect to nuclear spin. The beam component focused with the magnetic separator contains, in the case of a hydrogen beam, atoms of the spin states $m_{F=1} = 0; +1$ and this beam will be polarized in nuclear spin only if it is led from the strong field through an adiabatically changing magnetic field into a weak field where the ionization is taking place. This procedure can result in a nuclear spin polarization up to $P \leq 0.5$ for hydrogen and up to $P_3 \leq 0.33$, $P_{33} \leq -0.33$ for deuterium atoms.

In order to increase the degree of polarization a high frequency unit is placed between the magnetic separator and the ionizer to obtain high frequency field induced transitions (Paragraph 4.43 and Figs 4.61a and b) [503, 512, 549–552]. With induced transitions, a polarization degree of $P \leq 0.9$ has been achieved for hydrogen atoms and $P_3 \leq 0.55$, $P_{33} \leq \leq 0.74$ for deuterium atoms [512, 549–551].

A fraction of the nuclear spin polarized atom beam entering the ionizer placed in the weak magnetic field is ionized by electron bombardment. The types of ionizer most frequently used in polarized ion sources are shown in Figs 3.4, 3.5 and 3.6. The highest efficiency $\left(\eta = (1-3) \cdot 10^{-3} \right)$ has been achieved with the type in Fig. 3.6 in which the effective volume where the ionizing inelastic electron-atom collisions occur can be most expediently expanded.

If a nuclear spin polarized beam of negatively charged ions is required, the polarized, positively charged ions leaving the ionizer have to be accelerated to an energy determined by the given value of the charge transfer cross section (Paragraph 1.4, Fig. 1.64) and focused on the metal vapour or foil target in which the charge transfer takes place, as described in Paragraph 4.1.

Studies of depolarization during ionization have shown that in polarized atom beams ionized by the capture of electrons having energies below 1 keV, the depolarization due to ionizing collisions is practically negligible [262, 1163].

Depolarization during charge transfer is similarly of little importance if the thickness of the foil or metal vapour target is not more than a few $\mu g/cm^2$ and if the target is in a magnetic field with a properly chosen value [1194, 1195].

The intensity and polarization of the atomic beam are appreciably affected by the prevalent pressures generated in the source assembly. The intensity of the atomic beam decreases because of scattering on residual gas atoms and molecules, while the degree of polarization is reduced by the contribution from unpolarized ions formed from the residual gas in the ionizer. In order to avoid the loss of atoms due to collisions, the pressures in the different units of the assembly must be kept at values at which the mean free path of atoms is longer than the path that has to be covered by them in the given unit. To cope with this problem a differential vacuum system has to be applied with a special vacuum pump in each of the different units of the source assembly.

The decrease in polarization which occurs in the ionizer is due to the fact that the unpolarized ions produced from the residual gas in the ionizer leave the ionizer together with the polarized ions. To minimize the unpolarized contribution, the pressure in the ionizer must be such that the residual gas density is kept at a value lower by one–two orders of magnitude by comparison with the density of the polarized atom beam. In this case the contribution from unpolarized ions formed from residual gas can be reduced to less than 10 per cent.

Sources of positively charged polarized ion beams have been constructed in several laboratories [434, 441–444, 501–504, 509–513, 515, 516, 552]. One of these assemblies with a 6-pole magnetic separator is shown in Fig. 4.67 (Clausnitzer [516]). The atom beam is produced in this ion source with a high frequency dissociator of alternating electric field (Paragraph 2.2.5.3). The U-shaped dissociator is a pyrex glass tube 12 mm in diameter. The high frequency gas discharge is generated by a capacitively coupled oscillator of controllable frequency. The oscillator frequency is set to 30 MHz and the output power was 1 kW. The two cylindrical output coupling electrodes are spaced ∼50 cm apart; the gas pressure in the dissociator is ∼0.5 mmHg.

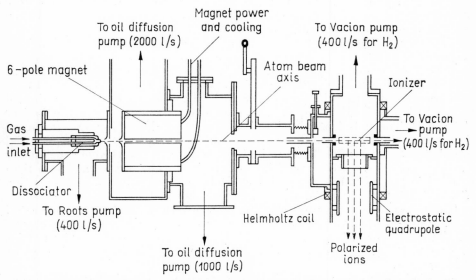

Fig. 4.67 Clausnitzer's version of ion source for producing nuclear spin polarized ion beams of positive charge with high frequency dissociator and separator in inhomogeneous sextupole magnetic field

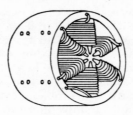

Sextupole magnet

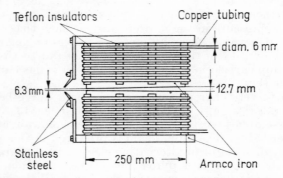

Fig. 4.68 Schematic diagram of magnetic separator with inhomogeneous sextupole magnetic field

The atoms flying out from the source are shaped to a beam of annular cross section at the input of the magnet because no separation can occur in the axis of the 6-pole magnet (Paragraph 4.4.3).

The 6.3 mm diameter input aperture of the 6-pole magnet (Fig. 4.68) expands gradually up to the output aperture of 12.7 mm of the 250 mm long magnet. The 6-pole magnet is excited by the d.c. current flowing through copper coils made of a 6 mm diameter copper tube. The coils are cooled with a distilled water flow in the copper tube. The output beam consisting of $\sim 5 \cdot 10^{14}$ polarized atoms/sec is driven from the magnet to the ionizer (Fig. 3.4). The polarized positive ions formed in the ionizer are extracted in the direction normal to that of the atomic beam by the negative potential applied to the electrode. The non-ionized atoms continue through the ionizer into the ion pump where they are pumped off. The weak magnetic field needed in the ionizer is generated by Helmholtz coils in a direction which determines that of the nuclear spin polarization.

The operational pressures required in the different units are generated by use of a differential vacuum system with pumps (Fig. 4.67) which can keep the pressure at the separator input at $5 \cdot 10^{-3}$ mmHg, at its output at $5 \cdot 10^{-6}$ mmHg until the ionizer input, while in the ionizer it is kept at 10^{-7} mmHg.

At this ionizer pressure, the density of the residual gas is higher than that of the polarized atom beam of velocity $v_H = 3 \cdot 10^5$ cm/sec and a current of $5 \cdot 10^{14}$ atoms/sec which gives $\varrho = 1.7 \cdot 10^9$ atoms/cm³. Consequently, the polarization degree of the proton beam obtained from this source was $P = 0.15$ with a polarized proton beam intensity of $3 \cdot 10^{-8}$ A.

A higher degree of polarization has been obtained with the assembly in which a 4-pole magnetic separator is applied, as shown in Fig. 4.69 (Adyase-vits, Antonenko, Polunin *et al.* [442]).

The atomic beam is produced in this source in a dissociator of alternating electric field and with internal electrodes. The discharge tube of the disso-ciator is 12 mm in diameter and ∼3 m long. Gas discharge is generated by applying an alternating voltage of 5 kV in amplitude with a frequency of 50 Hz. The outlet of the atom source is provided with a multi-collimator made of glass capillaries. The collimated atoms are separated under the action of the inhomogeneous magnetic field of a quadrupole magnet into components in different states. The aperture diameter of the magnet is 8 mm and the length of the magnet is 50 cm. The magnet is excited with a d.c. current of ∼225 A flowing through a coil made of a copper tube of 6 mm in diameter. The polarized beam current of $2 \cdot 10^{16}$ atoms/sec is driven adiabatically from the strong field of the separator magnet into weak field of the ionizer of the type seen in Fig. 3.5 where the atoms are ionized. The ionizer efficiency was measured between $(0.5$ and $1) \cdot 10^{-4}$.

The pressures needed in the differential vacuum system are produced by mercury diffusion and ion getter pumps in order to avoid contaminations

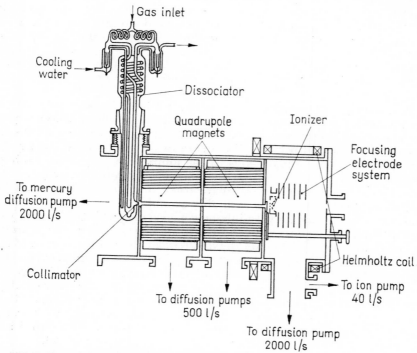

Fig. 4.69 Adjasevits–Antonenko–Polunyin–Femenko's version of an ion source producing positively charged, nuclear spin polarized ion beams with high frequency dissociator and separator in quadrupole inhomogeneous magnetic field

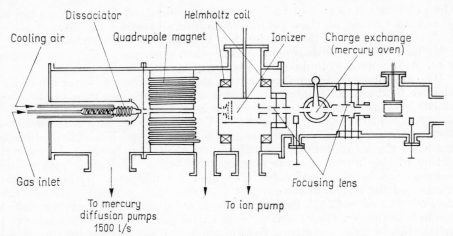

Fig. 4.70 Slabospickii–Kiselev–Karnauhov–Lopatko–Taranov's version of ion source producing negative ion beams with high frequency dissociator and separator in quadrupole inhomogeneous magnetic field

from oil. The residual gas pressure of 10^{-7} mmHg measured under these conditions in the ionizer was found to be less by an order of magnitude than the $\sim 2 \cdot 10^{-6}$ mmHg pressure of the polarized atomic beam. Thus, the contribution from unpolarized ions produced in the residual gas was kept below 10 per cent.

The proton and deuteron beam currents obtained from this source were found to be 0.15 μA and the degree of deuteron polarization was measured as $P_{33} = -0.318$.

A higher degree of polarization has been attained in positive ion beams by applying high frequency induced transitions (Collins, Glavish [503]; Thirion, Beurtey, Papineau [512]). The transitions are induced by the high frequency field applied between the separator output and the ionizer input as a result of which the input ions to the ionizer are polarized to a higher degree than the output ions from the magnet. A proton polarization to a degree $P \sim 0.90$ and deuteron polarization to $P_{33} \sim -0.40$ has been measured in beams generated with this assembly.

One of the arrangements with inhomogeneous magnetic field which yields beams of polarized negative ions by use of mercury vapour charge transfer target is shown in Fig. 4.70 (Slabospitskii, Kiselev, Karnaukhov *et al.* [444]).

The atom beam is obtained from a dissociator of alternating magnetic field. The gas discharge is generated with an oscillator operating at a frequency of 140 MHz with an output power of 100 W. A set of glass capillaries serves as a multi-collimator at the dissociator output. The collimated atom beam is led through a 250 mm long 4-pole magnetic separator with an aperture of 8 mm in diameter to be ionized with an efficiency of $2 \cdot 10^{-3}$ in the weak magnetic field of the ionizer of the type shown in Fig. 3.6.

The polarized beam of positive ions leaving the ionizer is focused onto the mercury vapour jet target of a density of 10^{15} atoms/cm³. The negative ions formed in the target are driven to the accelerator tube across a system of electrostatic plates correcting the direction of the beam. The different pressures in the different places of the vacuum system are generated by mercury diffusion and ion getter pumps. The pressure in the ionizer can be reduced to $\sim 10^{-7}$ mmHg, thus to less by an order of magnitude than the $\sim 2 \cdot 10^{-6}$ mmHg pressure of the polarized atom beam.

The polarized negative ion current yield from this source is $1.2 \cdot 10^{9}$ ions/sec with a polarization degree $P_{33} = -0.290 \pm 0.017$ in the case of a deuteron beam.

Higher degrees of polarization have been obtained in negative ion beams by means of high frequency field induced transitions [549–551].

An arrangement which induces transitions by high frequency field is shown in Fig. 4.71 (Grüebler, König, Schmelzbach [551]). The atom beam is produced by use of a high frequency dissociator with alternating magnetic field. The discharge chamber of the dissociator is made of a pyrex glass tube of 5 cm in diameter. The gas discharge is generated by an oscillator operating at a frequency of 13 MHz with a power output of 300 W. The atoms fly from the source through a glass capillary multi-collimator device into the inhomogeneous field of the 4-pole magnetic separator. The separated beam components are led from the magnet into a unit generating induced high frequency transitions (Fig. 4.72). If only transitions induced by weak magnetic field are used, theoretically, the degree of polarization is $P_3 = -1$ for hydrogen, and $P_3 = -2/3$ and $P_{33} = 0$ for deuterium atoms. Using polarization with only strong field induced transitions, the theoretically expected values for deuterium atoms are $P_3 = +1/3$, $P_{33} = -1$ calculating with "a" type transitions and $P_3 = +1/3$, $P_{33} = +1$ calculating with "b" type transitions in Fig. 4.61b.

If both weak and strong magnetic field induced transitions are applied, the calculated values are $P_3 = -1/3$, $P_{33} = +1$ for "a" transitions and $P_3 = -1/3$, $P_{33} = -1$ for "b" transitions.

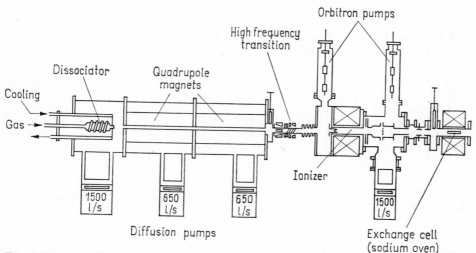

Fig. 4.71 Grüebler–König–Schmelzbach's version of ion source producing negatively charged, nuclear spin polarized ion beams with high frequency dissociator and separator in inhomogenous quadrupole magnetic field utilizing induced transitions

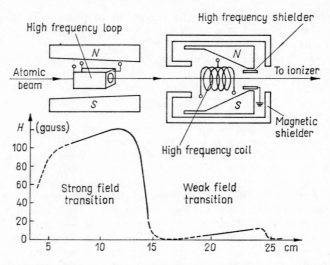

Fig. 4.72 Magnets and magnetic fields applied in radiofrequency induced transitions in strong and weak magnetic fields for the polarized ion source shown in Fig. 4.71. The first point of the scale is at the end of the quadrupole magnetic separator

The atom beam of an increased degree of polarization which leaves the high frequency unit is driven through the ionizer of the type shown in Fig. 3.6 where a fraction of the atoms is ionized. The ionization efficiency was measured as $\eta = 2.6 \cdot 10^{-3}$.

The polarized positive ion current of 1.1 μA obtained from the ionizer is led through the target chamber of 10 mm in diameter and 100 mg long. It is filled with sodium vapour in which the positive ions are transformed by charge transfer to negative ions.

A differential vacuum system is used to generate the required pressures in the ion source. The residual gas pressure of $\sim 10^{-7}$ mmHg in the ionizer is generated with the use of orbitron pumps and is thus kept below the pressure of the polarized beam.

This ion source has been reported to yield polarized negative ion currents in the range of $(1-2) \cdot 10^{-8}$ A. The degrees of polarization measured on deuterium beam upon weak magnetic field induced transitions are $P_3 = 0.50 \pm 0.01$, $P_{33} = 0$, and for strong magnetic field induced "a" transitions $P_3 = +0.25 \pm 0.1$, $P_{33} = -0.75 \pm 0.01$; for combined weak and strong field induced "a" transitions $P_3 = -0.24 \pm 0.01$, $P_{33} = +0.74 \pm 0.01$.

4.4.7.2 *Polarized ion sources using the Lamb shift of the* $2S_{1/2}$ *metastable level*
[445, 545–548, 1172, 1189, 1190, 1346]

The Lamb shift known to occur between the $2S_{1/2}$ metastable and the $2P_{1/2}$ excited levels of hydrogen and deuterium atoms can also be utilized in the design of polarized positive or negative ion sources.

This method is preferable to the inhomogeneous magnetic field technique because the possibility of selective ionization eliminates the contribution from ions formed in the residual gas in the ionizer which decreases the degree of ionization. Consequently, there is no need to generate as low pressures in the ionizer as required for the inhomogeneous field technique and the assembly is therefore simpler and easier to construct.

The ion source is again composed of three main units:

(a) source of excited state atoms
(b) separator
(c) selective ionizer.

Atoms excited to the level $n = 2$ can be produced by bombarding with electrons the atoms in the ground state [44, 445, 494] or by leading a H^+ ion beam with an energy of 1 keV ($v_H^+ = 4.4 \cdot 10^7$ cm/sec) through a caesium vapour charge transfer target, where the charge transfer process of the form

$$H^+ + Cs \rightarrow H(2S_{1/2}) + Cs^+$$

has a relatively large cross section ($\sigma \sim 3.4 \cdot 10^{-15}$ cm^2) for the production of atoms in $2S_{1/2}$ metastable state [1170, 1171, 1338].

The spin states of the beam with metastable atoms can be separated in a homogeneous field of 575 gauss at which the hyperfine level pair of the $2S_{1/2}$ state corresponding to quantum number $m_I = -1/2$ crosses the level pair of the $2P_{1/2}$ state corresponding to $m_I = +1/2$ (Fig. 4.54).

The separation can also occur without applying a magnetic field if transitions from the $2S_{1/2}$ ($F = 1$) to the $2P_{1/2}$ ($F = 0$) are induced in a high frequency field (Fig. 4.73); (Boyd, Lombardi, Robbins et al. [1190]).

The polarized atoms in metastable state are then selectively ionized by bombardment with electrons of energies in the range 3.4 eV $< E_e <$ 13.4 eV or in argon gas by the process

$$H(2S_{1/2}) + Ar \rightarrow H^- + Ar^+ \tag{4.38}$$

If the ionization is carried out by bombardment with electrons, the elec-

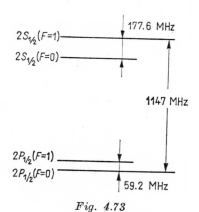

Fig. 4.73

tron energies are sufficient for the ionization of excited atoms only, and the atoms in the ground state remain unchanged. Since in the output atom beam of the separator only those atoms are excited which are in the polarized metastable state, the positive ions formed in the ionizer are also polarized with respect to nuclear spin.

In an alternative method for the production of negative ions the selectivity of the ionization is due to the fact that in a beam of atoms in $1S_{1/2}$ ground state moving at a rate $v \sim (2\text{--}4) \cdot 10^7$ cm/sec, the fraction of negative ions produced in the process

$$\mathrm{H}(1S_{1/2}) + \mathrm{Ar} \rightarrow \mathrm{H}^- + \mathrm{Ar}^+$$

is negligibly small compared with the fraction of negative ions formed from atoms in the $2S_{1/2}$ metastable state in the process of type (4.38) (Fig. 4.74; Donnally, Sawyer [1171]).

Thus, the negative ions leaving the argon gas target are formed almost entirely from polarized $2S_{1/2}$ state atoms.

An assembly built for the production of polarized positive ion beams by utilizing the Lamb shift of the $2S_{1/2}$ metastable state is shown in Fig. 4.75 (Vályi [445]).

The beam of atoms in ground state is produced in a high frequency dissociator with alternating electric field (Fig. 2.26). The U-shaped pyrex glass tube used as dissociator is 12 mm in diameter and the distance between the two coupling electrodes is 50 cm. The gas discharge is generated by an oscillator operating at a frequency of 23 MHz with an output power of 360 W. A multi-collimator made of glass capillaries is mounted at the dissociator output (Fig. 2.3). The atom beam current from the dissociator is $\sim 2 \cdot 10^{16}$ atoms/sec.

The collimated atom beam is led through a bombarding unit where it is excited by collisions of 12 eV electrons to energy levels $n = 2$.

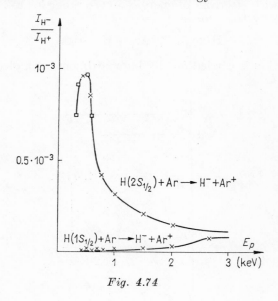

Fig. 4.74

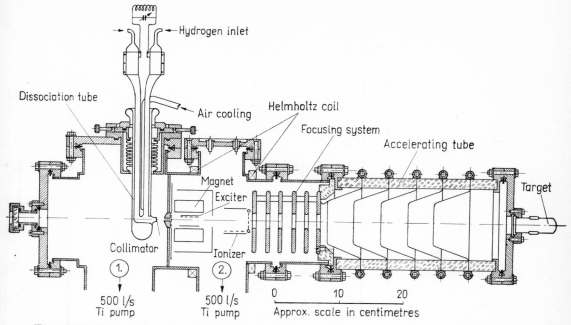

Fig. 4.75 Válvi's version of ion source producing positively charged, nuclear spin polarized ion beams with high frequency dissociator and separator utilizing the Lamb-shift of nuclear levels

The atoms excited to the $2S_{1/2}$ metastable state become polarized with respect to electron spin in the applied homogeneous magnetic field of 575 gauss and are led by adiabatic transition into the weak magnetic field of the ionizer where they are ionized by electron bombardment.

The required vacuum is generated in this source with two ion-getter pumps operating at a rate of 500 l/sec. The use of an ion-getter pump was found to considerably increase the lifetimes of the oxide cathodes in the bombarding and in the ionizer units as well as the lifetime of the tritium target used to measure the degree of polarization.

This ion source lends itself to the production of nuclear spin polarized positively charged proton or deuteron beams with a current of $4.2 \cdot 10^9$ ions/sec, polarized in the case of deuterons to a degree $P_{33} = -0.294 \pm 0.044$.

An arrangement for the production of polarized negative ion beams is shown in Fig. 4.76 (Cesati, Cristofori, Milazzo-Colli *et al.* [1172]).

In this assembly the metastable atoms are obtained from charge transfer processes. The ion beam is generated by high frequency gas discharge. The positive ion beam is focused and slowed down to 0.5–1 keV energy before the input to the 3 mm diameter aperture of the caesium vapour target. At a caesium vapour pressure of $3 \cdot 10^{-3}$ mmHg a fraction of ~ 2 per cent of the 500 eV energy ions transforms by charge exchange to metastable atoms. The metastable atoms are led through a magnetic field of 575 gauss into the argon gas target placed in a weak magnetic field.

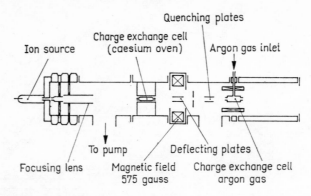

Fig. 4.76 Cesati–Cristofori–Milazzo–Colli–Sona's version of ion source producing negatively charged, nuclear spin polarized ion beams with high frequency dissociator, and separator utilizing caesium metal vapour excitation and the Lamb-shift of nuclear levels

The negatively charged polarized ion current output from the argon gas target is $5 \cdot 10^9$ ions/sec with a deuterium polarization of a degree $P_{33} = -0.296$ for a gas target in a weak magnetic field of 1.6 gauss.

In order to reduce contamination from caesium metal, vapour jets from caesium can also be applied as charge exchange targets in the polarized ion sources. A negative polarized ion beam source with caesium vapour jet exchange target is shown in Fig. 4.77 (Khirnii, Kotsemasova [1189]). The positive ion beam is obtained from a duoplasmatron type source. The focused ions slowed down to $\sim$1 keV energy are led through the caesium

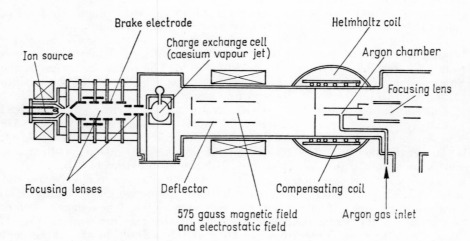

Fig. 4.77 Khirnii–Kotsemasova's version of an ion source for producing negatively charged, nuclear spin polarized ion beams with duoplasmatron type ion source and separator utilizing caesium jet excitation and the Lamb-shift of nuclear levels

vapour jet. The ions are separated by electrostatic deflector plates from the beam leaving the charge transfer target and thus only the neutral atom is taken to the magnetic field of 575 gauss. The atoms polarized to the spin state with $m_I = +1/2$ of the $2S_{1/2}$ level are led to the argon gas target in the weak magnetic field where a fraction of the polarized metastable atoms transforms to negative ions. The argon charge transfer target is 10 mm in diameter and 200 mm long with an argon gas pressure in the range of $(2-4) \cdot 10^{-5}$ mmHg.

The negative polarized ion current obtained from this source was measured as $4 \cdot 10^{-9}$ A with polarization degree $P_{33} = -(0.267 \pm 0.008)$ for deuterium.

The degree of polarization can be further increased by use of the "zero-field crossing" technique (Sona [1196]). The principle of this method is to apply two magnetic fields of 575 gauss for separation and an electrostatic field which generates the Stark effect. The two magnets generate longitudinal fields having opposite directions to each other. The metastable atomic beam passing from one into the other magnetic field of $H_0 = 575$ gauss crosses a point at which $H_0 = 0$ where the magnetic field changes its direction. If the crossing time of the atoms is shorter than the Larmor precession time in a weak magnetic field, the atoms remain in the states to be seen in Fig. 4.78.

If the atoms leaving the second magnetic field are ionized in a strong magnetic field, theoretically, the vector polarization of a deuterium beam can be as high as $P_3 = 2/3$ with a tensor polarization $P_{33} = 0$.

Now, if a transversal electrostatic field in the range of 10–20 V/cm is generated between the two plates placed in the second field of 575 gauss,

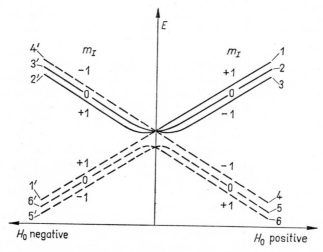

Fig. 4.78 Alteration of levels in the hyperfine structure of metastable 2S-state of deuterium atoms rapidly crossing the zero field between two opposite magnetic fields. States 4, 5 and 6 represented by a broken line are quenched passing the magnetic field of + direction and the state with quantum number $m_I = +1$ is quenched passing the magnetic field of − direction.

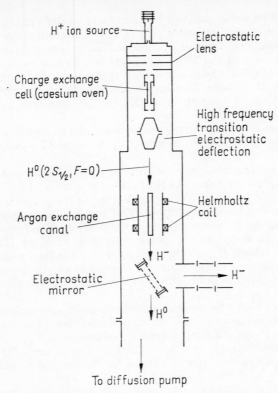

Fig. 4.79 Boyd–Lombardi–Robbins–Schecher's version of an ion source producing negatively charged, nuclear spin polarized ion beams with high frequency ion source, and separator utilizing caesium metal vapour excitation and the Lamb-shift of nuclear levels

the atomic state 1′ (Fig. 4.78) can be quenched, that is, the atoms are induced to decay to the ground state. If the atoms remaining in the two spin states of the metastable level are ionized in a strong magnetic field, the theoretically obtainable degrees of vector and tensor polarization in a deuterium beam are $P_3 = 1/2$ and $P_{33} = -1/2$, respectively.

The tensor polarization of the deuterium beam can be further increased [546] if the atoms remaining in the metastable state after extinction of the 1′ state are led adiabatically from the 575 gauss magnetic field into the weak magnetic field in the ionizer. Theoretically, the total tensor polarization of the deuterium beam ($P_{33} = -1$) can be achieved in this way, while the vector polarization disappears.

On applying the "zero-field crossing" method deuterium beams have been polarized by ionization in strong magnetic field to $P_{33} = -0.492 \pm 0.028$ and by ionization in weak magnetic field to $P_{33} = -0.829 \pm 0.031$ (Michel, Corrigan, Meiner *et al.* [548]).

An ion source assembly utilizing the Lamb shift of the $2S_{1/2}$ metastable level without applied magnetic field is shown in Fig. 4.79 (Boyd, Lombardi, Robbins *et al.* [1190]).

The positive ion beam is obtained from a high frequency gas-discharge unit. The extracted ion beam is slowed down from 1.5 keV to an energy of 550 eV before the input and focusing to the caesium vapour charge transfer target. The charge transfer target channel is 10 mm in diameter and 76 mm long. The metastable atoms formed by charge transfer are taken to a high frequency field generated by an oscillator operating at 1147 MHz and a power output of 0.5 W. The high frequency field induces the transition $2S_{1/2}\ (F = 1) \rightarrow 2P_{1/2}\ (F = 0)$.

The ions are removed by electrostatic deflectors from the direction of propagation of the beam containing atoms in state $2S_{1/2}\ (F = 0)$, consequently only the neutral atom beam reaches the argon charge transfer target. The here formed negative polarized ions are reflected by an electrostatic mirror made of tungsten filaments with a transparency of 92 per cent and deflected to 90° while the fraction of neutral atoms continues to fly in the direction of the vacuum system.

This type of ion source can produce nuclear spin polarized ion current of $5 \cdot 10^{-9}$ A and in the case of protons the polarization degree can be $P \leq 0.5$.

4.4.7.3 *Polarized ion sources using optical pumping*
[1179, 1182]

An arrangement for the production of polarized ions with optical pumping is shown in Fig. 4.80. The discharge chamber of the ion source is a thin-walled sphere made of pyrex glass. The extracting channel of the probe type extracting system is 0.343 mm in diameter and 2.03 mm long. The weak discharge in ^{3}He gas is generated by a high frequency field of 50 MHz. The light inducing the transitions is obtained from a lamp with high intensity light produced by a 100 MHz frequency electric field in ^{4}He gas filling the lamp to a pressure in the range of 10–20 mmHg. The output light from

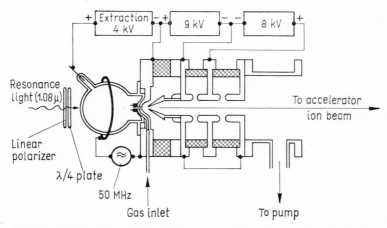

Fig. 4.80 Findley–Baker–Carter–Stock's version of ion source for producing nuclear spin polarized ion beams by optical pumping

the ^{4}He discharge is directed with a concave mirror mounted behind the lamp across a linear polarizer and the $\lambda/4$ plate into the ^{3}He gas discharge.

Owing to the optical pumping, the number of atoms in spin states with $m_F = \pm 1/2; +3/2$ of the 2^3S_1 level increases in the ^{3}He gas discharge exposed to the magnetic field applied parallel to the direction of the axis of the ion source. The degree of nuclear spin polarization depends on the value of the magnetic field.

Collisions occur in the discharge between the particles of the types: ^{3}He atoms in $1S_0$ ground state, unpolarized ^{3}He$^+$ ions and ^{3}He atoms in polarized metastable 2^3S_1 states. These collisions lead to excitation and charge exchange interactions. The ground state atoms and ions formed in these processes are also polarized.

The ^{3}He$^+$ ions extracted from this source are polarized to a small extent. The reported value of the ^{3}He$^+$ polarized ion current is 4 μA and the degree of polarization is 0.05.

CHAPTER V

DISTRIBUTION OF CHARGES AND MASSES
IN ION BEAMS

5.1 INTRODUCTION

The mass and charge distributions in ion beams extracted from ion sources are determined by the relative particle concentrations resulting from the different collision processes occurring in ion sources. The dynamical equilibrium concentration, which sets in around the extracting channel, is of particular importance in the composition of the extracted beam. This dynamical equilibrium value depends on the ratio of inflowing to outflowing particles in this region and on the number of particles extracted through the extracting channel.

In this state of dynamical equilibrium the concentrations of the different components of the beam to be extracted can be controlled within certain limits if the operational parameters and the design of the ion source assembly are appropriately chosen. Consequently, the mass and charge distributions in the ion beams extracted from different types of ion sources can vary with the operational parameters and the construction materials of the ion source.

5.2 RELATIVE CONCENTRATIONS OF THE DIFFERENT IONS
IN HYDROGEN ION SOURCES

The concentrations of the different particles appearing in low pressure hydrogen gas discharge are determined by the ionization and recombination processes occuring in the discharge. Ions are formed in hydrogen gas discharge in single collisions of types

$$H_2 + e \rightarrow H_2^+ + 2e \qquad (5.1)$$

$$H_2 + e \rightarrow H_1^+ + H_1 + 2e \qquad (5.2)$$

and by multiple collisions of types

$$H_2 + e \rightarrow H_1 + H_1 + e; \qquad H_1 + e \rightarrow H_1^+ + 2e \qquad (5.3)$$

$$H_2 + e \rightarrow H_2^+ + 2e; \qquad \begin{array}{l} H_2^+ + e \rightarrow H_1^+ + H_1 + e \\ H_2^+ + e \rightarrow H_1^+ + H_1^+ + 2e \end{array} \qquad (5.4)$$

$$H_2 + e \rightarrow H_2^+ + 2e; \qquad H_2^+ + e \rightarrow H_1 + H_1; \qquad H_1 + e \rightarrow H_1^+ + 2e \qquad (5.5)$$

$$H_2 + H_2^+ \rightarrow H_1 + H_1^+ + H_2; \qquad H_2 + H_1^+ \rightarrow H_3^+ \qquad (5.6)$$

It can be seen from these processes that in addition to the neutral particles

H_2 and H_1, the discharge population contains H_1^+, H_2^+ and H_3^+ atomic and molecular ions.

The processes leading to the formation of given types of particles occur with different probabilities at collision energies above the threshold energy of the collision in question. The threshold energies of the different types of collision are the following

$$H_2 \rightarrow 2H_1 (9 \text{ eV}), \qquad\qquad H_1 \rightarrow H_1^+ (13.6 \text{ eV})$$
$$H_1 \rightarrow H_2^+ (15.4 \text{ eV}), \qquad\qquad H_2 \rightarrow H_1 + H_1^+ (18 \text{ eV})$$
$$H_2 \rightarrow H_1^+ + H_1^+ \ (\sim 46 \text{ eV})$$

Theoretical and experimental studies of the values of cross sections for different processes have shown that at impact energies below 50 eV, molecular dissociation of H_2, ionization of H_2 molecules and ionization of H_1 atoms have the highest cross section values.

The dissociation cross section $\sigma_{(H_2 \rightarrow 2H_1)}$ has a maximum value at an impact energy of 15 eV. This maximum value exceeds by a factor of 6 the cross section for molecular ion formation $\sigma_{(H_2 \rightarrow H_2^+)}$ at the same impact energy [1199]. However, the ionization cross section of H_2 molecules is higher by an order of magnitude than that of H_1 atoms in type (5.2) single collisions [103].

It can be seen from these data that single collisions of type (5.2) do not lead to the production of that number of atomic ions which would appreciably affect the H_1^+ concentration. H_1^+ ions which determine the concentration of atomic ions are formed in the processes of types (5.3), (5.4), (5.5) induced by collisions with high cross sections values.

The most important source of atomic ions among these processes is the ionization resulting from collisions between H_1 atoms and electrons. The required concentration of H_1 atoms is obtained from the dissociation of H_2 molecules and H_2^+ ions (Paragraph 1.6) and from the dissociation recombination in the process of type (5.5) (Paragraph 1.5) which lead to the formation of hydrogen atoms in the ground state or in excited states.

H_2^+ molecular ions are formed in collisions with electrons in the process (5.1). In subsequent collisions the molecular ions transform to atomic ions in processes (5.4), (5.5) or to H_3^+ molecular ions (5.6). This last transformation of the type (5.6) occurs as follows. Because of the high concentration of H_2 molecules, the ion–molecule collision has a relatively high probability and the molecular ions dissociate at impact energies of a few electron volts. The H_1^+ ions formed in the dissociation combine in most cases with H_2 molecules and transform to H_3^+ molecular ions [1203].

The equilibrium concentration of atoms and ions depends not only on the above discussed ionization processes but also on the recombination of particles. Atomic ions can recombine to neutral atoms, while H_1 atoms to H_2 molecules.

These recombination processes can occur in the gas discharge volume and also on the surfaces of the discharge chamber and the electrodes.

The recombination of ions due to impact by electrons is negligible in the discharge since at electron energies of a few electron volts, the electron capture cross section of ions is very low and the probability of recombinations by double or triple collisions is similarly very low because of the low gas density (Paragraph 1.5).

In the relevant measurements the cross section for transformation of H_1 atoms to H_2 molecules proved to be very low as well so that volume recombination occurring in these processes can be ignored.

Thus, the transformations of ions to atoms and of atoms to molecules can be attributed above all to surface recombinations on the wall of the discharge tube and on the electrode surface. It is therefore important to know the surface recombination coefficient of the inside wall surface of the tube and of the surfaces of electrodes. The surface recombination coefficient of atomic hydrogen depends on the material, quality, contamination and temperature of the surface involved. The recombination coefficient of hydrogen atoms listed for a few materials in Appendix, Table 17 increases by more than an order of magnitude if the temperature increases by a few hundred degrees centigrade [1198].

For this reason the discharge tube wall has to be cooled by a forced air current to avoid the increase in the recombination coefficient.

As has been seen, the recombination of H_1 and H_1^+ particles is due mainly to surface recombination on solid surfaces in contact with the gas discharge. In the formation of H_2^+ ion concentration both the volume and the surface recombination processes have an important role.

The collisions which lead to dissociation processes in which H_2^+ ions have a high probability of being transformed to H_1^+ ions and H_1 neutral atoms depend on the gas pressure. Similarly, the number of H_2^+ ions decreases due to the gas pressure dependent surface recombination. However, since the number of volume collisions of H_2^+ ions increases with increasing gas pressure and that of surface collisions increases with decreasing gas pressure, the combination of the opposite effects allow only a very small change in H_2^+ ion concentration to be induced by varying the gas pressure. Consequently the combined effect of the two processes allow only very small changes in H_2^+ ion concentration to be induced by varying the gas pressure.

On the other hand, the concentration of H_3^+ ions is much more dependent on the gas pressure. The process of type (5.6) has a tendency to become predominant at high gas pressures so that the number of H_3^+ ions may exceed that of H_1^+ and H_2^+ ions.

5.3 MASS DISTRIBUTION OF HYDROGEN ION BEAMS EXTRACTED FROM DIFFERENT TYPES OF ION SOURCES

The ionization and recombination processes considered in the preceding section occur with different probabilities owing to the dissimilar design parameters in the different types of ion source. The relationship between the favourable and unfavourable conditions for given processes can cause appreciable changes in the percentual contributions from different particles to the ion beam extracted from the source. Though the composition of the ion beam is essentially determined by the design parameters, the percentages of the different components can be influenced by changes in the operational parameters.

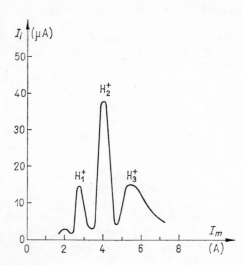

Fig. 5.1 Mass spectrum of ion beam extracted from the capillary arc-discharge ion source shown in Fig. 3.27 (Allison [625])

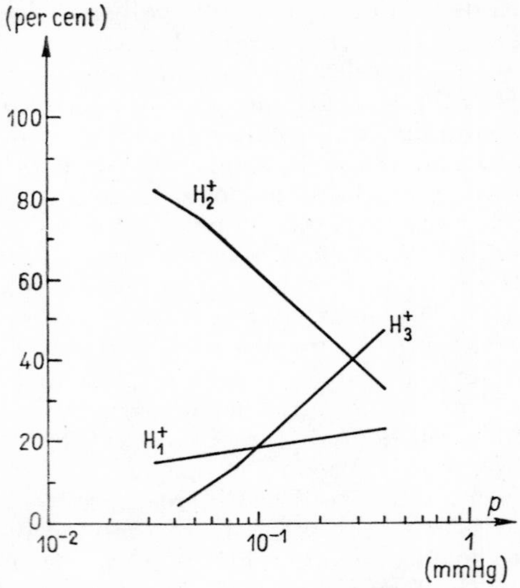

Fig. 5.2 Percentage ratio of H_1^+, H_2^+ and H_3^+ ions extracted from duoplasmatron-type ion source as a function of the gas pressure p in the source (Rose *et al.* [523])

In the following the composition of ion beams obtained from different types of ion source will be discussed and it will be shown how the percentual contributions from different components can be changed by modifying the operational parameters of the ion source.

5.3.1 MASS SPECTRUM OF ION BEAMS FROM CAPILLARY ARC-DISCHARGE ION SOURCES

In arc-discharge ion sources with metal capillary (Paragraph 3.6, Figs 3.26 and 3.27) the usually obtainable conditions favour above all the processes (5.2) and (5.6). Owing to the high recombination coefficient of the metal wall and the small diameter of the capillary tube, the formation of H_1^+ atomic ions is much less probable.

Consequently, the percentual contribution from H_1^+ atomic ions to the beams extracted from metal capillary arc-discharge ion sources is not more than 10 to 20 per cent and the beam is composed mainly of H_2^+ and H_3^+ ions. The relative concentration of the molecular ions depends on the gas pressure of the gas discharge. At higher gas pressures the H_3^+, while at lower gas pressure the H_2^+ ions are predominant.

Figure 5.1 shows the mass spectrum as a function of the analyser magnet current measured on the ion beam extracted from the capillary arc-discharge ion source to be seen in Fig. 3.27 (Allison [625]).

The percentage of H_1^+ atomic ions can be increased in the ion beams by increasing the discharge current I_a as this also increases the probability of multiple collision processes leading to the formation of atomic ions.

The percentage of atomic ions extracted from the source increase if quartz or glass instead of metal capillaries are applied. The lower surface recombination coefficient of the former allows the percentual contribution from H_1^+ ions to the beam to reach even 60 per cent [622].

5.3.2 MASS SPECTRUM OF THE ION BEAM EXTRACTED FROM A DUOPLASMATRON

The composition of ion beams extracted from duoplasmatron type ion sources (Paragraph 3.7.1.2) is markedly dependent on the operational parameters and the design of the source assembly.

The inherent features of duoplasmatrons and the applied inhomogeneous magnetic field restrict to a small volume the ionization and recombination processes which determine the relative ion concentrations. It follows that under the generally used operational conditions the probability of multiple collision processes generating H_1^+ atomic ions is low. Nevertheless, higher percentages of H_1^+ ions are obtained from this type of source than from arc-discharge ion sources with metal capillaries.

The reason for this is that the surface recombination probability is substantially diminished due to the contraction of the discharge plasma under the action of the applied magnetic field.

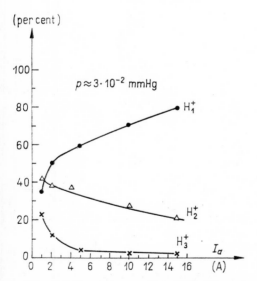

Fig. 5.3 Percentage ratio of H_1^+, H_2^+ and H_3^+ ions extracted from duoplasmatron-type ion source as a function of the discharge current I_a (Ardenne [936])

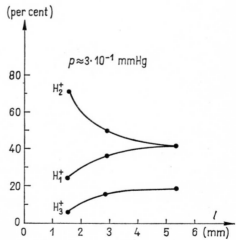

Fig. 5.4 Percentage ratio of H_1^+, H_2^+ and H_3^+ ions extracted from duoplasmatron-type ion source as a function of the distance l from electrode to anode (Kelly et al. [680])

The contributions from the different ionic species to the equilibrium concentration at the extraction slit can be influenced by a number of factors from which the chosen gas pressure, the discharge current and the distance between the intermediate electrode and the anode seem to be the most essential.

Experiments have shown that the concentration of H_2^+ and H_3^+ molecular ions varies appreciably while that of H_1^+ ions only to a small extent with the values of gas pressure (Fig. 5.2) (Rose, Bastide, Witkower [492]). As apparent from the figure, the H_2^+ species dominates at lower, the H_3^+ species at higher gas pressures.

The percentage of H_1^+ atomic ions can be substantially varied by changing the discharge current I_a as the probability of multiple collisions leading to the formation of H_1^+ atomic ions in the discharge plasma rapidly increases with increasing discharge current. The variation of the ion beam composition with the values of the discharge current I_a can be seen in Fig. 5.3 (Ardenne [936]).

The figure shows that for $I_a = 15$ A the contribution from H_1^+ atomic ions to the beam extracted from the duoplasmatron can be as high as 79 per cent while that from H_2^+ ions is simultaneously 19 per cent and that from H_3^+ ions is 1 per cent.

It has already been mentioned that the size of the ionization volume may also be responsible for the relative numbers of single and multiple collision processes in the given volume. Thus, it can be expected that the composition of the extracted beam can be changed by altering the distance between the anode and the intermediate electrode as this distance necessarily alters the size of the ionization volume. The results of experiments to this end are shown in Fig. 5.4 (Kelley, Lazar, Morgan [680]).

It can be seen from the figure that for small distances between the intermediate electrode and the anode in the duoplasmatron, the percentage of H_2^+ molecular ions from single collision processes may reach 71 per cent. As the distance from the intermediate electrode to the anode increases, the number of H_1^+ and H_3^+ ions also shows a gradual increase.

The cited examples prove that with a proper choice of the design and operational parameters of duoplasmatrons the extracted ion beam can contain more than 70 per cent of either H_1^+ atomic ions or H_2^+ molecular ions.

5.3.3 MASS SPECTRUM OF ION BEAMS EXTRACTED FROM ION SOURCES WITH OSCILLATING ELECTRONS

Similarly to the ion source types considered in the foregoing, the relative concentrations of the different ions in the ion beams extracted from Penning-type discharge with oscillating electrons (Paragraph 3.7.2) markedly vary with the design and operational parameters of the source assembly. The most important parameters which can induce changes in the ratio of single to multiple collision processes are the gas pressure, the discharge current, the magnetic field and the recombination coefficients of electrode materials in the environment of the discharge volume.

At low discharge currents $(I_a < 100 \text{ mA})$ and low pressures $(p < 10 \text{ } \mu\text{Hg})$ the H_2^+ ions predominate in the extracted beam (Fig. 5.5) (Nagy [675]),

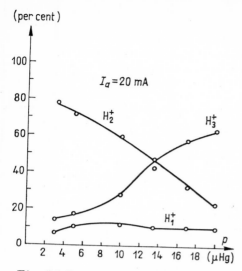

Fig. 5.5 Percentage ratio of H_1^+, H_2^+ and H_3^+ ions extracted from Penning-type discharge ion source with oscillating electrons as a function of the gas pressure p in the source (Nagy [674])

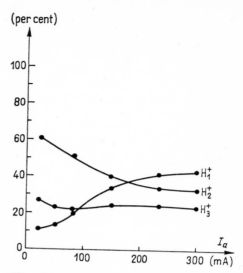

Fig. 5.6 Percentage ratio of H_1^+, H_2^+ and H_3^+ ions extracted from Penning-type discharge ion source with oscillating electrons as a function of the discharge current I_a (Nagy [674])

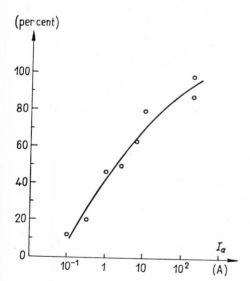

Fig. 5.7 Percentage ratio of H_1^+ ions extracted from a pulsed, Penning-type discharge ion source with oscillating electrons as a function of the discharge current I_a (Gabovits et al. [654])

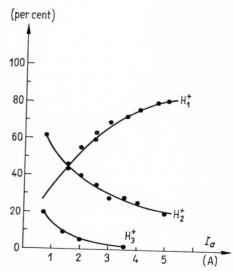

Fig. 5.8 Percentage ratio of H_1^+, H_2^+ and H_3^+ ions extracted from ion source with oscillating electrons and high magnetic field as a function of the discharge current I_a (Livingston et al. [964])

and it is apparent from the figure that as the gas pressure increases, the percentage of H_3^+ molecular ions resulting from the collisions of the abundant H_2^+ ions with H_2 molecules also gradually increases in the beam at the expense of the H_2^+ ions. At the same time the concentration of H_1^+ ions remains over the entire range of applied pressures at a value of about 10 per cent because of the recombinations occurring on the wall surfaces. At the discharge current $I_a > 100$ mA, at which the electron concentration is proportionally higher, the contribution from H_1^+ ions to the extracted beam exceeds the contributions from the H_2^+ and H_3^+ molecular ions (Fig. 5.6). With very high discharge currents in pulsed ion sources the percentage of H_1^+ ions can become almost 100 per cent (Fig. 5.7) (Gabovits, Nemec, Fedorus [654]).

The H_1^+ ion concentration also increases with increasing values of the applied magnetic field since the contraction of the discharge plasma and the reduced diffusion to the wall decreases the recombination probability on the electrode surface. Thus, if a magnetic field is applied, the H_1^+ ion concentration can increase in the discharge plasma even at a discharge current of a few amperes to such an extent that the extracted ion beam may contain ~ 80 per cent of H_1^+ ions (Fig. 5.8) (Livingston, Jones [964]).

It has been seen from the above that in the Penning-type gas discharge the H_1^+ ion concentration can be increased if high discharge currents are available and that with a suitable choice of the parameters these assemblies can be utilized also as a source of molecular and multiply charged ions.

5.3.4 MASS SPECTRUM OF ION BEAMS EXTRACTED FROM HIGH FREQUENCY GAS DISCHARGE ION SOURCES

In ion sources utilizing high frequency gas discharge (Paragraph 3.8) the multiple collision processes yielding H_1^+ atomic ions have the highest probability. The effect of surface recombination, which reduces the concentrations of H_1^+ ions and H_1 atoms, is much less important in high frequency ion sources than in the types already considered. The reason for this is that the electrodes in the discharge volume and the tube walls are shielded from the discharge by quartz or pyrex glass and the recombination coefficient of these materials is lower by orders of magnitude than that of the metals used in the previous types (Appendix, Table 17).

Since the recombination of H_1^+ ions and H_1 atoms is thus diminished, the concentrations of these species in the discharge volume are relatively high and also the extracted beam contains a high percentage of H_1^+ atomic ions.

Experimental investigations of the beams extracted from high frequency ion sources [683, 702, 705, 725, 730, 735] have shown that the contributions from H_2^+ atomic ions to the extracted beam can be as high as 80–90 per cent, if the gas pressure is low. At higher gas pressures this percentage decreases (Fig. 5.9) (Vályi, Gombos, Roosz [730, 735]).

The figure shows that the contribution from H_2^+ molecular ions is not appreciably affected by a change in the gas pressure, while that from H_3^+ molecular ions can reach 70 per cent, if the gas pressure is high enough.

The percentage of H_1^+ ions increases with increasing values of the high frequency power output generating the discharge (Fig. 5.10) since the thus obtained higher electron concentration promotes the formation of H_1^+ ions.

The effect of the static magnetic field applied to the high frequency discharge is apparent from the mass spectra measured as a function of the current of the analysing magnet in Figs 5.11 and 5.12. The spectra show the increase in the percentage of H_1^+ ions at the expense of that of H_3^+ ions for an essentially unchanged percentage of H_2^+ molecular ions. This can be explained by the increase in electron concentration under the action of the magnetic field which leads to a higher probability of multiple collisions enhancing the formation of H_1^+ ions with a simultaneous lower probability of collisions between H_2^+ ions and H_2 molecules with subsequent H_2^+ ion formation.

The relative ion concentration in high frequency gas discharge ion sources can be substantially modified if the gas discharge plasma is in contact with an unshielded metal surface. In this case surface recombination appreciably reduces the concentration of H_1^+ atomic ions. To avoid this inconvenience, the electrodes in the source should be prepared from a metal with low sputtering coefficient (e.g. pure aluminium) since the sputtered metal particles may form a metal layer on the wall of the discharge chamber and this may reduce the lifetime of the ion source.

It is of interest to note that if quartz or glass discharge tubes are washed with solutions of ∼3 per cent fluoric acid, the percentage of H_1^+ atomic ions can increase to more than 90 per cent. This can be explained by the fact that

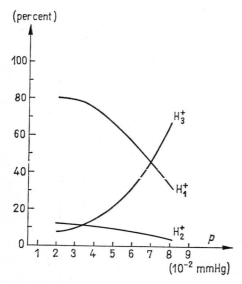

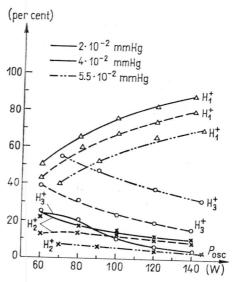

Fig. 5.9 Percentage ratio of H_1^+, H_2^+ and H_3^+ ions extracted from the inductively coupled high frequency ion source with diaphragm-type extraction system shown in Fig. 3.67 as a function of gas pressure p in the source (Vályi et al. [730, 735])

Fig. 5.10 Percentage ratio of H_1^+, H_2^+ and H_3^+ ions extracted from the inductively coupled high frequency ion source with diaphragm-type extraction system as a function of the oscillator output power for different values of gas consumption (Vályi et al. [730, 735])

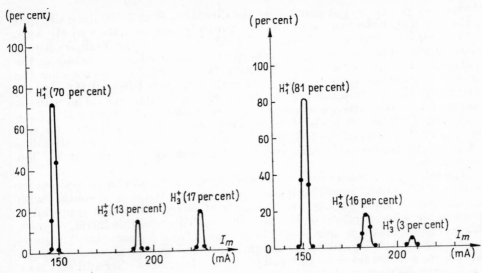

Fig. 5.11 Mass distribution of ion beams extracted from inductively coupled high frequency ion sources with diaphragm-type extraction system in the absence of an applied magnetic field

Fig. 5.12 Mass distribution of ions in the beams extracted from inductively coupled high frequency ion sources with diaphragm-type extraction system in the presence of an applied magnetic field

a molecular layer with low surface recombination coefficient is formed on the inside wall of the glass or quartz discharge tube. The effect of this layer asserts itself until it is removed at high temperatures of the wall of the discharge chamber.

5.4 CHARGE DISTRIBUTION OF MULTIPLY CHARGED ION BEAMS EXTRACTED FROM ION SOURCES

In the consideration of ion sources producing multiply charged ions (Paragraph 4.2), it has been shown that multiply charged ions are most probably produced in multiple collision ionization processes. Thus, ions ionized to higher degrees ($n = 3; \ldots; 10$) can be most conveniently obtained from sources with high electron and ion concentrations. Ions with lower number of charges ($n = 1; 2; 3$) can be obtained also from low current ion sources with lower electron and ion concentrations. Thus, for example He^{2+} ion currents of μA order have been obtained from capillary arc-discharge, Penning-type discharge and high frequency gas discharge ion sources. The composition of a helium ion beam obtained from capillary arc-discharge ion source is shown in Fig. 5.13 (Allison [625]) and that of a He ion beam extracted from a high frequency gas discharge ion source in Fig. 5.14 (Vályi [738]). The percentual contribution from He^{2+} ions to the extracted beam is strongly dependent on the gas pressure in the discharge. The highest percentage of He^{2+} ions is obtained at low gas pressures

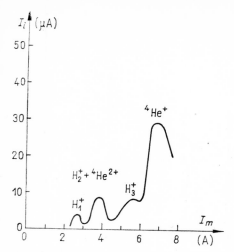

Fig. 5.13 Distribution of H_1^+, $H_2^+ + {}^4He^{2+}$, H_3^+ and ${}^4He^{2+}$ ions in the beam extracted from the capillary arc-discharge ion source shown in Fig. 3.27 (Allison [625])

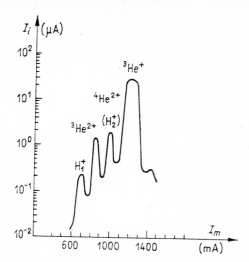

Fig. 5.14 Distribution of H_1^+, ${}^3He^{2+}$, $H_2^+ + {}^4He^{2+}$ and ${}^3He^+$ ions extracted from the capacitively coupled high frequency ion source with diaphragm-type extraction system shown in Fig. 4.20 (Vályi [738])

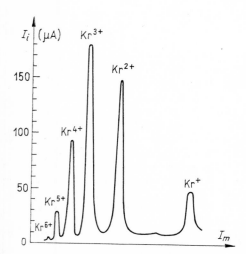

Fig. 5.15 Distribution of Kr^+, Kr^{2+}, Kr^{3+}, Kr^{4+}, Kr^{5+} and Kr^{6+} ions extracted from a high intensity duoplasmatron-type ion source (Krupp [1084])

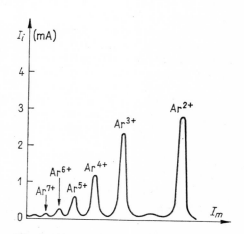

Fig. 5.16 Distribution of multiply charged argon ions extracted from cold cathode Penning-type discharge ion source with oscillating electrons (Anderson et al. [652])

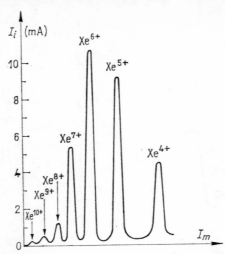

Fig. 5.17 Distribution of multiply charged xenon ions extracted from hot cathode Penning-type discharge ion source with oscillating electrons (Roos *et al.* [1061])

($\sim 10^{-3}$ mmHg) from each type of the sources mentioned above. The relative concentration of helium ion types varies with the partial pressure of hydrogen diffusing out from the tube wall and from the electrode surfaces. This partial pressure of hydrogen continuously decreases if the ion source is operated continuously.

Higher ion current intensities ($I_i > -100$ mA) and higher degrees of ionization ($n = 3; \ldots; 10$) can be obtained if the multiply charged ion beams are extracted from ion sources utilizing high arc-current duoplasmatron or Penning-type discharge. From a duoplasmatron with a discharge current $I_a = 37$ A in krypton gas at a discharge voltage $U_a = 265$ V, an ion beam is obtained with a particle distribution of the type shown in Fig. 5.15. The component of the highest degree of ionization consists of Kr^{6+} ions with maximum current intensities between 3 and 4 μA (Krupp [1084]). Multiply charged ion currents of higher intensities and higher degrees of ionization can be obtained from cold or hot cathode Penning-type gas discharge ion sources.

The composition of ion beams extracted from a cold cathode type source is shown in Fig. 5.16 (Anderson, Ehlers [652]), while that of the beams extracted from hot cathode type ion sources can be seen in Fig. 5.17 (Pasiok, Tretyakov, Gorbatsek [961]). The highest degree of ionization ($n = 10$) has been achieved in hot cathode ion sources because in this type of source the concentration and energy of the primary electrons emitted from the hot cathode can be set to optimum values by varying the operational parameters of the cathode.

Ion sources with low power consumption, yielding low ion currents can be applied at electrostatic accelerators, while the sources with higher ion currents of a high degree of ionization are suitable for use at linear accelerators and cyclotrons.

5.5 EXPERIMENTAL APPARATUS FOR MASS DISTRIBUTION ANALYSIS OF ION BEAMS

The mass distribution in ion beams extracted from ion sources is usually determined from analyses performed with electromagnetic separators, or mass spectrometers.

It has been shown [601, 1235, 1236] that charged particles are deflected by a homogeneous magnetic field H applied in the direction normal to their direction of motion and forced to move along a circular trajectory whose radius depends on the particle velocity v and on the value of $\dfrac{e}{m}$ as expressed by the formula

$$r = \sqrt{\frac{2V}{H^2 \dfrac{e}{m}}} \tag{5.7}$$

where V is the accelerating voltage by which the particle with mass m and charge e is accelerated to velocity $v = \sqrt{\dfrac{2eV}{m}}$.

It can be seen from expression (5.7) that particles which have the same velocity but differ in mass will move in an applied magnetic field of constant value along circular trajectories of different radiuses and thus strike different points of the receiver screen.

Usually, the ion beam extracted from the source consists of particles moving not in parallel but in slightly divergent directions. A special feature of the magnetic spectrometers is that the initially divergent beam of particles having the same velocity and the same value of $\dfrac{e}{m}$ are focused to a well defined small region after covering the first semicircular orbit. The magnitude of this region can be estimated from the drawing in Fig. 5.18.

Let us take an ion beam emitted through slit s in the solid angle α and composed of particles with mass m_1 or m_2 having charge e and the same velocity v. If semicircles are drawn with the extreme radiuses of the beam, we get for each side two semicircles, one with radius r_1, the other with

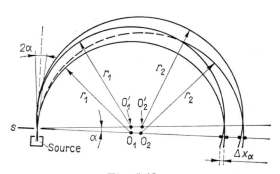

Fig. 5.18

radius r_2. The two types of semicircles cross the plane of the slit s at two points lying at a distance Δx_α from each other.

Simple calculations give for the value of this distance

$$\Delta x_\alpha = 2r_i(1 - \cos \alpha)$$

If the angle α is small, we can use the approximation

$$\Delta x_\alpha \approx r_i \alpha^2$$

Thus, the value of Δx_α determines the image width obtainable in mass spectrometers with $180°$ homogeneous magnetic field for particles having the same velocity v and the same value of $\dfrac{e}{m}$ which start from a given point and fly in the direction determined by the conical angle 2α.

The image width appearing in the plane of the slit s depends not only on the directional spread α. In the case of higher ion currents the image width becomes sensitive to the effect of space charge.

The contribution from this effect to the image width is given for a current I of particles with velocity v, mass m and charge e by the formula [1235] as

$$\Delta x_{\mathrm{ch}} = 2(r_+ - r_-) = \frac{8\pi e}{mv^3} I r_0^2$$

where r_0 is the radius of the trajectories of particles in the absence of space charge, and r_+ and r_- are the radiuses of the outside particle trajectories increased and decreased, respectively, due to the space charge effect.

Now, if, in addition, the particle velocity in the beam shows a velocity spread $\pm \Delta v$, this further increases the image width by the contribution of the form

$$\Delta x_v = 2\Delta r_v = \frac{\Delta v}{v}$$

It follows that for ion beams with a reasonably high current of ions having mass m and charge e with velocities in the range $v \pm \Delta v$, the broadening of the image is given by the sum formula of the three effects, as

$$\Delta x_k = \Delta x_\alpha + \Delta x_{\mathrm{ch}} + \Delta x_v$$

The distances between the focal points for ions having different masses m_i can be evaluated from the differences between the radiuses r_i of the mean trajectories of ions with mass m_i, as

$$\Delta r = 2(r_{i+1} - r_i)$$

or from the equation

$$\Delta x_m = \frac{|\, m_{i+1} - m_i\,|}{m_i} r$$

where r is the radius of the semicircular orbit running through the central points of the inlet and outlet slits of the mass spectrometer magnet.

Thus, separate images of ions differing in mass m_i can be obtained only if the distance x_m between the focal points exceeds the image width Δx_k, that is, if

$$\Delta x_m > \Delta x_k$$

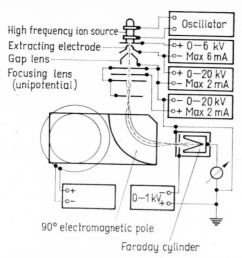

High frequency ion source
Extracting electrode
Gap lens
Focusing lens
(unipotential)

Oscillator

+ 0—6 kV
− Max 6 mA

+ 0—20 kV
− Max 2 mA

− 0—20 kV
+ Max 2 mA

+
−

0—1 kV

90° electromagnetic pole

Faraday cylinder

Fig. 5.19 Schematic drawing of apparatus for study of the mass spectrum of ions in ion beams

This is the requirement which has to be met by the apparatus suitable for the investigation of the mass distribution in ion beams extracted from ion sources.

It has to be noted that in mass spectrometers and magnetic separators sectional magnetic fields of 30°, 60° and 90° are also applied.

One of the possible experimental assemblies for ion mass distribution study and evaluation is schematically drawn in Fig. 5.19. In this arrangement the extracted ion beam is focused by gap and unipotential lenses on the input slit of the 90° mass spectrometer. The particles of the same velocity but of different masses m_i fly in the mass spectrometer along trajectories having different radiuses r_i and thus strike different points of the output slit of the magnet. By varying the magnetic field, the focused images of individual beam components consisting of particles with the same value of m_i can be separately brought to the output slit. The ion current crossing the slit can be detected with a high sensitivity current meter coupled to a shielded Faraday cylinder equipped with an opposite field diaphragm. In this way the mass spectrum of the ions extracted from the source can be determined as a function of the current generating the magnetic field.

5.6 ENERGY SPREAD OF IONS EMITTED FROM ION SOURCES

In experimental investigations carried out on assemblies with ion sources it is often necessary to work with ions having a small spread of energy. However, ions obtained from ion sources have more or less different energies and this spread of ion beam energy may vary to a large extent from one to another type of ion source.

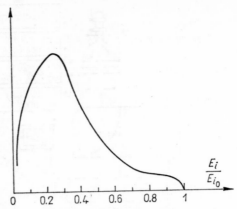

Fig. 5.20 Energy spectrum of ions extracted from high voltage arc-discharge ion source (Deutscher *et al.* [567])

A small energy spread of a few electron volts appears due to collision ionization processes and to the gas discharge plasma where the potential fall is low. The energy spread can be kept small only in ion sources where the ions leave the source by diffusion as in this case the beam is obtained by utilizing only the effect of the ion concentration gradient which does not contribute to the spread of ion energy.

The highest energy spread is observed on ions extracted from high voltage gas discharge (Paragraph 3.5) where the ions are driven towards the extracting electrode by the applied high voltage generating the gas discharge. In this type of ion sources the cathode of the discharge tube functions as the extracting electrode.

To maintain a high voltage gas discharge, the cathode potential fall must be as high as several hundreds or thousands of volts (Paragraph 3.5) and also the potential fall in the discharge plasma must be much steeper than in the low voltage arc or in the high frequency gas discharges. Ions can be thus produced by collisions with electrons in both the plasma and the cathode fall regions. Under these conditions ions can leave the source having discharge voltages U_a with energies in the range $E_{i_0} > E_i > 0$ as compared with $eU_a = E_{i_0}$. These facts have been confirmed by the observations made on high voltage arc-discharge ion sources (Deutscher, Kamke [567]). The energy spectrum in Fig. 5.20 measured on a high voltage arc-discharge ion source shows that ion energies can have any value in the range $0 < E_i < E_{i_0}$ and that most of the ions have an energy $E_i \approx E_{i_0}/4$.

Ions obtained from low voltage arc-discharge plasmatron and duoplasmatron type ion sources show a much lower spread of energy [805, 1138, 1208]. In this case the spread of the ion energy is determined by the negative potential fall before the anode and the value of the spread can be less than 10 eV. This energy spread increases if the channel aperture and the potential of the intermediate electrode are decreased since this causes an increased surface recombination of ions resulting in a higher fall of the negative potential.

The low spread of ion energy in the case of duoplasmatrons holds not only for the positive but also for the directly extracted negative ions. That is why duoplasmatron type ion sources are preferably applied if nanosecond ion current pulses have to be produced and also for any experimental study in physics where the measured values can be influenced by the energy spread of ions.

The energies of ions extracted in the axial direction from Penning-type gas discharge ion sources with oscillating electrons are usually equal to the energy corresponding to the cathode fall potential. This value depends on the material of the cathode and it is less by 16 to 30 per cent than the value of the energy corresponding to the anode potential U_a. The energy spread of ions extracted axially from Penning-type ion sources depends on the fluctuations in discharge current, the amplitude of ion oscillations taking place in the discharge and on the frequency of flash-overs in the cathode environment [645, 674, 1210].

Figure 5.21 (Nagy [674]) shows the spread of ion energy for axial ion extraction from a cold cathode ion source with an aluminium or iron cathode. The full width of the energy spectrum obtained with az aluminium cathode is between 45 and 50 eV with a half-width value of $\sim$16 eV. A similar energy spread has also been measured for magnesium cathodes. With these metal cathodes the broadening of the spectrum is due to the flash-overs on the cathode surface.

If iron cathodes are applied, the energy spread shows a substantial increase. The total width becomes $\sim$240 eV with a half-width value of $\sim$45 eV.

A large fraction of ions has energies exceeding the value corresponding to the anode potential U_a and the average ion energy is only a few electron volts less than the value of $eU_a = E_{i_0}$. Experiments have shown that in the case of iron cathodes ion oscillation sets in already at anode currents

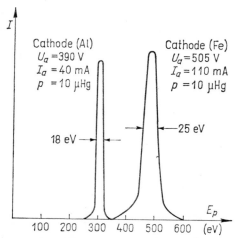

Fig. 5.21 Energy spread of ions extracted from cold-cathode Penning-type discharge ion source with oscillating electrons and axial ion extraction for aluminium or iron cathodes (Nagy [674])

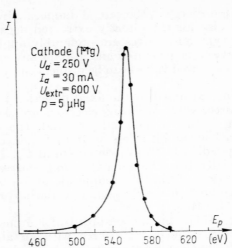

Fig. 5.22 Energy spread of ions extracted from cold cathode Penning-type discharge ion source with oscillating electrons and transversal ion extraction (Nagy [674])

above 30 mA with increasing frequencies and amplitudes as the anode current increases.

While taking the energy spread spectrum of Fig. 5.21 an ion oscillation frequency of $\sim$140 kc/s with a larger amplitude than 100 V has been observed at a discharge current $I_a = 110$ mA which contributed to the energy spread.

The presence of ions with energies above the value of eU_a can be attributed to the possible extinction of the gas discharge by ion oscillation as a result of which the higher ignition voltage causes at the time of ignition the emission of ions with higher energy than the value of eU_a. With tantalum cathodes, the spread of ion energy was measured as $\sim$100 eV for which also ion oscillation can be held responsible.

In tranversally extracted ion beams the energy of the ions is determined by the value of the extracting voltage U_{extr} applied to the extracting electrode and by the negative potential fall in the plasma. The spread of ion energy depends on the frequency of the flash-overs on the cathode surface, the amplitude of ion oscillation and on the processes occurring in the extracting system.

In Fig. 5.22 [674], the energy spread of ions extracted in transversal direction from a cold cathode ion source is shown. It can be seen that the potential fall of the plasma and the spread of ion energy are not essentially different from the values measured on the ion sources with axial ion extraction.

Thus, it seems that the ions extracted from Penning-type gas discharge ion sources with oscillating electrons and with the usually applied cathode material have energies which are from 10 to 30 per cent lower than the value eU_a corresponding to the cathode potential U_a. The average energy of ions approaches the value of eU_a with increasing values of the discharge current. I_a. The spread of ion energy is generally 10 to 15 per cent of the mean value of the ion energies.

The most extensive investigations have been carried out on the energy spread of ions extracted from high frequency gas discharge ion sources [658, 683, 685, 702, 704, 705, 725–727, 730, 732, 735, 738, 1207, 1211, 1212, 1215]. The spread of ion energy in high frequency gas discharge assemblies can be attributed to several effects. These include the potential fall in the gas discharge plasma, the charge transfer and ionization processes in the dark space, the flash-overs between the extracting electrode and the plasma, the fluctuations in extracting voltage and the superposition of a high frequency voltage on the extracting d.c. voltage. The contributions from these effect to the energy spread depend on the gas pressure, the high frequency power and on the extracting voltage.

Owing to the excess of positive ion formation in the discharge plasma caused by the electrons of high mobility diffusing from the plasma at a high rate, the potential fall in the gas volume can amount to a few times ten V. At higher values of the high frequency field and a decrease in the gas pressure of the discharge chamber, the potential fall can increase to a value of about 100 V [727].

In the dark space before the extracting electrode both elastic and inelastic atom-ion collisions may take place. Elastic collisions change the direction of the ions so that they cannot fly through the channel of the extracting electrode. In the charge transfer process during inelastic collisions slow ions and fast neutral atoms are formed. The ions are accelerated by the electric field in the dark space in the direction of the extracting channel. Since they are formed at different places, different energies are imparted to them, depending on the distance covered by the ions in the accelerating field.

The average energy of ions formed in this process is appreciably lower than the average energy of the ions in the extracted beam. Their contribution to the beam is less than 1 per cent [658].

Ions, which are produced in the dark space by secondary electrons ejected from the cathode by ions, contribute to the extracted beam similarly only $\sim$1 per cent of the total ion beam.

It follows from the above that the contribution from ions produced in the charge transfer and ionization processes occurring in the dark space can be considered negligible in the beams extracted from high frequency ion sources at the usually applied gas pressures from 10^{-3} to 10^{-2} mmHg.

The spread of ion energy may appreciably increase due to the fluctuation of the ectracting voltage or because of flash-overs between the gas discharge plasma and the extracting electrode. These effects can cause a spread of ion energy as high as 100 to 300 eV [704]. The flash-overs between the extracting electrode and the plasma can be detected by oscilloscopic observation of the extracting voltage. This is made possible by the fact that voltage pulses of about $\sim$30 μsec are generated by the potential fall on the protective resistor due to current pulses appearing simultaneously with the flash-overs in the current of the extracting circuit. The magnitude of the resistor output pulses depends slightly on the value of the extracting voltage but it is essentially independent of the discharge current. However, the repetition frequency of these pulses depends on the value of the discharge current, e.g. for $I_a = 0.5$ mA it varies from 1 to 200 pulses/sec while for $I_a \approx 1$ mA we find ~ 1000 pulses/sec and if I_a rises to values between 4 and 5 mA the pulses already become non-resolvable in time. Consequently, the energy spread caused by a few flash-overs at a low discharge current can be between 60 and 70 eV, while at a high number of flash-overs the energy spread can increase to values from 200 to 300 eV [704].

In the consideration of high frequency gas discharge ion sources (Paragraph 3.4.4) it has been shown that because of their large mass, ions cannot be energized by the high frequency field in the gas discharge plasma. On the other hand, the ion energies determined by the voltage U_{extr} applied to the extracting system of the ion source can be modulated under certain conditions by the high frequency field [704, 705, 1207, 1211, 1212].

Ion energies can be modulated in this way if their flight across the dark space before the extracting electrode takes less time than the period time of the high frequency oscillation, since then the high frequency field can increase or decrease the ion energies according to the time at which they enter the field. This procedure is similar to the velocity modulation discussed in Paragraph 4.3.2.

Owing to the high frequency modulation, the spread of ion energy can increase and the energy spectrum of the extracted ions may have one or two close lying peaks depending on the operational parameters of the ion source [705].

The occurrence of this high frequency modulation of ion energies was confirmed in a rapid analysis while applying a demodulating high frequency field of appropriately chosen phase and amplitude [1207].

In this case the two peaks of the energy spectrum were brought by the demodulating field not only closer to each other but also their coincidence

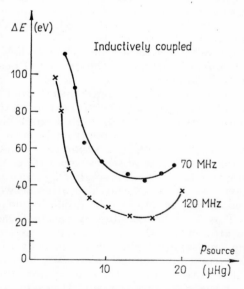

Fig. 5.23 Energy spread of ions extracted from inductively coupled high frequency ion source with diaphragm-type extraction as a function of the source pressure p for different frequencies of the high frequency field

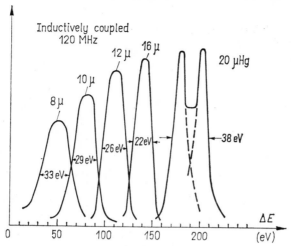

Fig. 5.24 Energy spread of ions extracted from inductively coupled high frequency ion source with diaphragm-type extraction system for various values of gas pressure

could be achieved by frequency demodulation so that the width of $\sim$50 eV observed on the energy spectrum with two peaks could be reduced to $\sim$15 eV in the modulated spectrum with a single peak. The modulation of ion energies causes the two peaks to reappear then to overlap again at a given ion energy.

Experiments concerning the spread of ion energy in beams extracted from high frequency gas discharge ion sources have been carried out on systems with alternating electric field (capacitively coupled), with alternating magnetic field (inductively coupled), and with different types of ion extracting channels.

The width and shape of the energy spectra of ions extracted from inductively coupled ion sources with probe extraction system (Paragraph 3.8.1) were found to vary with the current I_k flowing in the ion extracting circuit. For $I_k < 1$ mA the energy spectrum with a single peak showed widths between 40 and 70 eV, while for $I_k > 1$ mA energy spectra with two peaks and widths ranging from 80 to 250 eV were obtained [704, 705, 725, 732]. The distance between the two peaks, i.e. the total width of the energy spectrum with two peaks, was found to increase as the value of the applied high frequency field was being increased.

The spread of ion energy was studied on ion sources with probe extraction system and of the same design but with different types of coupling of the gas discharge generating oscillator [1138]. These experiments have shown that the energy spread of the extracted ions is less for the capacitively coupled than for the inductively coupled type.

The experiments carried out on high frequency assemblies with diaphragm type extraction system (Paragraph 3.8.2) proved the dependence of the energy spread on gas pressure, oscillator frequency and on the type of excitation used [1215]. The experimental results are shown for the inductively coupled ion sources in Figs 5.23 and 5.24, for the capacitively coupled assem-

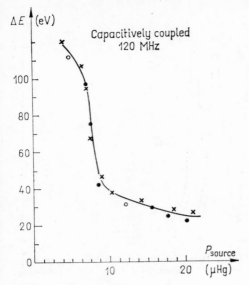

Fig. 5.25 Energy spread of ions extracted from capacitively coupled high frequency ion source with diaphragm-type extraction system as a function of gas pressure in the ion source

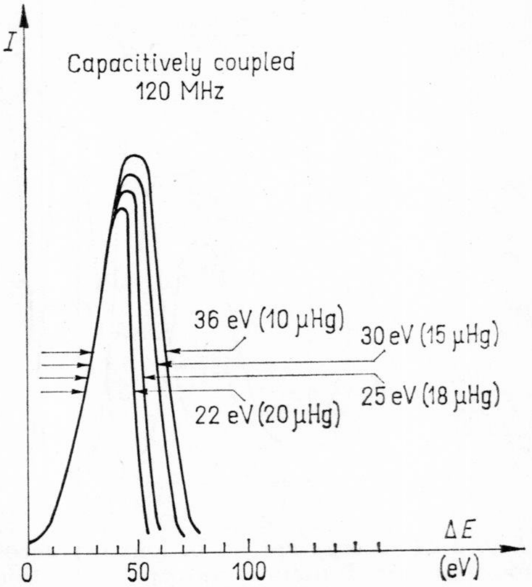

Fig. 5.26 Energy spread of ions extracted from capacitively coupled high frequency ion source with diaphragm-type extraction system for different values of gas pressure

blies in Figs 5.25 and 5.26. Figure 5.23 shows the energy spectra measured as a function of the gas pressure in the discharge tube for oscillator frequencies of 70 and 120 MHz in inductively coupled ion sources having diaphragm extraction system of the types to be seen in Figs 3.67 and 4.20. It can be seen from the figures that at pressure values $p_{source} < 10$ μHg the spread of ion energy appreciably increases as the gas pressure decreases. The decrease with pressure is less apparent at values of $p_{source} = (9-10)$ μHg.

The change in the energy spread can be attributed partly to the fall of the potential in the gas discharge plasma, partly to the effect of energy modulation. The increase in the width of the energy spectrum at a gas pressure of about 20 μHg can be explained by the appearance of a double peak (Fig. 5.24).

The analysis of the double spectrum revealed that the energies of the group of ions generating one of the two spectra lie in the range with a half width value below 15 eV. If the oscillator frequency which generates the high frequency gas discharge is decreased, the spread of ion energy increases and the double peaks caused by the high frequency modulation appear even at lower pressures.

The energy spectra measured on the ion sources with capacitive coupling, presented in Figs 5.25 and 5.26 do not show the double spectrum observed on inductively coupled assemblies at $\sim$20 μHg. Thus, at this value of pressure the spread of ion energy is 40 to 60 per cent lower if the 120 MHz frequency oscillator is not coupled inductively but capacitively in the high frequency gas discharge ion source.

The analysis of the energy spectra of ions extracted from different types of ion source furnishes information which suggests that the spread of ion energy can be regulated within some limits by the appropriate choice of the design and the operational parameters of the ion source.

In duoplasmatrons and in high frequency ion sources which can be used for physical investigations requiring ion beams with small energy spread it is the potential fall in the plasma which has to be first of all minimized. This can be achieved in duplasmatrons by suppressing surface ion recombinations if possible, and in high frequency ion sources by reducing the electron diffusion in the direction of the tube wall. Surface ion recombinations can be reduced in duoplasmatrons by increasing the channel length and the potential of the intermediate electrode [805] while in high frequency sources the gas pressure has to be increased (Figs 5.23 and 5.25) [1212, 1215].

High frequency energy modulation can also be minimized by a suitable choice of the design and operational parameters of the assembly.

The number of flash-overs at the extracting electrode in Penning-type sources can be strongly decreased by the addition of oxygen gas or by a suitably chosen cathode material, while in the high frequency ion sources a careful choice of the material of the extracting electrode, its precise machining, proper and accurate mounting can reduce flash-overs.

5.7 EXPERIMENTAL APPARATUS FOR MEASURING THE SPREAD OF ION ENERGY IN ION BEAMS EXTRACTED FROM ION SOURCES

The energy spread spectrum of ion beams emitted from ion sources is usually evaluated from data obtained with an electrostatic analyser of radial electric field.

It has been shown [601, 1235–1238] that in the radial electric field F of a cylindrical capacitor the particles with mass m, velocity v and charge e move along a circular orbit having a well defined radius r under the condition that

$$eF = \frac{mv^2}{r} \qquad (5.8)$$

if the direction of particle motion is normal to that of the electric field.

Now, if we use the relation between the particle velocity v and the accelerating voltage V_0 of the form $mv^2 = 2\,eV_0$ and the expression for the electric field of cylindrical symmetry given by

$$F = \frac{dV}{dr},$$

formula (5.8) transforms to

$$\frac{dV}{2V_0} = \frac{dr}{r}$$

Upon integration of both sides between the limits $-\frac{V}{2}$ to $+\frac{V}{2}$ and r_1

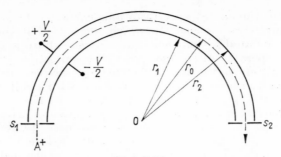

$$Fig.\ 5.27$$

to r_2, respectively, according to the notation in Fig. 5.27 we get the equation

$$\frac{V}{2V_0} = \ln \frac{r_2}{r_1}$$

where V is the direct voltage applied to the plates and r_1 and r_2 are the radiuses of the internal and external electrodes, respectively.

It can be read from this formula that the particles with charge e entering the field between the two cylindrical elctrodes with a velocity $v = \sqrt{\dfrac{2eV_0}{m}}$

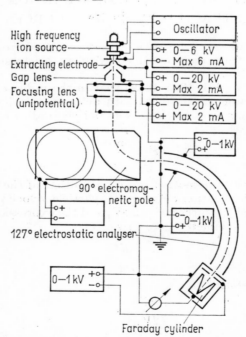

Fig. 5.28 Schematic drawing of apparatus for study of energy spread of ions extracted from the ion sources

in the direction normal to the field will move along an orbit with radius $r_2 > r > r_1$ under the condition that the direct voltage V applied to the electrodes satisfies the equation

$$V = 2 V_0 \ln \frac{r_2}{r_1}$$

The electrostatic energy measuring equipment with radial electric field focuses, similarly to the magnetic mass spectrometer, the ions entering the field in divergent directions.

It has been theoretically predicted and experimentally confirmed [1237, 1238] that a divergent ion beam emerging from the plane of the inlet slit is focused in the radial electric field of the cylindrical capacitor after the ions have completed the circular arch $\frac{\pi}{\sqrt{2}} = 127°17'$ of their orbital motion in the field.

The image of the line-shaped inlet slit on the plane of the outlet slit lying at a distance of $\frac{\pi}{\sqrt{2}}$ radian will have a width

$$\Delta x_k \approx \frac{4}{3r}$$

The schematic drawing of a possible experimental arrangement for the analysis of ion energies in beams extracted from ion sources is shown in Fig. 5.28.

In the assembly shown in the figure, the ion beam obtained from the ion source is separated according to masses so that only the beam component consisting of ions with velocity $v \pm \Delta v$, mass m_i and charge e can pass through the input slit of the electrostatic analyser.

By varying the electric field of the electrostatic analyser, the ion energy spectrum can be evaluated from the measured output currents of ions with velocity v_i. The current can be measured with a high sensitivity current meter coupled to a shielded Faraday cylinder equipped with a diaphragm of opposite field.

CHAPTER VI

EXTRACTION OF IONS FROM ION SOURCES AND FORMING OF THE EXTRACTED BEAM

The ions generated in different types of ion sources can be led out from the source, as has been already mentioned, either by diffusion or by means of an extracting system. Well-collimated intense ion beams with small diameter, which are required in the apparatuses utilizing an ion source, can be obtained only from sources equipped with an ion extracting system.

The utilization factor of the ions formed in the source depends on the type and design of the extracting system and on the features of the ion optical system forming in the environment of the extracting electrode. For this reason, the choice of the appropriate ion extraction method is as important as the generation of a high ion concentration in the environment of the extracting electrode.

In the present chapter some of the more important ion extraction methods will be reviewed and also the properties of the ion beams emitted from the sources will be discussed.

6.1 EXTRACTION OF IONS FORMED IN THE ION SOURCES

The properties (intensity, focusability, etc.) of ion beams extractable from ion sources depend on the ion concentration in the source, the area of the ion emitter surface and on the ion-optical properties of the extraction system.

In ion sources utilizing electron impact and surface ionization the ion concentration varies with the electron and ion currents, while the area of the emitter surface and its geometrical position are generally independent of the variation of operational parameters. In this type of source the value of the extracted ion current is determined by the ion concentration in the source and by the value of the applied extracting voltage in the extracting system.

In gas discharge sources, on the other hand, the position and the area of the ion emitter surface can be appreciably affected by a change in the operational parameters of the extracting system.

In the discussion of gas discharges (Paragraph 3.4) it has been shown that because of the higher mobility of electrons compared with that of the ions, the wall of the discharge tube is charged to a potential given by

$$U = \frac{kT_e}{e} \ln \frac{T_e m_i}{T_i m_e}$$

This potential causes the plasma boundary S_1 to move away from the wall surface S_2 (Fig. 6.1a). The measure of the plasma withdrawal is given by $2\lambda_D$, where λ_D is the Debye length (formula (3.17)). If an aperture of a larger size than $2\lambda_D$ is made on the discharge tube wall, the gas discharge plasma will balloon out through this aperture (Fig. 6.1b).

The geometry of the extracting electrode and the value of the negative potential applied to the electrode can determine the shape of the surface of the plasma fraction ballooning out into the expansion volume. Thus, at a given value of the negative potential the plasma surface coincides with the plane of the outlet aperture (Fig. 6.1c). If the negative potential is increased above this given value, the plasma surface can be forced back so that a curved surface forms which produces a depression in the plasma (Fig. 6.1d).

In the case shown in Fig. 6.1c we get from the ion source a beam of nearly parallel flying ions and the ion emitter surface area of the plasma is equal to the area of the outlet aperture.

Under the conditions shown in Fig. 6.1d, the ion emitter surface of the plasma increases and we get a convergent, focused beam of ions from the curved surface.

The shape and position of the ion emitter surface of the plasma are thus dependent on the value of the negative potential applied to the extracting electrode. This means that the ion-optical system composed of the extracting electrode and the plasma surface can have different geometries determined by the value of the negative potential applied to the extracting electrode. This fact imposes limitations on the variation of the extracting voltage, thus the beam intensity cannot be varied beyond a given measure by varying the extracting voltage.

This limitation can be overcome by keeping the shape and position of the

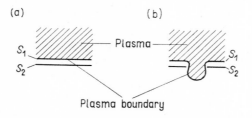

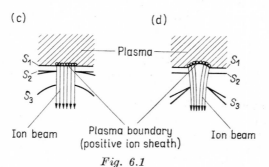

Fig. 6.1

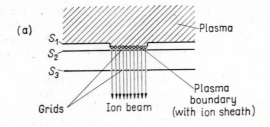

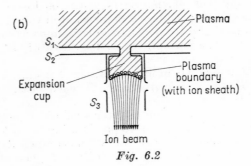

Fig. 6.2

ion emitter surface of the plasma of a fixed geometry (Fig. 6.2a). This has been achieved in the arrangement shown in the figure by placing a grid of high transparency into the plane of both the ion source outlet and the extracting electrode apertures. The grids ensure both the constant shape and position of the plasma boundary and the even distribution of the electric field in the optical system.

The ion emitter plasma surface can be increased to a multiple of the outlet aperture area of the ion source, if the plasma balloons out (Fig. 6.2b) into the expansion cup kept at the same potential as the outlet aperture.

In this case the negative potential applied to the extracting electrode S_3 causes the formation of a curved surface of positive ions on the plasma boundary and the ions emitted from this curved surface fly out from the electrode S_3 in a focused beam. With an extracting system of this type the ion emitter surface can be substantially increased without increasing the size of the outlet aperture in the ion source. Moreover, the thus increased emitter surface permits the angular divergency of the ion beam to be kept small.

6.2 ION-OPTICAL PROPERTIES
OF THE EXTRACTING SYSTEMS

It has been seen in the foregoing paragraph that the ion emitter plasma surface, which forms part of the ion-optical system in the environment of the extracting electrode, can be shaped by the choice of the geometry and the applied voltage of the ion extraction method. Consequently, the ion-optical properties can be different if different extraction systems are applied.

In ion sources with probe-type ion extraction and in those utilizing the expanded plasma surface, the ion-optical system is comparable with an immersion objective.

In ion sources with diaphragm-type ion extraction, where the electrodes S_2 and S_3 are shaped so as to reduce the space charge effect in the ion beam, the extracting system is comparable with a quasi-Pierce-type ion-optical system.

These comparisons, of course, do not account for all the features of the ion extraction system under consideration, but they are of help in the interpretation of the processes involved.

The precise calculation of the parameters of ion extraction systems is very difficult for apparatuses used in practice and therefore the optimum values of the operational parameters of a given ion source are usually determined experimentally.

6.2.1 ION-OPTICAL PROPERTIES OF PROBE-TYPE EXTRACTING SYSTEMS

In the probe-type extracting system (Fig. 6.3) the shielded tubular electrode, which is confined by the ion emitter plasma surface at the upper end of the insulating tube, can be regarded as an immersion objective (Fig. 6.4a) [1217].

The immersion objective lens consists, in the most simple case, of two electrodes both with diameter D; one of them is of length h and closed at one end while the other electrode (open at both ends) is placed at some distance from the former. The relations between the parameters of this simple immersion objective are given in Fig. 6.4b. It can be seen from the figure that for $h < 0.785\,D$, the distance K of the image is negative, thus we have a virtual image; for $h = 0.785\,D$ the distance $K = \pm\infty$, and for $h > 0.785\,D$, the image is real and the distance of the image decreases as h increases.

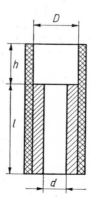

Fig. 6.3 Schematic drawing of the probe with surrounding isolator tube in probe-type extraction system

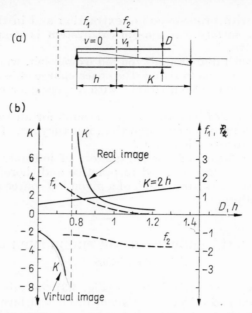

Fig. 6.4 Schematic drawing of electrostatic immersion lens and diagram of relationships between lens parameters

In the optimum case, the parameters of the extracting system are so chosen that the total cross section of the ion extracting channel in the electrode is fully utilized while the ion current to the channel wall is diminished. Therefore, the magnification of the immersion objective defined by

$$M = \frac{K}{2h}$$

should be chosen to have a value close to unity ($K \sim 2h$). At other values of this magnification the transmitted ion current decreases and at its low values the divergence angle of the ion beam increases on leaving the channel.

Under these conditions the parameters h and D of the extraction system (Fig. 6.3) are restricted, as to be seen in Fig. 6.4b, to values in the range

$$0.6\,D \le h \le 0.9\,D$$

The applicability of this relation has been confirmed by experimental studies [696, 699, 705, 710, 728, etc.].

This ion-optical analogy does not account for all the properties of the probe-type extraction system, since the formulation does not include the contribution from the space charge effect to the ion beam. A considerable ion current density can be obtained in the ion extracting channel of the electrode at the place where the diameter of the ion beam shows a minimum value. The effect of the space charge can be particularly important in ion beams of higher than the usual intensity since the electrostatic repulsing field generated between identical charged particles causes — due to the high charge density -- a substantial increase in the divergency of the beam.

Let us now consider the effect of space charge on an ion beam consisting of identically charged particles passing through the channel. The identically charged ions of a beam with circular cross section, moving parallel with the axis of the channel in the extracting electrode, have in the general case, input velocities composed of axial and radial components.

The axial velocity component is due to the accelerating electrostatic field $F_0 = \dfrac{U_0}{x}$ while the radial velocity component in the direction of the axis is due to focusing. Accordingly, the radius of the ion beam which at the input has the value $r = r_0$, reaches a minimum value ($r = r_{min}$) in the manner shown in Fig. 6.5. At this point the radius of the ion beam can be described by the expression [1221]

$$r_{min} = r_0 \exp\left(-\frac{8\pi \left(\dfrac{e}{m_i}\right) U_0^{3/2}}{I_i} \tan^2 \alpha\right)$$

If the radial contribution to the velocity is zero, that is, a beam of ions moving in parallel directions enters the channel of the extracting electrode we have $r_0 = r_{min}$. In this case the expansion of the beam due to charge effect extends over the whole channel length.

It follows from the above consideration that the loss of ions due to beam expansion because of the charge effect in an extracting channel of diameter d and length l can be less in the case of a focused beam than in the case of parallel moving ions.

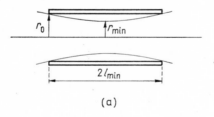

(a)

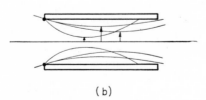

(b)

Fig. 6.5 Formation of minimum ion beam radius r_{min} in the ion extraction channel: (a) for optimum entrance angle α associated with maximum transmittable ion current of channel r_{min} is in the middle of the channel; (b) values of r_{min} for various values of entrance angle α

The maximum ion current through the channel can be obtained if $r = r_{\text{min}}$ is in the middle of the channel (Fig. 6.5a). However, ion beams can be transmitted in this manner only at given values of the applied voltage U_0 and of the input angle α [742, 1121, 1221]. The maximum transmissible ion current is given by the formula

$$I_{i(\text{max})} = 3.27\pi \left(\frac{e}{m_i}\right)^{1/2} U_0^{3/2} \left(\frac{d}{l}\right)^2$$

if the input angle is

$$\alpha = \text{arc tan} \left(\frac{d}{l}\right)$$

This formula shows that the maximum value of the ion current under the restriction imposed by the space charge effect varies with the applied extracting voltage and with the channel parameters. The latter are usually not free parameters as their values have to be chosen with respect to the ion-optical and vacuum technical features of the apparatus utilizing the ion source.

The maximum value of the ion current ($I_{i(\text{max})}$) transmissible through the extracting system should be chosen to be no higher than that of the current emitted from the plasma surface. The ion current emitted from the plasma surface is determined by the ion concentration on the ion emitter surface of the plasma and by the area and shape of this surface.

In ion sources with probe extraction, the probe is usually introduced into the volume of maximum ion concentration (Figs 3.56–3.58 and 3.61–3.64).

The ion current emitted from the plasma surface in the case of probe-type extraction can be given by use of formula (3.30) in the form

$$I_i = S_0 n_i e \sqrt{\frac{kT_i}{2\pi m_i}}$$

It has been shown by experimental studies [698, 699, 704, 705, 974, etc.] that the $U_{\text{extr}}^{3/2}$ type dependence of the transmissible ion current holds only for the ascendant portion of the $I_i(U_{\text{extr}})$ characteristic, then the slope of the characteristic decreases and it attains either a saturation value or shows a maximum with subsequent decrease (Figs 3.59, 3.60, 3.65). At increasing ion concentrations, obtained in these experiments by increasing the high frequency power input to the gas discharge, this deviation from the predicted characteristic appears at higher values of the extracting voltage and consequently of the ion current.

The maximum extractable ion current can be considerably increased also by introducing the tip of the extracting probe into the volume of maximum ion concentration [974]. It is possible to reduce the gas consumption without causing an appreciable decrease in the ion current by inserting a short diaphragm with an aperture equal to the minimum beam diameter into the extracting channel of the electrode [690, 699, 705] (Figs 3.58 and 3.62). However, if the extracting channel is equipped with a diaphragm, the ion-optical parameters have to be more carefully adjusted.

The characteristic for $I_i(U_{extr})$ and $I_d(U_{extr}) - I_d$ stand for probe current — plotted from the data measured on an ion source with the above type of extracting channel (Fig. 3.60) show that with the optimum ion-optical parameters, when the aperture of the diaphragm inserted in the channel is placed at the same point at which we have r_{min}, the value of the ion current extracted through the electrode suddenly increases while the probe current I_d decreases.

6.2.2 ION-OPTICAL PROPERTIES OF DIAPHRAGM-TYPE EXTRACTING SYSTEMS

With extracting systems using diaphragms, a much more precise geometry and a much better defined emitter plasma surface can be obtained than with probe type systems.

Extracting systems using either spherical (Fig. 3.68) or conical electrodes with the latter forming a given angle with the beam surface (Figs 3.31, 3.36, 3.38 and 3.74), can be regarded as a quasi-Pierce-type ion-optical system.

In the Pierce-type optical systems [741, 742, 1222] the electrodes S_2 and S_3 are so shaped as to prevent the divergency due to space charge effect in intense particle beams. This ion-optical system is based on the principle that the direction of a charged particle current between the surfaces of two concentric spheres with different radii is not affected by the space charge.

The potential distribution between two concentric spheres for a potential difference U in the case of space charge is given by the solution to the Poisson equation. With the thus obtained distribution, the particle current from the outer to the inner spherical surface can be approximated in the form

$$I_i \approx \frac{16\pi}{g} \sqrt{\frac{2e}{m_i}} \frac{U^{3/2}}{(-\xi)^2}$$

where ξ is a parameter depending on the value of the ratio r_a to r_k, that is, on the radii of the outer and inner spheres, respectively. The values of the parameter ξ are listed in Appendix, Table 18.

The above approximation to the particle current holds, of course, for the 4π solid angle of the total spherical surface. Thus, if the emitter surface is taken to be only that area of the outer spherical surface which is subtended by half conical-angle Θ, the ions will be transmitted only to the area of the half conical-angle Θ on the surface of the inner sphere and in the ideal case the particle current is given as

$$I_i(\Theta) = I_i \frac{1 - \cos \Theta}{2}$$

Since the outer and inner spherical surfaces are only partly utilized in this way, the effect of the remaining surface areas has to be compensated for by means of properly shaped electrodes kept at a suitable potential. Experiments performed in an electrolyte tank indicated that the electrode at the emitter surface has to form an angle of 67.5° to the boundary of the beam [742]. The shape and aperture of the equipotential electrode, which functions as extracting electrode, have to be chosen with respect to the value of the half-conical angle Θ and to that of the distance $r_a - r_k$. The aperture in the inner spherical surface must be chosen to be less than half the distance $r_a - r_k$ between the spherical surfaces in order to keep the potential distribution between the electrodes and the particle current through the aperture at values not substantially different from those calculated for the ideal case. Experimentally, it has been proved that with an aperture sized as

$$d = 0.7\,(r_a - r_k)$$

the field decreases by $\sim$5 per cent at the outer spherical surface [1121]. At higher values of d the distortion of the field can become important and it can lead to a decrease in the output current.

It is thus advisable to form in the inner sphere the aperture, which corresponds to the outlet slit of the extracting electrode, to a size given as

$$d \approx (0.5\text{--}0.7)(r_a - r_k)$$

The outlet aperture functions as a dispersing lens because its upper boundary faces a field, while its other side a field-free section. This dispersing effect of the aperture has to be minimized to ensure the full exploitation of the emitter surface. Thus, the focal distance of the lens must be kept at the minimum possible value by a suitable choice of the parameters. The minimum focus length of the lens is given by the parameter values through the relation

$$\frac{r_k}{r_a} = \frac{d}{D} \approx 0.5$$

The two last formulae furnish a number of useful data for the design of extraction systems with spherical electrodes.

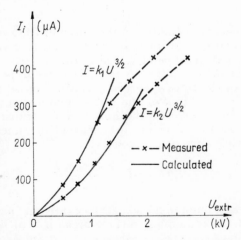

Fig. 6.6 Comparison of ion current I_i versus extraction voltage U_{extr} curve calculated from the relation $I_i = K_i U^{3/2}$ with that obtained from measured values for extraction from ion source with extracting system of spherical symmetry shown in Fig. 3.68

[735]

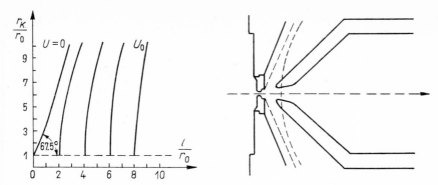

Fig. 6.7 Extracting system with quasi-Pierce-type ion-optical arrangement: Left figure: ideal Pierce-type electrode shapes where r_0 is the radius of aperture in input diaphragm and that of electrodes is r_k; Right figure: one of versions of quasi Pierce-type extracting system built with electrodes of approximately ideal shape

The parameters d, D and $(r_a - r_k) \approx h$ can be related by use of the two above formulae in the form

$$2d \approx D \approx 2(0.5\text{--}0.7)(r_a - r_k) \approx 2(0.5\text{--}0.7)h$$

The applicability of these relations was confirmed by experimental studies on ion sources with an extraction system of spherical geometry arranged as shown in Fig. 3.68 [730, 733, 735].

By appropriately setting the ratio r_a/r_k in the extraction systems with spherical geometry, it is possible to determine the point at which the diameter of the conically shaped particle beam emitted from the spherical surface has its minimum value. This point can lie at a distance either shorter or longer than the focus-length of the beam free of space charge. If long extraction channels are used, it is advisable to choose the ratio r_a/r_k between 2 and 3. The characteristic curve for $I_i(U_{\text{extr}})$ varies as $kU^{3/2}$ only in the initial rising part, similarly to the behaviour observed in probe extraction systems (Fig. 6.6), then it reaches either a saturation value or it exhibits a maximum with subsequent decrease depending on the values of the parameters l and h [735].

The ion-optically quasi-Pierce type extraction systems consisting of two conical electrodes (Fig. 6.7) are applied in duoplasmatrons and Penning-type ion sources with oscillating electrons (Paragraph 3.7). The ideal shape of electrodes is generally approached by the type of electrodes in the ion extraction system shown in Fig. 6.7 [680]. In this type of extraction system the emitter plasma surface is formed in the immediate environment of the small exit slit and the emitter surface changes its form as determined by the value of the negative potential applied to the extracting electrode. Owing to the high ion concentration in these ion sources, the current density can become so high at the exit slit that the effect of the thus generated space charge cannot be fully compensated by the Pierce-type electrodes. The ion beam extracted from the source through the exit slit shows, therefore, a relatively large angular divergence. This inconvenience can be eliminated by increasing the extracting voltage U_{extr} applied to the extracting electrode in order to obtain a beam of nearly parallel flying ions.

6.2.3 ION-OPTICAL PROPERTIES OF THE PLASMA SURFACE BALLOONED OUT FROM THE DISCHARGE VOLUME

In order to minimize the increase in the divergence of extracted beams caused by the high ion concentration at the exit slit, an expansion cap is mounted in front of the slit. The expansion cap is kept at the same potential as the extracting electrode and the plasma is left to partly expand out of the discharge volume into the cap [669, 670, 948–952].

The ion concentration in the plasma ballooning out into the expansion cap, as shown in Fig. 6.8, is changed in the presence of an applied magnetic field in a way which differs from its change in the absence of a magnetic field [949].

In the presence of a magnetic field the ion concentration changes due to the concentration of the plasma in the magnetic field [949]. On applying a negative potential to the electrode S_3 mounted before the expansion cap (Fig. 6.2b), the boundary surface of the expanded plasma transforms and takes the concave form to be seen in the figure. Investigations of so formed plasma boundary surfaces have shown [948–951] that in the major part of the expanded plasma fraction, the ion density distribution remains essentially unchanged if the applied negative potential is not too high ($\sim$5 keV), thus the ion density is not appreciably modified on the ion emitter surface of the plasma.

It can be estimated from the ion density distribution shown in Fig. 6.8 that the ion density on the emitter surface can, in this way, become lower

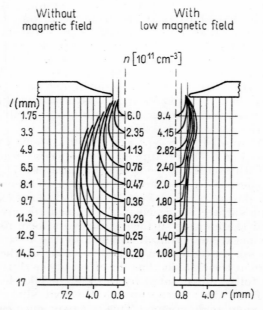

Fig. 6.8 Ion concentration distributions in the plasma ballooning out into the expansion cap of a quasi-Pierce extraction system in the absence and in the presence of an applied magnetic field

by about an order of magnitude while the emitter surface area can be increased by an order of magnitude if the electrode is appropriately designed.

The negative potential applied to the electrode S_3 can be chosen so as to set the emitting surface curvature needed for optimum focusing thereby permitting ion currents of high intensity to be obtained from the increased emitter surface in spite of the reduced ion concentration.

Expansion caps of this type are usually applied in the extracting system of duoplasmatrons or of Penning-type ion sources with oscillating electrons where the plasma with high ion concentration is the plane of or quite close to the extracting electrode and where the ion density can still be sufficiently high on the emitter surface of the plasma fraction expanded into the expansion cap.

6.3 PROPERTIES OF BEAMS EXTRACTED FROM ION SOURCES

Ion beams extracted from different types of sources have to be transmitted possibly without loss through the ion-optical system of the apparatuses utilizing the ions. It is very important that the particles striking the target of the experimental equipment should have the parameters required by the investigations. For a beam transmission without loss of particles, the cross section of the ion beam must not exceed a given maximum value at well defined points. In a well focused ion beam its contour has to be shaped according to the path of the rays in the beam. The beam contour is generally related to the emittance diagram of the ion beam leaving the source. Therefore, it is useful to know the emittance diagram of the ion source to be applied in order to construct a channel through which the ion beam can be transmitted without loss of ions.

6.3.1 THE EMITTANCE DIAGRAM OF ION SOURCES

The parameters of the beam extracted from ion sources can change during the passage of the beam through an ion-optical system. These changes can be of considerable importance as regards the loss of ions from the beam.

The ion beam properties can be well described in terms of the phase space known from mechanics [1219, 1223, 1276].

The movements of particles not interacting with one another and driven by an external force can be fully specified in the phase space by their x, y, z coordinates and the impulses p_x, p_y, p_z.

Any particle characterized by these six parameter values corresponds to a point in the six-dimensional phase space. Accordingly, the beam particles correspond to the population of a set of points in the phase space within a well defined phase volume. Since the particles form a conservative system, we can apply the Liouville theorem and this means that the phase volume populated by the points corresponding to the beam particles is constant. Thus, the phase volume representing the ion beam can be regarded as the motional constant of the system.

Now, if the direction of particle motion is normal to the plane of the x, y coordinates, the beam can be characterized at each point of the z axis

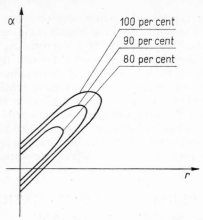

Fig. 6.9 Emittance diagram constructed from individual emittance diagrams of the beam fractions containing different fractions of the ion current extracted from ion source

by the phase volume given by the coordinates of the beam particles in the four-dimensional phase space specified by x, y, p_x and p_y.

Disregarding the constant factor, this phase volume is called emittance.

The beam is delimited in the x, y directions by the geometric cross section characterized by the values x_{max}, y_{max}, while in the p_x, p_y directions by the maximum angle of the beam path relative to the z-axis and characterized by the values $\alpha_{x(max)}$ and $\alpha_{y(max)}$ where $p_x = p_z \dfrac{dx}{dz} = p_z \operatorname{tg} \alpha_x \approx p_z \alpha_x$.

If we have a beam of cylindrical symmetry, it is sufficient to represent the beam in the two-dimensional phase space with parameters r, p_r where $p_r \approx p_z \alpha$.

Beams with ellipsoidal phase space are regarded as normal and the emittance can be described by the formula

$$\varepsilon = \frac{\oint p_r(r)dr}{\pi} = r\alpha \sqrt{2m_i eV} \ \ [\text{cm mrad (MeV)}^{1/2}]$$

The ion beam will be fully defined if, in addition to the phase volume, the percentual beam intensity is also given (Fig. 6.9).

The ion beam intensity has another parameter, called brilliance, which can be expressed as

$$B = \frac{IE}{\varepsilon^2} \left[\frac{\text{mA}}{\text{cm}^2}\right]$$

where I is the intensity of the ion beam, E is the ion energy and ε is the emittance.

The use of emittance as a beam parameter is advantageous because the volume populated by the beam particles in the phase space is not affected by the presence of electric or magnetic fields. The cross section and divergence of the beam can undergo any change while the phase volume remains the same.

The pair of parameter values $\alpha = \alpha(r)$ of the beam cross section can be determined at any point of the z-axis. The points defined by the pair of parameter values delimit a given area. The thus obtained graphical representation is the so-called emittance diagram. The calculated acceptance of the given ion-optical system (the phase volume of the beam that can be accepted by the optical system is called acceptance) is an ellipsoid given by $\alpha_0 r_0 \pi$. If the emittance area of the ion source is larger than the accelerator tube acceptance area represented in the same phase units, the beam cannot be transmitted through the tube without particle loss. On the other hand, if the emittance area of the source is smaller than the acceptance area of the accelerator tube, but cannot be accepted because of the beam shape, the beam can be adjusted with an appropriate ion-optical element to fit into the given accelerator tube [1223].

If the emittance of the beam leaving the source is known at a given point z_1 and the coupling matrix of the ion-optical system is evaluated in some other way, it is possible to evaluate the emittance after the ion-optical system at point z_2.

The determination of the emittance diagram of anion source is of essential importance in order to adjust the beam to the ion-optics of the apparatus which works with the ion source.

6.3.2 DETERMINATION OF THE EMITTANCE DIAGRAM OF THE ION SOURCE

Numerous methods are available for the determination of the emittance diagram [734, 735, 1213, 1214, 1216, 1218, 1220, 1224, 1227–1234, 1290, 1291].

The principle of each method is to single out elementary fractions of the ion beam by use of a diaphragm with holes or slits and to evaluate the angular deflection of the elementary fractions by a suitable detection procedure.

In most cases either static or dynamic detection technique is applied.

The principle of the static detection technique can be seen from the schematically drawn arrangement in Fig. 6.10. In the plane S_1 a diaphragm with holes or slits is placed in the path of the ion beam. The images of the ion beam fractions transmitted through the set of holes or slits are identified by the detectors placed in the plane S_2 at distance l from the plane S_1. The deflection angle α can be then evaluated from the sizes of aperture and image and the distance l between the planes S_1 and S_2 as

$\alpha \sim \dfrac{R_i - r_i}{l}$, where the symbols in Fig. 6.10 are used.

For the detection of the image in plane S_2 discs coated with ZnS layer [734, 735], or vacuum grease [1220], photographic paper [1229], photo or nuclear emulsions [1228], have been applied. These detectors show a discoloration under the bombarding ion beam and exhibit, with time, a permanent faded spot of an area corresponding to the cross section of the bombarding beam. The measure of discoloration is a function of the ion beam intensity and the irradiation time.

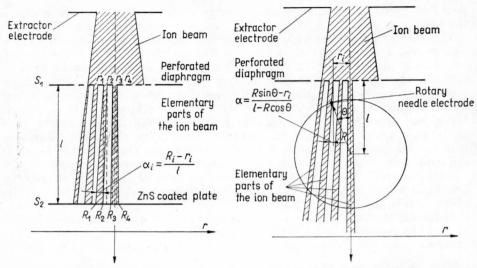

Fig. 6.10 Principle of detection method for determination of the emittance diagram of ion beams extracted from ion sources. The image of the beam appearing in plane S_2 can be detected with a ZnS coated aluminium disc

Fig. 6.11 Principle of a detection method for determination of the emittance diagram of ion beams extracted from ion sources. Parameters of beam fractions transmitted through the set of holes in plane S_1 are determined from the current measured with a needle-shaped electrode moving on a circular orbit of radius R

However, it has to be mentioned that with these detectors, the determination of the intensity distribution in the beam is rather inaccurate though the emittance for the total beam can be obtained to good accuracy.

In the dynamic detection method the detector, which is usually a Faraday

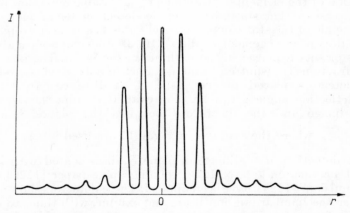

Fig. 6.12 The $I_i(r)$ diagram determined at the turn of 180° of the needle electrode moving on the circular orbit of radius R using detection method illustrated in Fig. 6.11

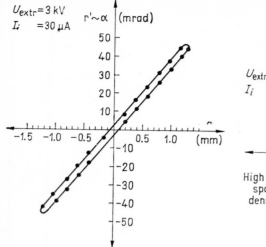

Fig. 6.13 Emittance diagram of ion beam extracted from low intensity high frequency ion current with diaphragm-type extraction system [734, 735]

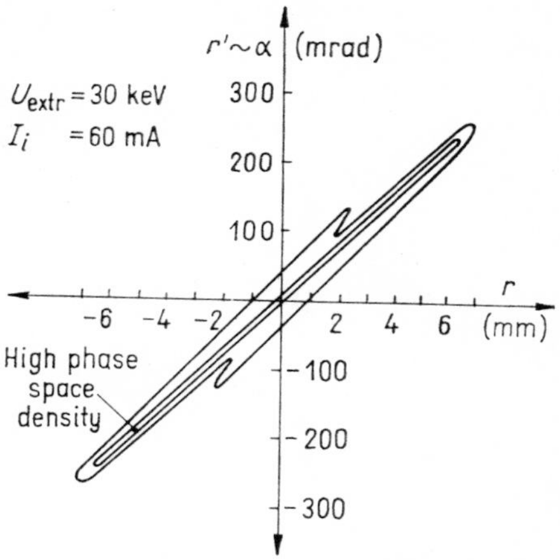

Fig. 6.14 Emittance diagram of ion beam extracted from a duoplasmatron-type ion source [1233]. The internal diagram corresponds to the higher intensity fraction of the higher intensity fraction of the ion beam

cylinder with a thin inlet slit, or a thin, needle-like electrode, is moved so as to cross the ion beam [1216, 1218, 1232, 1234].

The principle of the dynamic detection method can be seen from the schematic arrangement in Fig. 6.11. Here an electrode prepared from a thin tungsten filament placed normal to the direction of the ion beam is rotated on a circular orbit of radius R so that it crosses the ion beam fractions passing through the set of holes. The ion current striking the needle-like electrode can be registered with a high sensitivity line plotter (Fig. 6.12) and the angular deflections associated with the different coordinate values can be evaluated by elementary calculations. As is apparent from Fig. 6.11, the angular deflection α can be well approximated by the formula

$$\alpha \sim \frac{R \sin \Theta - r_i}{l - R \cos \Theta}$$

With this method of dynamic detection, as can also be seen from Fig. 6.12, the intensity distribution of the ion beam can be determined to a high degree of accuracy. Thus, the emittance diagram constructed from the data of dynamic measurement can also show the different percentual intensities of the ion beam in the manner seen in Fig. 6.9.

Figure 6.13 [734, 735] shows the emittance diagram of a high frequency ion source with diaphragm extraction, while Fig. 6.14 [1233] that of a duoplasmatron type ion source.

MEASUREMENT OF PARTICLE BEAM CURRENTS EXTRACTED FROM ATOM OR ION SOURCES

Numerous methods are available for the measurement of particle beam currents obtained from atom or ion sources.

Atom beam currents composed of neutral, non-charged particles can generally be assessed only from an indirect measurement. Ion beam currents which contain charged particles can be directly measured.

1. The current of beams consisting of neutral atoms is usually determined by using one of the following methods:

(a) high sensitivity manometry,
(b) surface ionization method,
(c) detection of conduction electron current induced by metastable atoms,
(d) condensation target method,
(e) chemical target method,
(f) calorimetric method.

(a) The method utilizing a high sensitivity manometer permits the number of atoms in the beam to be evaluated. For this purpose ionization or Pirani-type manometers are generally applied.

For either type of manometer the principle of the method is to introduce a beam of atoms flying in the direction ordered by collimation into the manometer, where the atoms start to move at random. These randomly moving atoms are prevented from leaving the manometer at the same rate as their input rate by the flow-obstructing slit of the manometer, thus the pressure in the manometer increases and exceeds that of the surrounding vacuum system.

Owing to the pressure increase in the manometer, eventually an equilibrium state sets in and the numbers of particles flying into and out from the detector become equal.

Taking a detector with an input slit of cross section A_d, at a distance L from the atom source, the number of input particles to the detector can be evaluated by the use of eq. (2.7) which gives the number of atoms obtainable from the atom source, as

$$N_b = \frac{1}{4\pi} \frac{A_d}{L^2} x n_s \bar{v}_s A_s$$

The number of atoms flying back through the flow resistant slit is formulated by the use of eq. (2.10) as

$$N_d = \frac{\xi}{4} x n_d \bar{v}_d A_d$$

The equality $N_b = N_d$ in the equilibrium state and the expression (2.3), which reads $p = nkT$, give for the equilibrium pressure p_d in the detector

$$p_d = \frac{kT_d}{\pi L^2 \xi \bar{v}_d} n_s \bar{v}_s A_s = \frac{4kT_d}{\xi x \bar{v}_d A_d} N_b$$

It is apparent from this equation that the equilibrium pressure formed in the detector is directly proportional to the input current of atoms to the detector and inversely proportional to the reduction factor ξ, the value of which can be calculated in terms of the geometry of the detector input slit with high flow resistance by use of eqs (2.11)–(2.16).

It has to be noted that the reduction factor ξ should not be chosen such that the outflow rate of particles is kept too low since in this case the heating filaments of both the ionization and the Pirani-type manometers would cause a change in the detector temperature T_d.

The particle intensity of the atom beam is evaluated from the difference between the detector pressures both in the presence and the absence of the atom beam. The difference between the two measured values of pressure is actually the equilibrium pressure of the manometer.

Using known values of p_d, the number of atoms flying out from the detector can be evaluated from the above equations.

In most of the cases an ionization manometer has been applied to evaluate the intensity of atom beams obtained from atom sources [433, 445, 1249–1251].

One of the arrangements using an ionization manometer, suitable for intensity measurement of atom beams, is schematically shown in Fig. A.1.

In the arrangement shown in the figure the beam is obtained from a high frequency dissociator atom source and shaped by the multi-collimator at the outlet of the source (see Paragraph 2.2.1.2). The beam is led through two collimator slits into the ionization manometer. The equilibrium pressure formed with or without exposure to the atom beam is measured by the manometer. The difference between the two pressure values permits the detector pressure p_d caused by the atom beam to be evaluated and this value is used for the calculation of the atom beam current.

This method is used above all for the determination of high atom beam intensities.

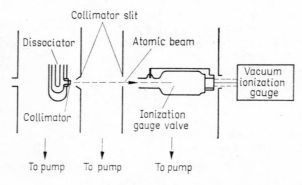

Fig. A.1

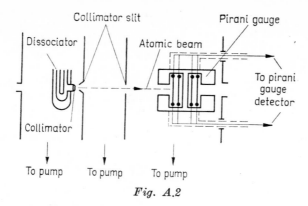

Fig. A.2

Pirani-type manometers are preferably used for the measurement of low currents of atom beams having a narrow, ribbon-like cross section [230, 334].

Figure A.2 shows a possible experimental arrangement for the measurement of atomic beam currents in which a Pirani-type manometer is applied.

The atom beam to be measured is taken through several collimating stages from the high frequency dissociator to the Pirani manometer (see Figs A.2 and A.3 [1252]). The atom beam intensity can be evaluated from the value of p_d obtained from the difference between the equilibrium pressures of the manometer in the presence and the absence of the atom beam — as in the case of ionization manometers.

(b) In atom beam detectors utilizing the surface ionization method, the surface ionization processes discussed in Paragraphs 1.2.5 and 3.3 are observed in order to evaluate the intensity of the neutral particle beam.

It has been shown in the paragraphs referred to that a fraction of the neutral atom beam striking a hot metal surface leaves the surface in the form of ions. The efficiency of the surface ionization is given by formula (3.4) as

$$\beta = \frac{n_i}{n_0} = \frac{A}{A + \exp\left[\dfrac{z_i(E^i - \varphi)}{kT}\right]}$$

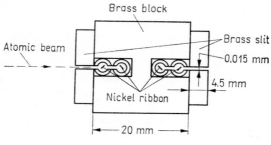

Fig. A.3

By use of this formula the number n_∂ of atoms incident on the hot metal surface can be evaluated from the data on the ionization energy E_i of the atoms in the beam, the work function φ of the ionizing metal surface and the detected ion current n_i, or from the efficiency β and the value of n_i [216, 1246, 1253–1255].

In this type of measurement care has to be taken of the surface purity of the metal used since even a small amount of impurity on the metal surface can strongly affect the value of the work function φ and thus that of the efficiency β.

(c) The current of metastable atoms can be evaluated from the measured current of electrons released from the cold metal surface struck by metastable atoms [418, 433, 445, 493, 495, 1256–1260].

The principle of this method is the following:

The metastable atoms excited to energy E_g transfer this excitation energy to the conduction electrons of the metal surface struck by the atom beam. If the work function of the metal surface for electrons

$$\varphi_e < E_g,$$

the conduction electrons leave the metal with an efficiency η.

The number N_e of electrons leaving the metal surface is proportional to the number N_{am} of metastable atoms incident on the surface, that is

$$N_e = \eta N_{am}$$

The measurement consists in counting the electrons released by the metastable atoms and the number N_{am} of incident metastable atoms can be evaluated if the efficiency η is known.

The purity of the cold metal surface is again of high importance as the surface layer formed from adsorbed gas atoms or molecules may considerably change the value of η. For this reason, it is advisable to degasing the metal surface by heating before the measurement.

(d) The most simple method for the detection of atom beam currents is the use of a condensation target [230, 334].

Thin, small-sized glass, quartz or metal plates kept at low temperature can be used as condensation detectors placed in the path of the atom beam. The atoms of the condensing solid element from which the beam has been obtained deposit in a stable form onto the plate [422].

However, stable deposition on the condensation detector can take place only up to a temperature limited both by the measured element and the material of the condensation target. It follows that the limit in temperature sets a limit also to the intensity of the measured atom beam as the condensation plate can reach a temperature exceeding the given limit only if the intensity of the investigated atom beam increases above a given value. In this case a fraction of atoms incident with the beam is scattered out of the detector. Owing to this scattering, the contour of the image of the atom beam becomes blurred and a fraction of atoms does not condense on the detector.

This detection method is suitable for the precise determination of the shape of the beam formed from the condensing atoms but the estimation of the atom beam intensity is difficult and inaccurate [1243, 1244].

(e) In the atom beam detectors utilizing chemical targets, the target material used reacts chemically with the element of the measured atom beam. Because of this chemical reaction, a change of colour is observable over a target area corresponding to the cross section of the atom beam.

For the detection of an oxygen atom beam a PbO target proved to be useful [422]. The target area struck by the atom beam changes from the bright yellow colour of PbO to brown due to the formation of PbO_2 in the reaction with the oxygen atoms of the beam.

For the detection of hydrogen atom beams, a bright yellow MoO_3 target can be used. In this case a blue MoO_2 spot appears on the target due to the reduction by atomic hydrogen over the area corresponding to the cross-section of the hydrogen atom beam [512–516, 1244].

Chemical targets are primarily useful for the determination of the shape of the atom beam, while the evaluation of the beam current by this method is difficult and inaccurate [230, 334].

(f) Calorimetry has proved to be useful mostly for the evaluation of fast atom or ion beam currents.

The principle of the method is to measure the increase in target temperature caused by collisions between target and beam particles [1260, 1261].

Assuming that the particles colliding with the target material lose their total energy and that this energy transforms to heat, then the heat generated by the number n of particles having energy $1/2 \, mv^2 = eV$, and incident per unit time is given by the formula

$$Q = n \, eV$$

If the heat capacity c of the target is known, and the change Δt^0 of the arget temperature is measured over time τ, then the beam current can be

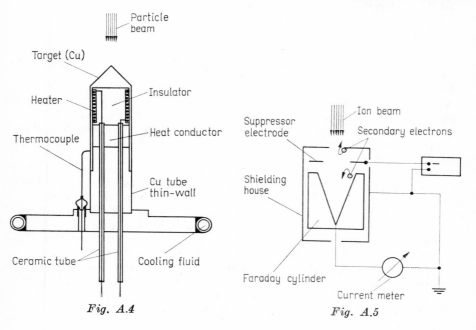

Fig. A.4   Fig. A.5

determined from the above expression and the equality 1 eV $= 3.827 \cdot 10^{-20}$ cal, since

$$Q = n \text{ eV } \tau = c \Delta t^0 \tau$$

An arrangement for particle intensity determination by calorimetry can be seen in Fig. A.4 [1261].

In the calorimeter shown in the figure the beam particles collide with a Cu target and the temperature of the target having heat capacity c is heated by the incident particles to Δt^0 C over the time τ. A thin-walled tube is the only heat conductor from the target and the opposite end of this tube is coupled to a water-cooled disc. The target temperature is measured with the thermocouple connected to the upper end of the thin-walled tube.

Two methods can be used for the calibration of the calorimeter. One of them is to generate a given quantity of heat Q with a heating spiral built into the target and to measure the change Δt of the target temperature. In the other method the target is bombarded for time τ with an electron beam of well defined energy and current in order to measure the temperature change Δt.

Particle intensity values obtained with calorimeters calibrated by either of the above methods can be of an accuracy from 1 to 2 per cent.

2. The intensity of beams composed of charged particles is usually determined by one of the following methods:

(a) Current measurement with Faraday cylinder,
(b) Current measurement with multiwire proportional or spark chamber,
(c) Measurement with secondary emission chamber,
(d) Measurement with electromagnetic detector,
(e) Fluorescent targets,
(f) Calorimetry for current measurement.

(a) Ion currents, particularly continuous beams, are usually measured by the use of a Faraday cylinder.

In current measurements with a Faraday cylinder the measured values may be uncertain because of contributions from secondary electron emission induced by ions striking the surface and the border of metal cylinders.

In order to avoid this effect, it is advisable to use a Faraday cylinder shielded in the manner schematically shown in Fig. A.5 and equipped with a diaphragm generating an opposite field.

It can be seen in the figure that the secondary electrons released both in the measuring cylinder and on the edges of the input diaphragm are driven back by the opposite field of the diaphragm and thus they do not contribute to the measured current, while the current measuring cylinder is protected from the ions scattered to its sides by the earthed shielding.

A detailed drawing of this Faraday cylinder protected from the effect of secondary electron emission is presented in Fig. A.6 [735].

A similar shielding and opposite field have to be provided if a movable Faraday cylinder with a thin input slit is used for the study of beam parameters, or if detectors with vibrating or rotating needle-like electrodes are used for current measurement (see Paragraph 6.3.2).

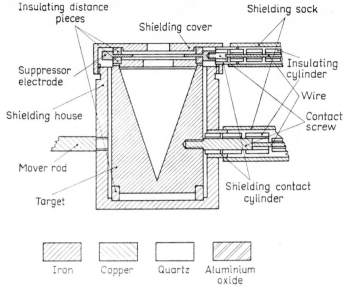

Fig. A.6

(b) The intensity, shape and position of ion beams can be determined to high accuracy by using multiwire proportional [1264–1266] or spark [1263] chambers.

The signal and high voltage electrodes are replaced in multiwire chambers by planes prepared from thin metal wires. The planes of signal wires, one of which is stretched in the vertical, the other in the horizontal direction, are mounted each between two high voltage electrodes.

The input pulses can be observed separately on each of the signal wires. The wires mounted in directions normal to each other permit the shape, intensity distribution and the geometry of the beam to be determined to an accuracy from 1 to 2 mm.

(c) Secondary emission chambers are also suitable for measuring ion beam currents [1267, 1268].

The principle of the ion current measurement is the following:

The ions of a beam striking a metal target cause the emission of secondary electrons from the metal surface. If the secondary electron emission coefficient δ_i of the metal surface is known (see Paragraph 1.8), the number n_i of incident ions can be evaluated from the measured secondary electron current by use of the formula

$$I_{se} = \delta_i n_i e$$

It should be noted that in this type of current measurement great care has to be taken of target surface purity and of the pressure of residual gases in the secondary electron emission chamber. This is necessary because the value of the secondary electron emission coefficient δ_i remains stable only if the metal surface is free from impurities and because a residual gas pres-

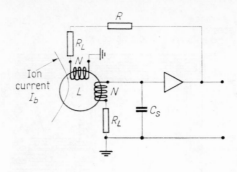

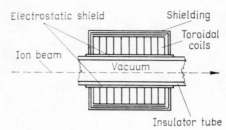

Fig. A.7

sure between 10^{-7} and 10^{-8} mmHg keeps the contribution from electrons due to the ionization of residual gas at the low values of between 2 and 3 per cent.

(d) The number of ions in ion beam pulses can also be evaluated from measurements made with electromagnetic [1289–1271, 1340] or electrostatic [1272] detectors.

The electromagnetic detector operates similarly to a transformer in which the ion beam acts as a primary coil with a single turn while the secondary coil has many turns with an output coupled to the input of a low resistance amplifier (Fig. A.7a).

Hereward-type transformers have been found to increase the sensitivity of electromagnetic detectors (Fig. A.7b) [1272, 1273].

The electromagnetic detector shown in the figure is composed of 10 toroids and the iron core is made of "Ultraperm" trade mark permalloy ($\mu = 120\,000$). The secondary coil, which generates the output pulses has 15 000 turns. An ion beam pulse of 200 μsec length, containing $n_i = 10^{12}$ protons, generates in the detector an output pulse of ~ 7.5 μV. This means that amplifiers having gains between 10^6 and 10^7 need to be applied.

(e) Fluorescent or luminescent materials or targets coated with such material are frequently used for observing the shape of ion beams. Generally, ZnS, $MgWO_4$ or quartz are the materials chosen for this purpose.

ZnS is mixed with some carrier lacquer and spread in this way on an Al foil to form a thin layer from which the lacquer is burnt out at the temperature used for activation while the ZnS coating remains on the metal surface.

A luminous greenish spot is observable on this target over an area corresponding to the cross section of the incident ion beam. ZnS coated targets are useful for the study of both low and high energy ion beams.

Targets made of quartz, which exhibit bluish luminous spots if struck by an ion beam, are less sensitive but of a longer useful lifetime than the ZnS coated detectors.

(f) The calorimetric detection method employed for the measurement of fast neutral atom beams can also be used for the measurement of ion beam currents.

However, in this case the ions have to be separated from the neutral particles admixed to the beam extracted from the ion source since the presence of neutral particles impairs the accuracy of the ion beam measurement.

Neutral particles can be separated from the charged ions by applying a magnetic or electrostatic field.

The isolated ion beam is introduced into the calorimeter which measures the change Δt of the temperature taking place over time τ. The intensity of the ion beam is then evaluated from the experimental data in the same way as in the case of neutral atom beams.

PHYSICAL QUANTITIES

Velocity of light $c = 2.9979 \cdot 10^{10}$ cm/sec

Electron charge $e = 4.8029 \cdot 10^{-10}$ abs.esu

Electron mass $m_e = 9.109 \cdot 10^{-28}$ g $= 0.51098$ MeV

Specific electron charge $\dfrac{e}{m_e} = 5.273 \cdot 10^{17}$ esu/g $= 1.759 \cdot 10^8$ Coulomb/g

Proton mass $m_p = 1.6726 \cdot 10^{-24}$ g $= 938.232$ MeV

Specific proton mass $\dfrac{e}{m_p} = 2.8716 \cdot 10^{14}$ esu/g

Proton to electron mass ratio $\dfrac{m_p}{m_e} = 1836.15$

Classical electron radius $r_0 = \dfrac{e^2}{m_e c^2} = 2.818 \cdot 10^{-13}$ cm

Radius of the electron orbit $a_0 = \dfrac{\hbar^2}{m_e c^2} = 5.2917 \cdot 10^{-9}$ cm

Electron velocity on the first Bohr electron orbit $v_0 = \dfrac{e^2}{\hbar} = 2.1877 \cdot 10^8$ cm/sec

Electron Compton wavelength $\lambda_e = 2.4265 \cdot 10^{-10}$ cm

$$\lambda_e = \frac{\lambda_e}{2\pi} = 3.863 \cdot 10^{-11} \text{ cm}$$

Borh magneton $\mu_0 = \dfrac{e\hbar}{2m_e c} = 0.9273 \cdot 10^{-20}$ erg $\cdot$ gauss^{-1}

Fine structure constant $\alpha = \dfrac{e^2}{\hbar c} = \dfrac{1}{137.039} = 7.2973 \cdot 10^{-3}$

Planck constant $h = 6.626 \cdot 10^{-27}$ erg $\cdot$ sec

$$\hbar = \frac{h}{2\pi} = 1.0546 \cdot 10^{-27} \text{ erg} \cdot \text{sec}$$

Rydberg constant $R = \dfrac{m_e e^4}{4\pi \hbar^3 c} = 109\ 737.3$ cm^{-1}

Boltzmann constant $k = 1.380 \cdot 10^{-16}$ erg $\cdot$ deg^{-1}

Avogadro number $N_0 = 6.022 \cdot 10^{23}$ (g $\cdot$ mol)$^{-1}$

Loschmidt number $L_0 = 2.687 \cdot 10^{19}$ cm^{-3}

1 eV $= 1.602 \cdot 10^{-12}$ erg $= 8066$ cm$^{-1} = 1.1605 \cdot 10^4\,^{\circ}$K

1 erg $= 6.2420 \cdot 10^{11}$ eV $= 5.0348$ cm$^{-1} = 7.2438 \cdot 10^{15}\,^{\circ}$K

1 cm$^{-1} = 1.2398 \cdot 10^{-4}$ eV $= 1.9862 \cdot 10^{-16}$ erg $= 1.4388\,^{\circ}$K

$1\,^{\circ}$K $= 8.6165 \cdot 10^{-5}$ eV $= 1.3804 \cdot 10^{-16} = 0.695$ cm^{-1}

$kT = 8.6165 \cdot 10^{-5}$ eV $= 1.3804 \cdot 10^{-16} = 0.695$ cm^{-1}

Thermal velocity (T, in °K)

$$\text{of electrons } v_e(T) = \sqrt{\frac{2kT}{m_e}} = 5.506 \cdot 10^5 \, T^{1/2} \text{ cm/sec}$$

$$\text{of protons } v_p(T) = \sqrt{\frac{2kT}{m_p}} = 1.29 \cdot 10^4 \, T^{1/2} \text{ cm/sec}$$

1 μA $= 6.2419 \cdot 10^{12}$ electrons/sec

1 Curie $= 3.7 \cdot 10^{10}$ decays/sec

1 mmHg $= 10^3 \, \mu$Hg $= 1333$ bar $= 1.315 \cdot 10^{-3}$ phys.atm. $= 1.359 \cdot 10^{-3}$ techn.atm.

1 bar $= 0.749 \cdot 10^{-3}$ mmHg $= 0.749 \cdot 10^{-6} \, \mu$Hg $= 9.86 \cdot 10^{-7}$ phys.atm $=$ $= 1.02 \cdot 10^{-6}$ techn.atm.

1 phys.atm. $= 760$ mmHg $= 760 \cdot 10^3 \, \mu$Hg $= 1.014 \cdot 10^6$ bar $= 1.0335$ techn.atm.

1 techn.atm. $= 1$ kg/cm$^2 = 735$ mmHg $= 735 \cdot 10^3 \, \mu$Hg $= 9.81 \cdot 10^5$ bar$=$ $= 0.9676$ phys.atm.

Composition of air [1239, 1240];

N$_2$	78.03 per cent	N$_3$	0.0012 per cent
O$_2$	20.99 per cent	He	0.0004 per cent
Ar	0.94 per cent	Ne	0.00161 per cent
CO$_2$	0.03 per cent	Kr	0.000108 per cent
H$_2$	0.01 per cent	Xe	0.000008 per cent

BOILING, MELTING AND SUBLIMATION TEMPERATURES
AT $p = 760$ mmHg OF A FEW MATERIALS OF INTEREST
[1240]

Melting ice	0°C			
Solid glycerine	−22°C	Liquid	O$_2$	−183°C
Liquid NH$_3$	−33.5°C	Liquid	Ar	−185°C
Solid CO$_2$	−78.5°C	Liquid	N$_2$	−195.7°C
Solid CO$_2$ + alcohol	−70.0°C	Liquid	H$_2$	−252.8°C
Solid CO$_2^+$ + acetone	−86.0°C	Liquid	He	−268.4°C

PERIODIC SYSTEM

1 1.008 H Hydrogen								

Atomic number ─ 2 4.003 ─ Atomic weight
Symbol ─ He Helium ─ Name

3 6.939 Li Lithium	4 9.012 Be Beryllium							
11 22.99 Na Sodium	12 24.31 Mg Magnesium							
19 39.10 K Potassium	20 40.08 Ca Calcium	21 44.96 Sc Scandium	22 47.90 Ti Titanium	23 50.94 V Vanadium	24 51.996 Cr Chromium	25 54.938 Mn Manganese	26 55.847 Fe Iron	27 58.933 Co Cobalt
37 85.47 Rb Rubidium	38 87.62 Sr Strontium	39 88.905 Y Yttrium	40 91.22 Zr Zirconium	41 92.91 Nb Niobium	42 95.94 Mo Molybdenum	43 99 Tc Technetium	44 101.1 Ru Ruthenium	45 102.91 Rh Rhodium
55 132.905 Cs Cesium	56 137.36 Ba Barium	57 Lanthanides	72 178.5 Hf Hafnium	73 180.95 Ta Tantalum	74 183.85 W Tungsten	75 186.2 Re Rhenium	76 190.2 Os Osmium	77 192.2 Ir Iridium
87 223 Fr Francium	88 226 Ra Radium	89 Actinides	104 260 Bo Bohrium					

Lanthanides	57 138.91 La Lanthanum	58 140.12 Ce Cerium	59 140.91 Pr Praseody-mium	60 144.24 Nd Neodymium	61 145 Pm Promethium	62 150.35 Sm Samarium	63 151.96 Eu Europium

64 157.25 Gd Gadolinium	65 158.9 Tb Terbium	66 162.5 Dy Dysprosium	67 164.9 Ho Holmium	68 167.3 Er Erbium	69 168.9 Tm Thulium	70 173.04 Yb Ytterbium	71 174.97 Lu Lutetium

OF ELEMENTS

								2 4.003 He Helium
			5 10.81 B Boron	6 12.011 C Carbon	7 14.007 N Nitrogen	8 15.999 O Oxygen	9 18.998 F Fluorine	10 20.183 Ne Neon
			13 26.98 Al Aluminium	14 28.09 Si Silicon	15 30.97 P Phosphorus	16 32.06 S Sulphur	17 35.453 Cl Chlorine	18 39.95 Ar Argon
28 58.71 Ni Nickel	29 63.54 Cu Copper	30 65.37 Zn Zinc	31 69.72 Ga Gallium	32 72.59 Ge Germanium	33 74.922 As Arsenic	34 78.96 Se Selenium	35 79.91 Br Bromine	36 83.80 Kr Krypton
46 106.4 Pd Palladium	47 107.87 Ag Silver	48 112.40 Cd Cadmium	49 114.82 In Indium	50 118.69 Sn Tin	51 121.75 Sb Antimony	52 127.60 Te Tellurium	53 126.90 I Iodine	54 131.30 Xe Xenon
78 195.1 Pt Platinum	79 196.97 Au Gold	80 200.6 Hg Mercury	81 204.37 Tl Thallium	82 207.2 Pb Lead	83 208.98 Bi Bismuth	84 210 Po Polonium	85 210 At Astatine	86 222 Em Radon

Actinides	89 227 Ac Actinium	90 232.04 Th Thorium	91 231 Pa Protactinium	92 238.03 U Uranium	93 237 Np Neptunium	94 244 Pu Plutonium	95 246 Am Americium

96 247 Cm Curium	97 247 Bk Berkelium	98 249 Cf Californium	99 254 E Einsteinium	100 253 Fm Fermium	101 256 Mv Mendele-vium	102 258 No Nobelium	103 257 Lw Lawrencium

TABLE 1

Low lying atomic levels and the corresponding values of term and energy [1245]

Atomic term	Term value (cm^{-1})	Energy value (eV)	Atomic term	Term value (cm^{-1})	Energy value (eV)
H ($^2S_{1/2}$)	0	0	$^2D_{3/2}$	19231.00	2.3842
$2^2P_{1/2}$	82258.91	10.19822	^2P	28840.00	3.5755
$2^2S_{1/2}$	82258.94	10.19823	$^4P_{1/2}$	83285.00	10.3254
$2^2P_{3/2}$	82259.27	10.19827	$^4P_{3/2}$	83319.00	10.3297
$3^2P_{1/2}$	97492.00	11.8388			
			O ($^3P_2$)	0	0
He ($^1S_0$)	0	0	$^3P_1$	158.50	0.0197
2^3S_1	159850.00	19.8177	$^3P_0$	226.50	0.0281
2^1S_0	166972.00	20.7001	$^1D_2$	15867.70	1.9671
2^3P	169081.00	20.9621	$^1S_0$	33792.40	4.1894
2^1P_1	171129.00	21.2160	^5S	73767.80	9.1455
3^3S_1	183231.00	22.7164	^3S	76794.70	9.5208
Li ($^2S_{1/2}$)	0	0			
$2^2P_{1/2}$	14903.00	1.8476	F ($^2P_{3/2}$)	0	0
$2^2P_{3/2}$	14904.00	1.8477	$^2P_{1/2}$	404.00	0.0501
$3^2S_{1/2}$	27206.00	3.3729	$^4P_{5/2}$	102406.50	12.6960
3^2P	30925.00	3.8340	$^4P_{3/2}$	102681.20	12.7301
3^2D	31283.00	3.8784	$^4P_{1/2}$	102841.20	12.7499
Be ($^1S_0$)	0	0	Ne ($^1S_0$)	0	0
$^3P_0$	21979.00	2.7249	($^2P_{1/2}$)^3S	134043.80	16.6182
$^3P_1$	21980.00	2.7250		134461.20	16.6700
$^3P_2$	21982.00	2.7253		134820.60	16.7146
$^1P_1$	42565.00	5.2771		135890.70	16.8473
$^3S_1$	52082.00	6.4570	($^2P_{1/2}$)^3p	148259.80	18.3807
B ($^2P_{1/2}$)	0	0		149659.00	18.5543
$^4P_{1/2}$	28805.00	3.5712		149826.20	18.5750
$^4P_{3/2}$	28810.00	3.5718			
$^4P_{5/2}$	28816.00	3.5725	Na ($^2S_{1/2}$)	0	0
$^2S_{1/2}$	40040.00	4.9640	$^2P_{1/2}$	16956.20	2.1021
			$^2P_{3/2}$	16973.40	2.1043
			$^2S_{1/2}$	25739.90	3.1910
C ($^3P_0$)	0	0	$^2D_{5/2}$	29172.86	3.61676
$^3P_1$	16.40		$^2D_{3/2}$	29172.91	3.61677
$^3P_2$	43.50				
$^1D_2$	10193.70	1.2630	Mg ($^1S_0$)	0	0
$^1S_0$	21648.40	2.6839	$^3P_0$	21850.40	2.7089
^5S	33735.20	4.1824	$^3P_1$	21870.40	2.7114
			$^3P_2$	21911.10	2.7165
N ($^4S_{3/2}$)	0	0	^1P	35051.40	4.3455
$^2D_{5/2}$	19223.00	2.3832	^3S	41197.40	5.1075

TABLE 1 (cont.)

Atomic term	Term value (cm^{-1})	Energy value (eV)	Atomic term	Term value (cm^{-1})	Energy value (eV)
Al ($^2P_{1/2}$)	0	0	$^3P_1$	15210.00	1.8857
$^2P_{3/2}$	112.00	0.0139	$^3P_2$	15316.00	1.8988
$^2S_{1/2}$	25347.70	3.1425	$^3D_1$	20335.30	2.5211
$^4P_{1/2}$	29020.30	3.5978	$^3D_2$	20349.20	2.5228
$^4P_{3/2}$	29066.90	3.6036	$^3D_3$	20370.90	2.5255
$^4P_{5/2}$	29142.70	3.6130	Ti ($^3F_2$)	0	0
Si ($^3P_0$)	0	0	$^3F_3$	170.10	0.0211
$^3P_1$	77.10	0.0096	$^3F_4$	216.70	0.0269
$^3P_2$	223.30	0.0277	$^5F_1$	6556.80	0.8129
$^1D_1$	6299.00	0.7809	$^5F_2$	6598.80	0.8181
$^1S_0$	15394.00	1.9085	$^5F_3$	6661.00	0.8258
P ($^4S_{3/2}$)	0	0	$^5F_4$	6742.80	0.8359
$^2D_{3/2}$	11361.70	1.4086	Cr ($^7S_3$)	0	0
$^2D_{5/2}$	11376.50	1.4104	$^5S_2$	7593.10	0.9414
$^2P_{1/2}$	18722.40	2.3212	$^5D_0$	7750.80	0.9609
$^2P_{3/2}$	18748.10	2.3243	$^5D_1$	7810.80	0.9684
Cl ($^2P_{3/2}$)	0		$^5D_2$	7927.50	0.9828
$^2P_{1/2}$	881.00	0.1092	$^5D_3$	8095.20	1.0036
$^4P_{5/2}$	71954.00	8.9207	Mn ($^6S_{5/2}$)	0	0
$^4P_{3/2}$	72484.20	8.9864	$^6D_{9/2}$	17052.30	2.1141
$^4P_{1/2}$	72822.60	9.0283	Fe ($^5D_4$)	0	0
Ar ($^1S_0$)	0	0	$^5D_3$	415.90	0.0516
($^2P_{1/2}$)4s	93143.80	11.5477	$^5D_2$	704.00	0.0873
	93750.60	11.6229	$^5D_1$	888.10	0.1101
	94553.70	11.7225	$^5D_0$	978.00	0.1213
	95399.90	11.8273	$^5F_5$	6928.30	0.8589
($^2P_{1/2}$)4p	104102.10	12.9062	$^5F_4$	7376.80	0.9146
	105462.80	13.0749	Co ($^4F_{9/2}$)	0	0
	106087.30	13.1523	$^4F_{7/2}$	816.00	0.1012
	106237.60	13.1710	$^4F_{5/2}$	1406.80	0.1744
K ($^1S_0$)	0	0	$^4F_{3/2}$	1809.30	0.2243
$^2P_{1/2}$	12985.20	1.6099			
$^2P_{3/2}$	13042.90	1.6170	Ni ($^3F_4$)	0	0
$^2S_{1/2}$	21026.80	2.6068	$^3F_3$	1332.00	0.1651
$^2D_{5/2}$	21534.40	2.6698	$^3F_2$	2216.50	0.2748
$^2D_{3/2}$	21536.70	2.6701	Cu ($^2S_{1/2}$)	0	0
Ca ($^1S_0$)	0	0	$^2D_{5/2}$	11202.50	1.3888
$^3P_0$	15157.90	1.8792	$^2D_{3/2}$	13245.40	1.6421

TABLE 1 (cont.)

Atomic term	Term value (cm^{-1})	Energy value (eV)
$^2P_{1/2}$	30525.30	3.7857
$^2P_{3/2}$	30783.70	3.8165
Zn $(^1S_0)$	0	0
$^3P_0$	32311.30	4.0058
$^3P_1$	32501.40	4.0294
$^3P_2$	32890.30	4.0776
Ge $(^3P_0)$	0	0
$^3P_1$	557.10	0.0691
$^3P_2$	1409.90	0.1748
$^1D_2$	7125.30	0.8834
$^1S_0$	16367.10	2.0291
As $(^4S_{3/2})$	0	0
$^2D_{3/2}$	10592.50	1.3132
$^2D_{5/2}$	10914.60	1.3532
$^2P_{1/2}$	18186.10	2.2546
$^2P_{3/2}$	18647.50	2.3119
Se $(^3P_2)$	0	0
$^3P_1$	1989.50	0.2467
$^3P_0$	2534.50	0.3142
$^1D_2$	9576.10	1.1872
$^1S_0$	22446.00	2.7828
Br $(^2P_{1/2})$	0	0
$^2P_{3/2}$	3685.00	0.4569
$^4P_{5/2}$	63429.80	7.8637
$^4P_{3/2}$	64900.50	8.0461
$^4P_{1/2}$	66877.20	8.2912
Kr $(^1S_0)$	0	0
$(^2P_{1/2})5s$	79972.50	9.9147
	80917.60	10.0319
	85192.40	10.5619
	85847.50	10.6431
$(^2P_{1/2})5p$	91169.30	11.3029
	92295.20	11.4424
	92308.20	11.4441
	92965.20	11.5255
Rb $(^2S_{1/2})$	0	0
$^2P_{1/2}$	12578.90	1.5595
$^2P_{3/2}$	12816.60	1.5890
$^2D_{5/2}$	19355.01	2.39957
$^2D_{3/2}$	19355.45	2.39963

Atomic term	Term value (cm^{-1})	Energy value (eV)
Sr $(^1S_0)$	0	0
$^3P_0$	14317.50	1.7750
$^3P_1$	14504.40	1.7982
$^3P_2$	14898.50	1.8471
Y $(^2D_{3/2})$	0	0
$^2D_{5/2}$	530.40	0.0657
$^2P_{1/2}$	10529.20	1.3054
$^2P_{3/2}$	11359.70	1.4083
Zr $(^3F_2)$	0	0
$^3F_3$	570.40	0.0707
$^3F_4$	1240.80	0.1538
$^3P_2$	4186.10	0.5190
$^3P_1$	4376.30	0.5425
Mo $(^7S_3)$	0	0
$^5S_2$	10768.30	1.3350
$^5D_0$	10966.00	1.3595
$^5D_1$	11142.80	1.3815
$^5D_2$	11454.40	1.4201
$^5D_3$	11858.50	1.4702
Ag $(^2S_{1/2})$	0	0
$^2P_{1/2}$	29552.00	3.6638
$^2D_{5/2}$	30242.70	3.7493
$^2P_{3/2}$	30472.30	3.7779
$^2D_{3/2}$	34714.20	4.3037
Cd $(^1S_0)$	0	0
$^3P_0$	30114.00	3.7334
$^3P_1$	30656.00	3.8006
$^3P_2$	31827.00	3.9458
In $(^2P_{1/2})$	0	0
$^2P_{3/2}$	2212.60	0.2743
$^2S_{1/2}$	24372.80	3.0216
Sn $(^3P_0)$	0	0
$^3P_1$	1691.80	0.2097
$^3P_2$	3427.70	0.4250
Te $(^3P_2)$	0	0
$^3P_1$	4751.00	0.5890
$^3P_2$	4757.00	0.5898
I $(^2P_{3/2})$	0	0
$^2P_{1/2}$	7603.20	0.9426
$^4P_{5/2}$	54633.50	6.7733

TABLE 1 (cont.)

Atomic term	Term value (cm^{-1})	Energy value (eV)	Atomic term	Term value (cm^{-1})	Energy value (eV)
Xe ($^1S_0$)	0	0	Pt ($^3P_3$)	0	0
($^2P_{1/2}$)6s	67068.00	8.3149	$^3D_2$	775.90	0.0962
	68045.70	8.4361	$^3D_1$	823.30	0.1021
($^2P_{1/2}$)6p	77269.60	9.5797			
	78120.30	9.6851	Au ($^2S_{1/2}$)	0	0
			$^2D_{5/2}$	9161.30	1.1358
Cs ($^2S_{1/2}$)	0	0	$^2D_{3/2}$	21435.30	2.6575
$^2P_{1/2}$	11178.20	1.3858			
$^2P_{3/2}$	11732.40	1.4545	Hg ($^1S_0$)	0	0
$^2D_{3/2}$	14499.50	1.7976	$^3P_0$	37645.00	4.6671
$^2D_{5/2}$	14597.00	1.8097	$^3P_1$	39412.00	4.8862
Ba ($^1S_0$)	0	0	$^3P_2$	44043.00	5.4603
$^3D_1$	9034.00	1.1200			
$^3D_2$	9215.50	1.1425	Pb ($^3P_0$)	0	0
$^3D_3$	9596.50	1.1897	$^3P_1$	7819.30	0.9694
			$^3P_2$	10650.50	1.3204
W ($^5D_0$)	0	0			
$^5D_1$	1670.30	0.2071	Bi ($^4S_{3/2}$)	0	0
$^5D_2$	3325.30	0.4123	$^2D_{3/2}$	11419.00	1.4157
$^5D_3$	4830.00	0.5988	$^2D_{5/2}$	15437.70	1.9139

TABLE 2

Metastable excited states [1245]

Atom	Term	Excitation potential (eV)	Atom	Term	Excitation potential (eV)
H	$^2S_{1/2}$	11.80	Zn	$^3P_0$	4.00
He	$^3S_1$	19.80	Ce	$^1D_2$	0.88
	$^1S_0$	20.61	As	$^2D_{3/2}$	1.31
C	$^1D_2$	1.26	Kr	$^2P_{1/2}$	9.91
N	$^2D_{5/2}$	2.38	Sr	$^3P_0$	1.77
O	$^1D_2$	1.97	Zr	$^3P_0$	0.52
Ne	$^2P_{1/2}$	16.62	Mo	$^5S_2$	1.34
Mg	$^3P_0$	2.71	Cd	$^3P_0$	3.73
Si	$^1D_1$	0.78	Sn	$^1D_2$	1.07
P	$^2D_{3/2}$	1.41	Sb	$^2D_{3/2}$	1.06
Ar	$^2P_{1/2}$	11.55	Te	$^1D_2$	1.31
Ca	$^3P_0$	1.88	Xe	$^2P_{1/2}$	8.32
Ti	$^5F_1$	0.81	Ba	$^3D_2$	1.13
V	$^6D_{3/2}$	0.26	W	$^5D_2$	0.37
Cr	$^5S_2$	0.94	Pt	$^3D_1$	0.10
Fe	$^5F_5$	0.85	Au	$^2D_{5/2}$	1.14
Co	$^4F_{9/2}$	0.43	Hg	$^3P_0$	4.67
Ni	$^3D_2$	0.42	Pb	$^1D_2$	2.66
Cu	$^2D_{5/2}$	1.38	Bi	$^2D_{3/2}$	1.42

374

APPENDIX

TABLE 3

The first ionization potentials and the corresponding values of λ_i [1245–1248]

Atom or molecule	E_i' (eV)	cm^{-1}	Å	Atom or molecule	E_i' (eV)	cm^{-1}	Å
H	13.597	109 678.758	911.96	Mo	7.099	57 260.00	1746.7
D	13.601	109 708.596	911.69	Ag	7.576	61 106.50	1636.7
T	13.603	109 718.526	911.56	Cd	8.993	72 538.80	1378.8
He	24.586	198 310.800	504.35	In	5.786	46 669.93	2143.1
Li	5.391	43 487.190	2300.10	Sn	7.343	59 231.80	1688.6
Be	9.322	75 192.290	1330.10	Sb	7.848	63 300.00	1580.0
B	8.297	66 930.000	1537.30	Te	9.009	72 667.00	1376.4
C	11.269	90 820.420	1100.36	I	10.456	84 340.00	1185.9
N	14.548	117 345.000	852.35	Xe	12.129	97 834.00	1022.3
O	13.617	109 837.030	910.62.	Cs	3.894	31 406.45	3184.3
F	17.422	140 524.500	711.74	Ba	5.211	42 032.40	2379.5
Ne	21.564	173 931.700	575.03	W	7.984	64 400.00	1553.1
Na	5.139	41 449.650	2412.90	Pt	8.964	72 300.00	1383.3
Mg	7.646	61 671.020	1621.70	Au	9.225	74 410.00	1344.1
Al	5.985	48 279.160	2071.80	Hg	10.437	84 184.00	1188.6
Si	8.151	65 747.500	1521.20	Pb	7.415	59 809.00	1672.2
P	10.484	84 531.000	1183.20	Bi	7.287	58 778.00	1701.6
S	10.359	83 559.300	1197.00	U	6.2		2000.0
Cl	13.017	107 995.460	952.60	H$_2$	15.42		804.2
Ar	15.759	127 109.900	786.85	HBr	11.7		1059.8
K	4.340	35 009.78	2857.1	HCl	12.8		968.7
Ca	6.113	49 304.80	2028.4	HF	15.77		786.3
Sc	6.561	52 920.00	1889.9	HI	10.4		1192.3
Ti	6.836	55 138.00	1813.9	HN$_3$	10.3		1203.9
V	6.739	54 361.00	1840.0	HO$_2$	11.5		1078.3
Cr	6.765	54 565.00	1832.9	H$_2$O	12.6		984.1
Mn	7.434	59 960.00	1668.0	LiI	8.5		1458.8
Fe	7.897	63 700.00	1570.2	BH	10.1		1227.7
Co	7.865	63 438.00	1576.6	BH$_2$	8.1		1530.8
Ni	7.634	61 579.00	1624.3	C$_2$	12.0		1033.3
Cu	7.726	62 317.20	1604.9	CBr	11.43		1084.7
Zn	9.393	75 766.80	1320.1	CCl	12.9		961.2
Ga	5.998	48 380.00	2067.3	CClF	13.1		946.6
Ge	7.885	63 600.00	1572.6	CF	13.8		898.6
As	9.815	79 165.00	1263.3	CF$_2$	13.3		932.3
Se	9.752	78 658.20	1271.5	CF$_3$Cl	12.8		968.8
Br	11.846	95 550.00	1046.8	CF$_3$I	10.2		1215.7
Kr	13.999	112 915.20	885.8	CH	11.1		1117.1
Rb	4.177	33 691.00	2968.6	CH$_2$	11.8		1050.8
Sr	5.694	45 925.00	2177.7	CH$_3$	9.86		1257.6
Y	6.527	52 650.00	1899.8	C$_2$H	11.3		1097.3
Zr	6.952	56 077.00	1783.6	C$_2$H$_2$	11.4		1087.7

TABLE 3 (cont.)

Atom or molecule	E'_i (eV)	cm^{-1}	Å	Atom or molecule	E'_i (eV)	cm^{-1}	Å
H_2H_3	9.45		1312.2	F_2O	13.7		905.1
CN	15.1		821.2	NaI	8.8		1409.1
C_2N	12.8		968.7	Al_2O	7.7		1610.3
CO	14.01		885.1	$SiCl_4$	11.6		1069.0
CO_2	13.8		898.5	SiF	7.3		1698.6
CS	10.7		1158.9	SiO	10.8		1148.1
N_2	15.6		794.9	P_2	11.8		1050.8
NF	12.0		1033.3	S_2	8.3		1493.9
NH	13.1		946.6	SO_2	12.34		1004.1
NH_2	10.4		1192.3	Cl_2	11.5		1078.3
NO	9.25		1340.5	ClF_3	13.0		953.8
N_2O	12.7		976.4	ClO	10.4		1192.3
NO_2	9.8		1265.3	KI	8.3		1493.9
O_2	12.2		1016.4	Br_2	12.0		1033.3
O_3	11.7		1059.8	BrCl	11.1		1117.1
OH	13.2		939.4	I_2	9.35		1326.2
F_2	15.8		784.8	IBr	10.3		1203.9
FO	13.0		953.8	ICl	10.4		1192.3

TABLE 4

Values of the ionization energy $E_i^{(n+i)} - E_i^n = E$ of atoms and ions in eV
[1245–1247]

Atom	Degree of ionization (n)					
	1	2	3	4	5	6
H	13.597					
He	24.586	54.414				
Li	5.391	75.635	122.446			
Be	9.322	18.206	153.850	217.657		
B	8.297	25.149	37.920	259.298	340.127	
C	11.259	24.381	47.881	64.489	392.067	489.946
N	14.548	29.611	47.436	77.466	97.881	552.04
O	13.617	35.146	54.934	77.394	113.873	138.080
F	17.422	34.985	62.659	87.157	114.237	157.151
Ne	21.564	41.079	63.742	97.044	126.287	157.94
Na	5.139	47.30	71.66	98.902	138.627	172.36
Mg	7.646	15.034	80.134	109.318	141.262	186.898
Al	5.985	18.827	28.447	119.983	153.806	190.466
Si	8.151	16.342	33.466	45.140	166.762	205.157
P	10.484	19.72	30.156	51.354	65.007	
S	10.359	23.4	35.0	47.29	72.5	
Cl	13.017	23.80	39.912	53.462	67.81	96.70
Ar	15.759	27.628	40.908	59.806	75.02	91.32
K	4.340	31.817	45.747	60.909	82.6	90.741
Ca	6.113	11.870	51.218	67.196	84.39	109,0
Sc	6.561	12.80	24.75	73.9	92.0	
Ti	6.836	13.637	28.143	43.245	98.8	119.762
V	6.739	14.65	29.31	48.0	65.0	
Cr	6.765	16.49	30.95	50.0	73.0	
Mn	7.434	15.636	33.69	53.0	76.0	
Fe	7.897	16.182	30.647	57.1	78.0	102.0
Co	7.865	17.05	33.49			
Ni	7.634	18.15	35.16			
Cu	7.726	20.291	36.834			
Zn	9.393	17.96	39.70			
Ga	5.998	20.51	30.70	64.2		
Ge	7.885	15.93	34.21	45.7	93.4	
As	9.815	18.63	28.34	50.1	62.6	
Se	9.752	21.5	32.0	43.0	68.0	
Br	11.846	21.6	35.9	47.3	59.7	
Kr	13.999	24.57	36.947			
Rb	4.177	27.5	40.0			
Sr	5.694	11.027	42.8			
Y	6.527	12.23	20.5			
Zr	6.952	13.13	22.98	33.33		

TABLE 4 (cont.)

Atom	Degree of ionization (n)					
	1	2	3	4	5	6
Nb	6.88	14.0	25.04	38.3	50.0	
Mo	7.099	16.15	27.13	46.4	61.2	
Tc	7.28	15.26				
Ru	7.364	16.76	28.46			
Rh	7.45	18.07	31.05			
Pd	8.33	19.42	32.92			
Ag	7.576	21.48	34.82			
Cd	8.993	16.904	37.47			
In	5.786	18.86	28.03	54.4		
Sn	7.343	14.628	30.49	40.72	72.3	
Sb	7.848	16.5	25.3	44.1	56.0	
Te	9.009	18.6	31.0	38.0	60.0	
I	10.456	19.09				
Xe	12.129	21.208	32.121			
Cs	3.894	25.076	34.0			
Ba	5.211	10.001	35.5			
La	5.61	11.06	19.17			
Ce	5.57	10.85	19.70	36.715		
Pr	5.42	10.55				
Nd	5.45	10.73				
Pm	5.55	10.90				
Sm	5.6	11.07				
Eu	5.64	11.25				
Gd	6.16	12.1				
Tb	5.98	11.52				
Dy	5.93	11.67				
Ho	6.02	11.80				
Er	6.10	11.93				
Tu	6.18	12.05				
Yb	6.22	12.17				
Lu	6.15	13.9				
Hf	6.8	14.9				
Ta	7.88	16.2				
W	7.984	17.7				
Re	7.87	16.6				
Os	8.73	17.0				
Ir	9					
Pt	8.964	18.56				
Au	9.225	20.5				
Hg	10.437	18.751	34.2			
Tl	6.106	20.42	29.8	50.7		
Pb	7.415	15.028	31.93	42.31	68.8	

APPENDIX

TABLE 4 (cont.)

Atom	Degree of ionization (n)					
	1	2	3	4	5	6
Bi	7.287	16.68	25.56	43.3	56.0	
Po	8.43					
Rn	10.746					
Ra	5.277	10.144				
Ac	6.9	12.1	20.0			
Th	6.2		20.2	28.6		
U	6.2					
Am	6.0					

TABLE 5

Experimental and theoretical values of the multiple ionization cross section for He, Ne, Ar and Hg

Process	Maximum cross section σ_n cm^2	
	Experimental [103, 111, 112, 118]	Calculated from expression (1.30)
$He + e \rightarrow He^{n+} + (n+1)e$	σ_1 3.6 · 10^{-17}	
	σ_2 1.4 · 10^{-19}	1.14 · 10^{-19}
$Ne + e \rightarrow Ne^{n+} + (n+1)e$	σ_1 7.9 · 10^{-17}	
	σ_2 4.5 · 10^{-18}	1.6 · 10^{-18}
	σ_3 3.4 · 10^{-19}	3.25 · 10^{-19}
$Ar + e \rightarrow Ar^{n+} + (n+1)e$	σ_1 3.5 · 10^{-16}	
	σ_2 3.0 · 10^{-17}	1.3 · 10^{-17}
	σ_3 1.0 · 10^{-18}	3.3 · 10^{-18}
	σ_4 1.5 · 10^{-19}	1.0 · 10^{-18}
$Hg + e \rightarrow Hg^{n+} + (n+1)e$	σ_1 6.0 · 10^{-16}	
	σ_2 8.0 · 10^{-17}	6.0 · 10^{-17}
	σ_3 2.0 · 10^{-17}	2.5 · 10^{-17}
	σ_4 4.0 · 10^{-18}	1.2 · 10^{-17}
	σ_5 1.4 · 10^{-18}	6.3 · 10^{-18}

TABLE 6

Ionization cross section values for metastable atom impact

Process	$\sigma \; [\pi a_0^2]$	References
$He \; (2^3S) + Ne \rightarrow He + Ne^+$	0.32	[141]
$He \; (2^3S) + Ar \rightarrow He + Ar^+$	1.06	[136]
	1.1	[137]
	0.34	[138]
	11.37	[139]
	8.64	[140]
	7.5	[141]
$He \; (2^3S) + Kr \rightarrow He + Kr^+$	10.23	[140]
	11.71	[141]
$He \; (2^3S) + Xe \rightarrow He + Xe^+$	11.37	[139]
	13.65	[140]
	15.81	[141]
$He \; (2^3S) + H_2 \rightarrow He + H_2^+$	2.96	[140]
	6.82	[141]
$He \; (2^3S) + N_2 \rightarrow He + N_2^+$	7.28	[141]
	7.96	[140]
$He \; (2^3S) + O_2 \rightarrow He + O_2^+$	15.92	[140]
$He \; (2^3S) + CO \rightarrow He + CO^+$	7.96	[140]
$He \; (2^3S) + Hg \rightarrow He + Hg^+$	159.19	[137]
$He \; (2^1S) + Ne \rightarrow He + Ne^+$	4.66	[141]
$He \; (2^1S) + Ar \rightarrow He + Ar^+$	8.53	[140]
	62.54	[141]
$He \; (2^1S) + Kr \rightarrow He + Kr^+$	10.23	[140]
	72.77	[141]
$He \; (2^1S) + Xe \rightarrow He + Xe^+$	13.65	[140]
	117.12	[141]
$He \; (2^1S) + H_2 \rightarrow He + H_2^+$	1.93	[140]
$He \; (2^1S) + N_2 \rightarrow He + N_2^+$	7.96	[140]
$He \; (2^1S) + O_2 \rightarrow He + O_2^+$	15.92	[140]
$He \; (2^1S) + CO \rightarrow He + CO^+$	7.96	[140]
$Ne \; (^3P_2) + Ar \rightarrow Ne + Ar^+$	2.27	[138]
	2.96	[137]
	3.3	[136]
$Ar \; (^3P_2) + Kr \rightarrow Ar + Kr^+$	1.14	[138]
$Ar \; (^3P_2) + Hg \rightarrow Ar + Hg^+$	34.11	[138]
$Ar \; (^3P_2) + H_2 \rightarrow Ar + H_2^+$	0.34	[138]

TABLE 7

State of autoionization

State of auto-ionization	Excitation energy (eV)	
	calculated [175]	measured [176]
Li $(1s2s2p)^4P$	55.47	56 ± 1
Li $(1s2p^2)^4P$	59.67	60
Li $(1s2s3s)^4S$	61.11	
Li $(1s2s3p)^4P$	61.75	—
Li $(1s2s3d)^4D$	61.80	—
Li $(1s2p3s)^4P$	63.17	—
Li $(1s2p3p)^4P$	63.99	—
Li $(1s2p3d)$	64.40	—
Li $(1s2sns)^4S$	—	65

TABLE 8

Autoionization states of H^- *ions* [198]

State	Energy (eV)	State	Energy (eV)
^1S	9.559	^1P	10.178
	10.1668		10.2029
	10.2016		10.2037
^3S	10.149	^3P	9.727
	10.2016		10.1875
	10.2036		10.2032

TABLE 9

Excitation energies and level widths of autoionization states in He *atoms*

State	Excitation energy (eV)		Level width (eV)	
	Experimental [191, 192]	Calculated [194, 195]	Experimental [191, 192]	Calculated [194, 195]
¹S	57.86	58.0	0.141	0.23
	62.91	62.75	0.061	0.0037
	64.18	—	0.0194	—
	64.67	—	0.0092	—
¹P	60.26	60.35	0.043	0.025
	62.77	—	—	—
	63.87	—	0.0087	—
	64.19	—	0.0037	—
	64.42	—	0.0019	—
³P	58.35	58.4	0.0105	0.033
	63.13	59.8	0.0027	—
	63.28	—	—	—
	64.12	—	—	—
	64.25	—	0.0012	—
	64.70	—	0.0006	—

TABLE 10

Binding energies of electrons in negative ions

Negative ion	Experimental	References	Calculated	References	Semi-empirical and isoelectronic extrapolation	References
H⁻ (^1S)	0.77 ±0.02	[270]	0.75406	[236]		
H⁻ (^1S)	0.8 ±0.1	[258]		[274]		
He⁻ (^4P)	0.06 ±0.005	[281]	0.075			
Li⁻			0.616	[268]	0.76	[217]
			0.6	[237]		
Be⁻					0.19	[246]
B⁻ (^3P)					0.30±0.06	[238–240]
					0.33	[246, 247]
C⁻ (^4S)	1.25 ±0.03	[251]			1.24, 1.21	[245, 246]
					1.17±0.06	[238–240]
					1.38, 1.33	[247, 299]
N⁻	0	[283]			0.05, 0.15	[246, 298]
O⁻ (^2P)	1.465±0.005	[301]			1.47	[245, 246]
	1.461	[295]			1.46	[247]
F⁻ (^1S)	3.448±0.005	[293]	3.48	[259, 260]	1.22±0.14	[238–240]
	3.48 ±0.17	[253–255]			3.45	[299]
	3.56 ±0.1	[257]			3.47, 3.50	[246, 247]
	3.39	[288]			3.50, 3.62	[245, 246]
					3.37±0.08	[238–240]
Na⁻ (^1S)	0.35	[280, 281]	0.84	[241]	0.47	[246]
Mg⁻	not measured	[282]			0.78±0.06	[238–240]
					−0.32	[246]
Al⁻ (^3P)					0.49±0.14	[238–240]
					0.52	[246]
Si⁻					1.36, 1.39	[238–240, 247]

Ion	Value	Ref	Value	Ref	Value	Ref
P⁻					1.40, 1.46	[246, 299]
					0.72, 0.77	[246, 247]
					0.78	[238, 240]
					2.08, 2.03	[247, 299]
					2.12, 2.15	[238–240, 246]
					3.61	[299]
					3.56	[238–240]
S⁻ (^2P)	2.07 ±0.07	[249]				
	2.14 ±0.05	[296]				
Cl⁻ (^1S)	3.613±0.003	[293]	3.68	[259, 260]		
	3.62	[288]				
		[253–255]				
	3.62 ±0.04	[257]				
	3.75 ±0.09	[280]				
K⁻ (^1S)	0.30	[282]			3.69, 3.70, 3.84	[245, 246, 247]
Ca⁻	not measured					
Sc⁻			not measured	[240]		
Ti⁻ (^4F)			0.39±0.2	[240]		
V⁻					0.65	[248]
Cr⁻ (^6S)			0.98±0.35	[240]		
Mn⁻			not measured	[240]		
Fe⁻ (^4F)			0.58±0.2	[240]		
Co⁻ (^3F)			0.94±0.15	[240]		
Ni⁻ (^2D)			1.28±0.2	[240]		
Cu⁻ (^1S)	1.5 ±0.5	[285]	1.80±0.1	[240]		
As⁻	< 2	[288]				
Se⁻	~ 2	[288]	3.7 ±2	[286]		
Br⁻ (^1S)	3.363±0.003	[293]	3.40±0.04	[259, 260]	3.50	[302]
		[253–255]				
	3.49 ±0.02	[257]				
	3.50 ±0.07	[280]				
Rb⁻ (^1S)	0.27	[287]				
	0.2					
Y⁻					0.3	[248]
Zr⁻					1.0	[248]
Nb⁻					1.3	
Mo⁻					1.3	
Tc⁻					1.0	

TABLE 10 (cont.)

Negative ion	Experimental	References	Calculated	References	Semi-empirical and isoelectronic extrapolation	References
Ru⁻					1.4	
Rh⁻					1.3	
Pd⁻					1.4	
Ag⁻	2.0 ±0.2	[285]				
Sb⁻	2	[288]				
	2	[289]				
Te⁻	2	[288]	3.6 ±1.7	[286]		
	2.3	[289]				
I⁻	3.063±0.003	[293]	3.14	[259, 260]		
	3.076±0.005	[294]				
	3.13 ±0.12	[303]				
	3.13 ±0.07	[253–255]				
	3.17 ±0.05	[291]				
	3.17 ±0.08	[233]				
	3.23	[302]				
Cs⁻ (¹S)	0.23	[280]				
Au⁻ (¹S)	2.8 ±0.1	[295]				
Au⁻ (¹S)	1					
Tl⁻			2.1	[290]		
Bi⁻	0.7	[289]				

TABLE 11

Molecular dissociation energies [1246–1248]

Molecule	Dissociation energy		Molecule	Dissociation energy	
	(eV)	(Kcal/mol)		(eV)	(Kcal/mol)
Ag_2	1.65	38	BiF	3.21	74
AgAu	2.08	48	BiH	2.6	60
AgBr	3.04	70	BiO	3.69	85
AgCl	3.29	76	Br_2	1.97	46
AgH	2.3	53	BrCl	2.23	52
AgI	2.6	60	BrF	2.34	54
AgO	2.47	57	BrO	2.4	55
Al_2	1.95	45	C_2	4.9	113
AlBr	4.55	105	CCl	3.47	80
AlCl	5.03	116	CF	4.68	108
AlF	6.76	156	CH	3.47	80
AlH	2.91	67	CN	8.15	188
AlI	3.77	87	CO	11.14	257
AlN	3.04	70	CP	5.98	138
AlO	4.99	115	Ca_2	0.23	5.5
As_2	3.94	91	CaBr	2.86	66
Au_2	2.25	52	CaCl	2.78	64
AuCl	3.04	70	CaF	3.17	73
AuCr	2.17	50	CaH	1.69	39
AuCu	2.34	54	CaI	2.78	64
AuH	3.08	71	CaO	4.87	115
AuPd	1.43	33	Cd_2	0.09	2
B_2	3.12	72	Cl_2	2.48	57.1
BBr	4.51	104	ClF	2.56	59
BCl	5.12	118	Cr_2	1.73	40
BF	7.59	175	CrO	4.38	101
BH	3.38	78	Cs_2	0.45	10.4
BO	7.98	184	CsBr	4.34	100
BaBr	2.78	64	CsCl	4.42	102
BaCl	2.69	62	CsF	4.99	115
BaF	3.77	87	CsH	1.91	44
BaH	1.78	41	CsI	3.64	84
BaO	4.64	107	Cu_2	1.99	46
Be_2	0.69	16	CuBr	3.38	78
BeCl	4.77	110	CuCl	3.69	85
BeF	6.93	160	CuF	3.04	70
BeH	2.3	53	CuH	2.86	66
BeO	4.59	106	CuI	2.99	69
Bi_2	2.04	47	CuO	4.78	110
BiCl	3.04	70	F_2	1.65	38

TABLE 11 (cont.)

Molecule	Dissociation energy		Molecule	Dissociation energy	
	(eV)	(Kcal/mol)		(eV)	(Kcal/mol)
FO	1.91	44	LiCl	4.86	112
FeCl	3.6	83	LiF	5.94	137
FeO	4.25	98	LiH	2.43	56.1
Ga_2	1.52	35	LiI	3.64	84
Ge_2	2.76	64	LiO	3.38	78
GeBr	2.60	60	Mg_2	0.31	7.2
GeC	4.73	109	MgO	4.34	100
GeCl	3.51	81	MnO	4.16	96
GeF	4.94	114	MoO	5.20	120
GeH	3.30	76	N_2	9.76	225
GeO	6.89	159	NH	3.6	83
GeSi	3.08	71	NO	6.59	152
H_2	4.48	103.26	Na_2	0.75	17.3
HD	4.51	104.08	NaBr	3.82	88
D_2	4.56	105.06	NaCl	4.25	98
HT	4.53	104.38	NaF	4.94	114
DT	4.57	105.44	NaH	2.04	47
T_2	4.59	105.86	NaI	3.06	70.6
HBr	3.73	86.7	NiBr	?.69	85
HCl	4.43	102.2	NiCl	3.82	88
HF	5.86	135.1	NiH	2.6	60
HI	3.05	70.4	NiI	3.04	70
Hg_2	0.06	1.4	NiO	4.21	97
HgBr	0.71	16.4	O_2	5.12	117.9
HgCl	0.99	23	OH	4.4	101.4
HgF	1.73	40	P_2	5.03	116
HgH	0.37	8.6	PCl	2.95	68
HgI	0.36	8.2	PF	4.55	105
In_2	0.95	22	PH	3.04	70
I_2	1.54	35.57	PO	6.11	141
IBr	1.82	41.92	Pb_2	0.99	23
ICl	2.16	49.8	PbBr	2.47	57
IF	2.86	66	PbCl	3.04	70
IO	1.91	44	PbF	3.21	74
K_2	0.51	11.8	PbH	1.78	41
KBr	3.95	91	PbI	1.99	46
KCl	4.38	101	PbO	4.06	93.6
KF	5.03	117	Po_2	1.56	36
KH	1.86	43	Rb_2	0.47	10.8
KO	3.34	77	RbBr	3.99	92
Li_2	1.11	25.5	RbCl	4.73	109
LiBr	4.34	100	RbF	4.94	114

TABLE 11 (cont.)

Molecule	Dissociation energy		Molecule	Dissociation energy	
	(eV)	(Kcal/mol)		(eV)	(Kcal/mol)
RbH	1.78	41	TiCl	4.34	100
S_2	4.27	98.5	TiO	6.76	156
SO	5.36	123.6	Tl_2	0.61	14
Sb_2	3.06	70.6	TlBr	3.38	78
SbO	4.42	102	TlCl	4.03	93
Se_2	2.82	65	TlF	4.77	110
SeO	6.94	160	TlH	1.91	44
Si_2	3.21	74	TlI	2.82	65
SiBr	3.69	85	UO	7.8	180
SiCl	3.99	92	VO	6.07	140
SiF	5.42	125	WO	6.68	154
SiH	3.21	74	YO	7.24	167
SiN	4.51	104	Zn_2	0.26	6
SiO	8.33	192	ZnBr	2.3	53
Sn_2	1.99	46	ZnCl	2.6	60
SnO	5.81	134	ZnF	3.34	77
SrO	4.86	112	ZnH	0.87	20
TaO	8.41	194	ZnI	1.39	32
Te_2	2.25	52	ZnO	3.99	92
TeO	2.72	62.8	ZnS	4.21	97
ThO	8.5	196	ZrO	7.85	181
TiC	5.51	127			

TABLE 12

Values of the gas kinetical collision cross section σ_k and of the atomic diameter δ

Atom	Diameter σ (10^{-8}cm)	$\sigma_k = \dfrac{\pi}{4}\delta^2$ (10^{-16}cm^2)	Atom	Diameter σ (10^{-8}cm)	$\sigma_k = \dfrac{\pi}{4}\delta^2$ (10^{-16}cm^2)
H	3.0	7.07	Ru	2.64	5.47
He	2.44	4.67	Rh	2.69	5.68
Li	3.12	7.64	Pd	2.74	5.89
Be	2.16	3.66	Ag	2.85	6.38
B	1.76	2.43	Cd	3.0	7.07
C	1.54	1.86	In	3.08	7.45
N	1.26	1.25	Sn	3.12	7.64
O	1.20	1.13	Sb	3.18	7.94
F	1.36	1.45	Te	2.79	6.11
Ne	3.22	8.14	I	2.71	5.77
Na	3.72	10.86	Xe	4.26	14.25
Mg	3.20	8.04	Cs	5.25	21.63
Al	2.85	6.38	Ba	4.42	15.34
Si	2.32	4.22	La	3.75	11.04
P	2.16	3.66	Ce	3.65	10.38
S	2.11	3.49	Pr	3.63	10.35
Cl	2.04	3.27	Nd	3.61	10.23
Ar	3.83	11.52	Eu	4.08	13.07
K	4.68	17.19	Gd	3.48	9.51
Ca	3.92	12.07	Tb	3.54	9.84
Sc	3.15	7.79	Dy	3.53	9.78
Ti	2.96	6.88	Ho	3.51	9.67
V	2.68	5.64	Er	3.50	9.61
Cr	2.52	4.98	Tm	3.49	9.56
Mn	2.54	5.06	Yb	3.87	11.76
Fe	2.52	4.98	Lu	3.43	9.23
Co	2.52	4.98	Hf	3.24	8.24
Ni	2.48	4.83	Ta	2.89	6.56
Cu	2.54	5.06	W	2.76	5.98
Zn	2.71	5.77	Re	2.73	5.92
Ga	2.70	5.72	Os	2.68	5.64
Ge	2.76	5.98	Ir	2.70	5.72
As	2.41	4.56	Pt	2.76	5.98
Se	2.29	4.12	Au	2.88	6.51
Br	2.34	4.30	Hg	3.05	7.30
Kr	3.98	12.43	Tl	3.68	10.63
Rb	5.02	19.78	Pb	3.48	9.51
Sr	4.26	14.25	Bi	3.78	11.21
Zr	3.19	7.99	Th	3.64	10.40
Nb	2.90	6.60	U	3.0	7.07
Mo	2.78	6.07			

TABLE 13

Melting and boiling points of metals.
Temperatures at given pressures [1033, 1240]

Ele-ment	Melting Temperature t_m °C	Melting Pressure p mmHg	Boiling Temperature t_b °C	10^{-5}	10^{-4}	10^{-3}	10^{-2}	10^{-1}	1
Ac	1050	—	—	—	—	—	—	—	—
Ag	960.8	$1.78 \cdot 10^{-3}$	2163	767	848	936	1047	1184	1353
Al	660.1	$1.8 \cdot 10^{-8}$	2348	843	929	1030	1148	1291	1465
Am	1200	—	2606	—	—	—	—	—	—
Au	1063	$6 \cdot 10^{-6}$	2710	1083	1190	1316	1465	1646	1867
B	2075	—	3860	1052	1140	1239	1355	1489	1648
Ba	717	$7.6 \cdot 10^{-2}$	1634	418	476	546	629	730	858
Be	1280	$1.9 \cdot 10^{-2}$	2471	942	1029	1130	1246	1395	1582
Bi	271.3	$1.2 \cdot 10^{-10}$	1560	474	536	609	698	802	934
C	3700	—	4800	2129	2288	2471	2681	2926	3214
Ca	850	$8.9 \cdot 10^{-1}$	1439	480	463	538	625	730	887
Cd	321.03	$1.0 \cdot 10^{-1}$	767	148	180	220	264	321	379
Ce	785	$5.5 \cdot 10^{-9}$	2530	1004	1091	1190	1305	1439	1599
Co	1492	$7.6 \cdot 10^{-4}$	2250	1249	1362	1524	1649	1833	2056
Cr	1890	$6.4 \cdot 10^{-1}$	2570	907	992	1090	1205	1342	1504
Cs	28.5	$1.5 \cdot 10^{-6}$	688	45	74	110	153	207	277
Cu	1083	$3.1 \cdot 10^{-4}$	2580	946	1035	1141	1273	1432	1628
Dy	1380	—	2230	—	—	—	—	—	—
Er	1525	—	2400	—	—	—	—	—	—
Eu	1150	—	1470	—	—	—	—	—	—
Fe	1539	$3.72 \cdot 10^{-2}$	2770	1094	1195	1310	1447	1602	1783
Ga	29.7	0	2230	771	859	965	1093	1248	1443
Gd	1350	—	1500	—	—	—	—	—	—
Ge	958.5	$4.5 \cdot 10^{-5}$	2850	897	996	1112	1251	1421	1635
Hf	2222	—	5400	—	—	—	—	—	—
Hg	−38.87	$2.5 \cdot 10^{-6}$	356.58	−23.9	−5.5	18	48	82	126
Ho	1500	—	2380	—	—	—	—	—	—
In	156.4	0	2000	667	746	840	952	1088	1260
Ir	2443	$3.5 \cdot 10^{-3}$	4180	1993	2154	2340	2556	2811	3118
K	63.6	$9.8 \cdot 10^{-7}$	765	91	123	161	207	265	338
La	887	$2.3 \cdot 10^{-7}$	3370	1023	1125	1242	1381	1549	1754
Li	179	$9 \cdot 10^{-10}$	1350	325	377	439	514	607	725
Lu	1675	—	2260	—	—	—	—	—	—
Mg	651	2.2	1103	287	331	383	443	515	605
Mn	1244	$9 \cdot 10^{-1}$	2120	717	791	878	980	1103	1251
Mo	2620	$2.2 \cdot 10^{-2}$	4800	1923	2095	2295	2533	—	—
Na	97.8	$8.2 \cdot 10^{-8}$	877	158	195	238	291	356	437

TABLE 13 (cont.)

Element	Melting Temperature t_m °C	Melting Pressure p mmHg	Boiling Temperature t_b °C	10^{-5}	10^{-4}	10^{-3}	10^{-2}	10^{-1}	1
Nb	2500	—	4840	—	—	—	—	—	—
Nd	1024	—	3110	—	—	—	—	—	—
Ni	1453	$4.4 \cdot 10^{-3}$	2140	1157	1257	1371	1510	1679	1884
Np	640	—	—	—	—	—	—	—	—
Os	2700	$1.3 \cdot 10^{-2}$	4610	2101	2264	2451	2667	2920	3221
Pa	1430	—	4230	—	—	—	—	—	—
Pb	327.3	$5.4 \cdot 10^{-9}$	1751	483	548	625	718	832	975
Pd	1552	$8.7 \cdot 10^{-3}$	3110	1156	1271	1405	1566	1759	2000
Pm	1035	—	—	—	—	—	—	—	—
Po	252	—	962	—	—	—	—	—	—
Pr	935	—	3017	—	—	—	—	—	—
Pt	1769	$1.6 \cdot 10^{-4}$	3710	1606	1744	1904	2090	2313	2582
Pn	639	—	3235	—	—	—	—	—	—
Ra	960	—	1140	—	—	—	—	—	—
Rb	38.8	$1.5 \cdot 10^{-6}$	705	59	88	123	165	217	283
Re	3167	—	5500	—	—	—	—	—	—
Rh	1960	$1 \cdot 10^{-3}$	3670	1681	1815	1971	2149	2358	2607
Ru	2430	$9.8 \cdot 10^{-3}$	4200	1913	2058	2230	2431	2666	2946
S	112.8	—	444.6	—	—	—	—	—	—
Sb	630.5	$2.8 \cdot 10^{-3}$	1625	466	525	595	678	779	904
Sc	1538	$6.6 \cdot 10^{-3}$	2430	1098	1221	1352	1505	1695	1906
Si	1410	$3.2 \cdot 10^{-2}$	2600	1024	1116	1223	1343	1485	1670
Sn	231.91	0	2720	823	922	1042	1189	1373	1609
Sm	1050	—	1600	—	—	—	—	—	—
Sr	777	1.4	1383	361	413	475	549	639	750
Ta	3000	$5 \cdot 10^{-3}$	5300	2407	2599	2820	2950	—	—
Tb	1368	—	2500	—	—	—	—	—	—
Tc	2140	—	4700	—	—	—	—	—	—
Te	452	—	1012	—	—	—	—	—	—
Th	1830	$9.3 \cdot 10^{-5}$	4200	1686	1831	1999	2196	2431	2717
Ti	1725	$8.4 \cdot 10^{-2}$	3170	1134	1249	1384	1546	1742	1965
Tl	303.5	$5.0 \cdot 10^{-8}$	1472	405	461	527	606	702	821
Tu	1600	—	2400	—	—	—	—	—	—
U	1133	$3.2 \cdot 10^{-9}$	3490	1461	1585	1730	1898	2098	2338
V	1900	$6.5 \cdot 10^{-4}$	3330	1662	1783	1996	2196	2298	2487
W	3380	$1.75 \cdot 10^{-2}$	5370	2554	2767	3016	3309	—	—
Y	1525	$7.6 \cdot 10^{-4}$	2780	1279	1404	1544	1698	1888	2106
Zn	419	$1.6 \cdot 10^{-1}$	—	211	248	292	343	405	—
Zr	1852	$4.1 \cdot 10^{-2}$	4500	1380	1516	1652	1798	2056	2316

TABLE 14

Melting and boiling points of metal oxides [1247]

Oxides	Melting point t_m °C	Boiling point t_b °C
Ag_2O	300 (decomp.)	
Al_2O_3	2010–2050	2700
B_2O_3	450	1700
BaO	1920	2000
BeO	2550	3900
Bi_2O_3	820	1890–1900
CaO	2580	2850
CeO_2	2500	—
CrO	1875	· 2480
Cr_2O_3	1990	—
Cs_2O	360–400	decomp.
CsO_2	600	decomp.
Cu_2O	1229	1800
CuO	1026 (decomp.)	—
FeO	1360	—
Fe_2O_3	1565	—
Ga_2O_3	1740	—
GeO_2	1116	—
HgO	500 (decomp.)	—
In_2O_3	850 (decomp.)	—
Ir_2O_3	400 (decomp.)	—
La_2O_3	2320	4200
MgO	2640	3600
MnO	1650	—
Mn_3O_4	1560	—
Mn_2O_3	1080 (decomp.)	—
MnO_2	535 (decomp.)	—
MoO_3	795	1155
Na_2O	1275	—
Nb_2O_3	1780	—
Nb_2O_5	1512	—
Nd_2O_3	1900	—
NiO	1950	—
OsO_2	650 (decomp.)	—
PbO	890	1473
PdO	750 (decomp.)	—
Re_2O_7	300	360
Sb_2O_3	656	1425
SeO_3	120 (decomp.)	—
SiO_2	1670	—
SrO	2430	—
Ta_2O_5	1470 (decomp.)	—

TABLE 14 (cont.)

Oxides	Melting point t_m °C	Boiling point t_b °C
ThO_2	3050	—
TiO_2	1560	—
UO_2	2800	—
VO	2000	—
V_2O_3	1970	—
WO_2	1270	1700
Y_2O_3	2410	4300
ZnO	419	913

TABLE 15

Values of the work function φ for elements and compounds under consideration [873]

Element or compound	φ (eV)	Element or compound	φ (eV)
Li	2.39	Sr	2.35
Be	3.37	Y	3.3
B	4.5	Zr	3.84
C	4.39	Nb	3.99
Na	2.27	Mo	4.27
Mg	3.46	Tc	4.4
Al	3.74	Ru	4.52
Si	4.1	Rh	4.65
K	2.15	Pd	4.82
Ca	2.76	Ag	4.28
Sc	3.3	Cd	3.92
Ti	4.09	In	3.75
V	4.11	Sn	4.11
Cr	4.51	Sb	4.08
Mn	3.95	Te	4.73
Fe	4.36	I	2.7
Co	4.18	Cs	1.89
Ni	4.84	Ba	2.29
Cu	4.47	La	3.3
Zn	3.74	Ce	2.7
Ga	3.96	Pr	2.7
Ge	4.56	Nd	3.3
As	5.15	Pm	3.05
Se	4.72	Sm	3.2
Rb	2.13	Eu	2.55

TABLE 15 (cont.)

Element or compound	φ (eV)	Element or compound		φ (eV)
Gd	3.05	U		3.74
Tb	3.05	CsO		0.99–1.17
Dy	3.1	MgO		3.1–4.4
Ho	3.05	BeO		3.8–4.7
Er	3.1	Al_2O_3		4.7
Tu	3.1	B_2O_3		4.7
Yb	2.6	NiO		5.55
Lu	3.15	FeO		3.85
Hf	3.53	CuO		4.35
Ta	4.12	(Ba, Sr) O		1.1
W	4.50	BaO		1.3
Re	5.1	W + molecular layer	Th	2.63
Os	4.83	W + molecular layer	Zr	3.14
Ir	5.27	W + molecular layer	Ba	1.6
Pt	5.32	W + molecular layer	Cs	1.36
Au	4.58	W + molecular layer	O	6.28
Hg	4.52	W + molecular layer	H	5.8
Tl	3.76	W + molecular layer	O + Cs	1.1
Pb	4.02	Mo+ molecular layer	Th	2.58
Bi	4.28	Ta + molecular layer	Th	2.52
Po	4.8	Pt + molecular layer	O	6.55
Fr	1.8	Pt + molecular layer	H	4.24
Ra	3.25	Ni + molecular layer	O	6.34
Ac	2.65	Au + molecular layer	O	6.46
Th	3.41	Mn+ molecular layer	O	6.5

TABLE 16

Values of the secondary electron emission coefficient δ_{em} and those of the associated primary electron energy E_{em}

Element or compound	δ_{em}	E_{em} (eV)	References
Li	0.5	85	[831, 833]
LiF (crystal)	8.5	—	[871]
LiF (layer)	5.6	700	[843]
Be	0.5	200	[835, 837, 839]
BeO	3.4	2000	[848]
B	1.2	150	[853]
C (diamond)	2.8	750	[860]
C (graphite)	1.0	300	[838]
C (carbon black)	0.45	500	[838]
Na	0.82	300	[831, 834]
NaF (crystal)	14.0	1200	[871]
NaF (layer)	5.7	—	[843]
NaCl (crystal)	14.0	1200	[858, 866, 871]
NaCl (layer)	6.8	600	[842, 843]
NaBr (crystal)	24.0	1800	[868]
NaBr (layer)	6.3	—	[843]
NaI (crystal)	19.0	1300	[871]
NaI (layer)	5.5	—	[843]
Mg	0.95	300	[831, 835]
MgO (crystal)	20–25	1500	[859, 867, 870, 872]
MgO (layer)	3–15	400–1500	[848, 865, 869, 872]
Al	1.0	300	[835]
Al_2O_3 (layer)	2–9	—	[844, 848, 858]
Si	1.1	250	[853]
SiO_2 (quartz)	2.1–4	400	[844, 848]
K	0.7	200	[832, 846]
KCl (crystal)	12.0	1600	[843, 851]
KCl (layer)	7.5	1200	[843, 851]
KBr (crystal)	14.0	1800	[866]
KI (crystal)	10.0	1600	[866]
KI (layer)	5.6	—	[843]
Ti	0.9	250	[838]
Cr	1.1	400	[859]
Mn	1.35	200	[837]
Fe	1.3	350	[828, 833, 836]
Co	1.35	600	[836, 841]
Ni	1.3	550	[828, 831, 836] [856]
Cu	1.3	600	[831, 835, 849]

TABLE 16 (cont.)

Element or compound	δ_{em}	E_{em} (eV)	References
Cu_2O	1.25	400	[863]
Zn	1.1	200	[836]
ZnS	1.8	350	[864]
Ga	1.55	500	[855]
Ge	1.15	500	[853, 857]
GeCs	7.0	700	[866]
Rb	0.9	350	[832]
RbCl (layer)	5.8	—	[835]
Zr	1.1	350	[838]
Nb	1.2	375	[831, 833]
Mo	1.25	375	[862]
MoO_2	1.25	350	[830]
MoS_2	1.10	—	[843]
Pd	1.3	250	[827]
Ag	1.5	800	[831, 833]
Cd	1.1	450	[849, 850]
Sn	1.35	500	[847, 862]
SnO_2	3.2	640	[862]
Sb	1.3	600	[833]
Cs	0.7	400	[834, 835]
CsCl	6.5	—	[834, 835]
Ba	0.8	400	[835]
BaO (layer)	2.3–4.8	400	[843, 848]
BaF_2 (layer)	4.5	—	[843]
Ta	1.25	600	[831]
W	1.45	650	[827, 831, 833]
WSz	0.46–1.04	—	[837]
Pt	1.8	700	[833]
Au	1.4	800	[831, 833]
Hg	1.3	600	[855]
Tl	1.7	650	[857]
Pb	1.1	500	[847, 855]
Bi	1.2	550	[847, 857]
BiCs	1.9	1000	[837]
Th	1.1	800	[835]
Tungsten with thorium	1.2	350	[858]
Pyrex glass	2–3	300–400	[844, 852]
Sodium glass	~2	300	[844, 852]
Ground glass	~3	400	[844, 852]
Mica	2.4	350	[844]

TABLE 17

Values of the surface recombination coefficient of materials used for the inside wall of discharge chambers for hydrogen atoms

Material	Surface recombination coefficient	
	[1197]	[1198]
Pyrex glass	$7.5 \cdot 10^{-4}$	$(0.75-4.6) \cdot 10^{-4}$
Quartz	—	$7 \cdot 10^{-4}$
Aluminium	10^{-3}	—
Iron	0.17	—
Copper	0.19	—
Platinum	0.25	1

TABLE 18

$$\xi = l_n \frac{r_a}{r_k} - 0.3 \left(l_n \frac{r_a}{r_k} \right)^2 + \ldots$$

r_a/r_k; r_k/r_a	$\xi^2 (r_k > r_a)$	$(-\xi)^2 (r_a > r_k)$
1.0	0.0000	0.0000
1.1	0.0086	0.0096
1.2	0.0299	0.0373
1.3	0.0592	0.0812
1.4	0.0933	0.1393
1.5	0.1306	0.2118
1.6	0.1692	0.2969
1.7	0.2090	0.3960
1.8	0.2480	0.5040
1.9	0.2870	0.6210
2.0	0.3290	0.7520
2.1	0.3650	0.8890
2.2	0.4030	1.0390
2.3	0.4390	1.1930
2.4	0.4740	1.3600
2.5	0.5100	1.5320
2.6	0.5450	1.7150
2.7	0.5780	1.9030
2.8	0.6090	2.0990
2.9	0.6400	2.3030
3.0	0.6700	2.5150
4.0	0.9360	4.9690
5.0	1.1420	7.9780
6.0	1.3120	11.4600

REFERENCES

1. H. A. Bethe, E. E. Salpeter, *Quantum Mechanics of One and Two Electron Atoms*. Springer Verlag Berlin (1957).
2. N. F. Mott, H. S. W. Massey, *The Theory of Atomic Collisions*. Oxford, Clarendon Press (1965).
3. L. D. Landau, E. M. Lifsitz, *Quantum Mechanics*. Moscow, Fizmatgiz (1963).
4. W. L. Fite, R. T. Brackmann, *Phys. Rev.*, **112**, 1151 (1958).
5. R. McFahrland, *Phys. Rev. Lett.*, **10**, 397 (1963).
6. W. L. Fite, R. F. Stebbings, R. T. Brackmann, *Phys. Rev.*, **116**, 356 (1959).
7. W. Lichten, G. J. Schultz, *Phys. Rev.*, **116**, 1132 (1959).
8. V. Burke, M. Seaton, *Proc. Phys. Soc.*, **77**, 199 (1961).
9. O. Thieme, *Z. f. Phys.*, **86**, 646 (1933).
10. J. J. Thomson, *Phil. Mag.*, **23**, 419 (1912).
11. M. Gryzinski, *Phys. Rev.*, **115**, 374 (1959).
12. M. Gryzinski, *JNR Swierk, Poland Rep.* 448/XVIII 1963.
13. M. Gryzinski, *Phys. Rev.*, **138A**, 322 (1965).
14. L. Thomas, *Proc. Camb. Phil. Soc.*, **28**, 713 (1927).
15. D. Webster, V. Hansen, F. Duveneck, *Phys. Rev.*, **43**, 833 (1933)
16. P. Burke, K. Smith, *Rev. Mod. Phys.*, **34**, 458 (1962).
17. E. Erskine, H. S. W. Massey, *Proc. Roy. Soc.*, **212A**, 521 (1952)
18. R. Mariott, *Proc. Phys. Soc.* **72**, 121 (1958).
19. P. K. Peterkop, R. J. Damburg, *JETP*, **43**, 1763 (1962).
20. N. Bohr, *Phys. Rev.*, **58**, 654 (1940).
21. R. Stebbings, W. Fite, D. Hummer, R. Brackmann, *Phys. Rev.*, **124**, 2051 (1962).
22. S. Khaschaba, H. S. W. Massey, *Proc. Phys. Soc.*, **71**, 574 (1958).
23. H. S. W. Massey, *Rev. Mod. Phys.*, **28**, 199 (1956).
24. K. Smith, *Phys. Rev.*, **120**, 845 (1960).
25. J. Leech, D. Bates, A. Fundaminsky, H. S. W. Massey, *Phil. Trans. Roy. Soc.*, **243A**, 93 (1950).
26. A. V. Eleckii, B. M. Smirnov, *JETP*, **38**, 3 (1968).
27. I. P. Zapesotskii, L. P. Simon, *Optika i spektroskopiya*, **19**, 480, 864 (1965).
28. S. Silverman, E. Lassetre, *Sci. Rep. Dep. of Chem., University* No. 9, Cont. No AF 19 (122) 642 Columbia Ohio 1967.
29. R. McCarroll, *Proc. Phys. Soc.*, **83**, 409 (1964).
30. M. Bailitis, R. Damberg, *Proc. Phys. Soc.*, **82**, 192 (1963).
31. G. J. Schultz, *Phys. Rev. Lett.*, **13**, 583 (1964).
32. V. J. Otskur, *JETP*, **45**, 734 (1963).
33. V. J. Otskur, *JETP*, **47**, 1746 (1964).
34. H. S. W. Massey, B. L. Moiseiwitsch, *Proc. Phys. Soc.*, **66A**, 406 (1953).
35. P. G. Burke, H. M. Schey, *Phys. Rev.*, **126**, 147 (1962).
36. V. Burke, R. McCarroll, *Proc. Phys. Soc.*, **80**, 422 (1962).
37. J. R. Oppenheimer, *Phys. Rev.*, **32**, 361 (1928).
38. H. Eyiring, J. O. Hirschfelder, H. S. Taylor, *Journ. Chem. Phys.*, **4**, 479 (1936).
39. H. S. W. Massey, B. L. Moiseiwitsch, *Proc. Roy. Soc.*, **227A**, 38 (1954).
40. H. Moier-Liebniz, *Z. f. Phys.*, **95**, 489 (1935).
41. G. J. Schulz, R. E. Fox, *Phys. Rev.*, **108**, 1179 (1957).
42. G. Breit, E. Teller, *Astrophys. J.* **91**, 215 (1940).
43. J. Shapiro, G. Breit, *Phys. Rev.*, **113**, 179 (1959).
44. N. E. Lamb, R. C. Rutherford, *Phys. Rev.*, **72**, 241 (1947).
45. J. M. Heberle, H. A. Reich, P. Kusch, *Phys. Rev.*, **101**, 612 (1956).
46. J. L. Snoch, S. Orstein, F. Zernike, *Z. f. Phys.*, **47**, 627 (1928).

47. W. L. Fite, *Phys. Rev.*, **116**, 363 (1959).
48. E. Commins, *Bull. Amer. Phys. Soc.*, **7**, 258 (1962).
49. A. G. Engergardt, A. V. Phelps, C. G. Risk, *Phys. Rev.*, **135A**, 1566 (1964).
50. H. F. Winters, *J. Chem. Phys.*, **43**, 926 (1965).
51. E. Gerjouy, S. Stein, *Phys. Rev.*, **97**, 1671 (1955).
52. A. Dalgarno, W. Henry, *Proc. Phys. Soc.*, **85**, 679 (1965).
53. A. F. Forst, A. V. Phelps, *Phys. Rev.*, **127**, 1621 (1962).
54. A. G. Engelgart, A. V. Phelps, *Phys. Rev.*, **131**, 2115 (1963).
55. T. R. Carson, *Proc. Phys. Soc.*, **67A**, 909 (1954).
56. P. M. Morse, *Phys. Rev.*, **90**, 51 (1953).
57. G. J. Schulz, J. T. Dowell, *Phys. Rev.*, **128**, 174 (1962).
58. J. Franck, *Trans. Farad. Soc.*, **21**, 536 (1925).
59. E. U. Condon, *Phys. Rev.*, **32**, 858 (1928).
60. G. J. Schulz, *Phys. Rev.*, **112**, 150 (1958).
61. G. J. Schulz, *Phys. Rev.*, **116**, 1141 (1959).
62. G. J. Schulz, *Phys. Rev.*, **125**, 229 (1962).
63. G. J. Schulz, *Phys. Rev.*, **135A**, 988 (1964).
64. D. R. Bates, G. W. Griffing, *Proc. Phys. Soc.*, **66A**, 613 (1953).
65. D. R. Bates, G. W. Griffing, *Proc. Phys. Soc.*, **67A**, 663 (1954).
66. D. R. Bates, G. W. Griffing, *Proc. Phys. Soc. A.*, **68**, 90 (1955).
67. D. R. Bates, H. S. W. Massey, A. L. Stewart, *Proc. Roy. Soc.*, **216**, 437 (1953).
68. L. D. Landau, *Phys. J. Sov.*, **2**, 41, 88 (1932).
69. C. Zener, *Proc. Roy. Soc. A.*, **137**, 696 (1932).
70. S. Geltman, *Phys. Rev.*, **102**, 171 (1956).
71. G. H. Wannier, *Phys. Rev.*, **90**, 873 (1953).
72. K. T. Dolder, M. F. A. Harrison, P. C. Thonemann, *Proc. Roy. Soc.*, **264**, 367 (1961).
73. J. W. McGowan, L. Kerwin, *Can. J. Phys.*, **41**, 1535 (1963).
74. M. F. A. Harrison, K. T. Dolder, P. C. Thonemann, *Proc. Phys. Soc.*, **82**, 368 (1963).
75. K. T. Dolder, M. F. A. Harrison, P. C. Thonemann, *Proc. Roy. Soc.*, **274**, 546 (1963).
76. V. H. Dibeler, R. M. Reese, *J. Chem. Phys.*, **31**, 282 (1959).
77. R. E. Fox, W. M. Hickam, T. Kjeldass, *Phys. Rev.*, **89**, 555 (1953).
78. F. H. Dorman, J. D. Morrison, A. J. C. Nicholson, *J. Chem. Phys.*, **31**, 1335 (1959).
79. J. D. Morrison, A. J. C. Nicholson, *J. Chem. Phys.*, **31**, 1320 (1959).
80. D. P. Stevenson, J. A. Hipple, *Phys. Rev.*, **49**, 237 (1942).
81. G. Elwert, *Z. f. Naturf.*, **7A**, 432 (1952).
82. T. Killian, *Phys. Rev.*, **35**, 1238 (1930).
83. N. Morgulis, *Sov. Phys.*, **5**, 410 (1934).
84. E. Kononovits, *Astron. Journ.*, **30**, 1045 (1962).
85. F. de La-Ripelle, *J. Phys. Radium*, **10**, 318 (1949).
86. H. W. Drawin, *Z. f. Phys.*, **164**, 513 (1961).
87. L. Vriens, *Phys. Lett.*, **8**, 260 (1964).
88. H. A. Bethe, *Ann. der Phys.*, **5**, 325 (1930).
89. H. S. W. Massey, C. Mohr, *Proc. Roy. Soc.*, **144A**, 613 (1933).
90. R. Peterkop, *Izv. AN SSSR Phys.*, **24**, 946 (1960).
91. G. Peack, *Proc. Phys. Soc.*, **85**, 709 (1965).
92. R. Peterkop, *JETP*, **41**, 1938 (1961).
93. L. Biberman, J. Toropkin, K. Uljanova, *JETP*, **32**, 827 (1962).
94. L. D. Landau, E. Lifsitz, *Mechanics*, Moscow, Nauka 1938.
95. R. Stabler, *Phys. Rev.*, **133A**, 1268 (1964).
96. V. J. Ofskur, *Opt. and Spectr.*, **14**, 457 (1963).
97. G. F. Drukarev, *JETP*, **25**, 129 (1953).
98. V. Veldre, R. Peterkop, *Opt. and Spectr.*, **13**, 461 (1962).
99. G. S. Ivanov, G. M. Nikolskii, R. A. Guljaev, *Astron. Journ.*, **37**, 799 (1960).
100. L. Vriens, *Phys. Lett.*, **9**, 295 (1964).
101. L. Vriens, *Phys. Lett.*, **10**, 170 (1964).
102. P. G. Burke, H. M. Schey, K. Schmith, *Phys. Rev.*, **129**, 1258 (1963).
103. W. Bleakney, *Phys. Rev.*, **35**, 1180 (1930).
104. P. T. Smyth, *Mod. Phys.*, **3**, 347 (1931).
105. H. Ramien, *Z. f. Phys.*, **70**, 353 (1931).

106. L. S. Frost, A. V. Phelps, *Westinghouse Res. Rep.*, 62-908-113-P1 (1962).
107. P. Swan, *Proc. Phys. Soc.*, **68A**, 1157 (1955).
108. F. Mond, *AERE Rep. T/R* 1006 *H. M. Stat. off.*, (1952) .
109. C. F. Barnet, W. B. Gauster, J. A. Roy, *Atomic and Molecular Collision Cross Section* ORNL 3113
110. B. Yavorski, *Dokl. AN, SSSR*, **49**, 256 (1945).
111. W. Bleakney, *Phys. Rev.*, **35**, 139 (1930).
112. W. Bleakney, *Phys. Rev.*, **36**, 1303 (1930).
113. P. T. Smith, *Phys. Rev.*, **36**, 1293 (1930).
114. D. R. Bates, D. A. Williams, *Proc. Phys. Soc.*, **83**, 425 (1964).
115. D. F. Gallaker, L. Wiles, *Phys. Rev.*, **169**, 139 (1968).
116. G. Ryding, A. B. Witthower, H. B. Gilbody, *Proc. Phys. Soc.*, **89**, 547 (1966).
117. S. E. Lovell, M. B. McElroy, *Proc. Roy. Soc.*, **283A**, 100 (1965).
118. W. Bleakney, P. T. Smith, *Phys. Rev.*, **49**, 402 (1936).
119. V. Beloselski, *Trud. Voronezki Univ.*, **55**, 69 (1961).
120. A. Burgess, *Proc. Roy. Soc.*, **273A**, 372 (1963).
121. E. Trefftz, M. Rudge, *Proc. Roy. Soc.*, **271A**, 379 (1963).
122. M. McDowell, G. Peack, *Phys. Rev.*, **121**, 1383 (1961).
123. M. McDowell, V. Myerscong, G. Peack, *Proc. Phys. Soc.*, **85**, 703 (1965).
124. E. Burkop, *Proc. Camb. Phil. Soc.*, **36**, 43 (1940).
125. M. Seaton, *Phys. Rev.*, **113**, 814 (1959).
126. W. Bleakney, J. T. Tate, *Phys. Rev.*, **35**, 658 (1930).
127. W. Lozier, *Phys. Rev.*, **36**, 1285 (1930).
128. W. Lozier, *Phys. Rev.*, **36**, 1417 (1930).
129. J. T. Tate, P. T. Smith, *Phys. Rev.*, **39**, 270 (1932).
130. J. T. Tate, P. T. Smith, A. L. Vaughan, *Phys. Rev.*, **48**, 525 (1935).
131. H. D. Hagstrum, J. T. Tate, *Phys. Rev.*, **59**, 354 (1941).
132. J. P. Hooper, *Phys. Rev.*, **125**, 2000 (1962).
133. F. Horton, D. M. Millest, *Proc. Roy. Soc. A.*, **185**, 381 (1946).
134. N. G. Utterback, *Bull. Amer. Phys. Soc.*, II, **7**, 7 (1962).
135. W. Weizol, *Z. f. Phys.*, **76**, 250 (1932).
136. M. A. Biondi, *Phys. Rev.*, **83**, 653 (1951).
137. M. A. Biondi, *Phys. Rev.*, **88**, 660 (1952).
138. A. V. Phelps, J. P. Molnár, *Phys. Rev.*, **89**, 1202 (1953).
139. F. Colegrove, P. A. Franken, *Phys. Rev.*, **119**, 680 (1960).
140. W. P. Sholette, E. E. Muschlitz, *J. Chem. Phys.*, **36**, 3368 (1962).
141. E. E. Benton, *Phys. Rev.*, **128**, 206 (1962).
142. E. V. Soloviev, R. N. Ilin, V. A. Oparin, N. V. Fedorenko, *JETP*, **42**, 659 (1962).
143. E. V. Soloviev, R. N. Ilin, V. A. Oparin, N. V. Fedorenko, *JETP*, **45**, 496 (1963).
144. R. A. Mapleton, *Phys. Rev.*, **122**, 528 (1961).
145. H. B. Gilbody, J. B. Hasted, *Proc. Roy. Soc.*, **240A**, 382 (1957).
146. A. R. Lee, H. B. Gilbody, *Proc. Roy. Soc.*, **274A**, 365 (1963).
147. D. R. Bates, G. W. Griffing, *Proc. Phys. Soc.*, **66A**, 961 (1953).
148. R. A. Langley, D. W. Martin, D. S. Harmer, J. W. Hooper, E. W. McDaniel, *Phys. Rev.*, **136A**, 379 (1964).
149. R. A. Langley, D. W. Martin, D. S. Harmer, J. W. Hooper, E. W. McDaniel, *Phys. Rev. A.*, **136**, 385 (1964).
150. M. Stobbe, *Ann. d. Phys.*, **7**, 661 (1930).
151. F. Sautter, *Ann. d. Phys.*, **9**, 217 (1931).
152. F. Sautter, *Ann. d. Phys.*, **11**, 454 (1931).
153. H. R. Hulme, J. McDougall, R. A. Budkingham, R. H. Fowler, *Proc. Roy. Soc.*, **149**, 131 (1935).
154. J. M. Allen, *Phys. Rev.*, **27**, 266 (1926).
155. J. M. Allen, *Phys. Rev.*, **28**, 907 (1926).
156. L. H. Gray, *Proc. Camb. Phil. Soc.*, **27**, 103 (1931).
157. J. A. Gaunt, *Phil. Trans.*, **229A**, 163 (1930).
158. D. H. Menzel, C. L. Pekeris, *Roy. Astron. Soc.*, **96**, 77 (1935).
159. G. V. Marr, *Proc. Phys. Soc.*, **81**, 9 (1963).
160. E. Schonheit, *Z. f. Naturf.*, **16A**, 1094 (1961).
161. J. W. Cooper, *Phys. Rev.*, **128**, 681 (1962).
162. R. E. Huffmann, Y. Tanaka, J. C. Larrabee, *J. Chem. Phys.*, **39**, 902 (1963).
163. A. Dalgerno, *Proc. Phys. Soc.*, **65A**, 663 (1952).
164. N. Wainfan, W. C. Walker, G. L. Weissler, *Phys. Rev.*, **99**, 542 (1955).

165. Lee Po, G. L. Weissler, *Proc. Roy. Soc.*, **219A**, 71 (1953).
166. Lee Po, G. L. Weissler, *Phys. Rev.*, **99**, 540 (1955).
167. Lee Po, G. L. Weissler, *Journ. Opt. Soc. Amer.*, **42**, 200 (1952).
168. Lee Po, *Journ. Opt. Soc. Amer.*, **45**, 703 (1955).
169. K. Watanabe, F. F. Marmo, *Journ. Chem. Phys.*, **25**, 965 (1956).
170. H. Beutler, H. O. Junger, *Z. f. Phys.*, **101**, 285 (1936).
171. Lee Po, G. L. Weissler, *Astrophys. Journ.*, **115**, 570 (1952).
172. C. D. Mannsell, *Phys. Rev.*, **98**, 703 (1955).
173. N. Astoin, J. G. *Compt. Rend.*, **244**, 1350 (1957).
174. E. Holoien, T. Midtel, *Proc. Phys. Soc.*, **88A**, 815 (1955).
175. J. D. Gracia, J. E. Mack, *Phys. Rev.*, **138A**, 987 (1965).
176. P. Feldman, R. Novick, *Phys. Rev. Lett.*, **11**, 278 (1963).
177. G. Herzberg, H. R. Moohre, *Can. J. Phys.*, **37**, 1239 (1959).
178. P. G. Kruger, *Phys. Rev.*, **36**, 853 (1930).
179. J. Pitenpol. *Phys. Rev. Lett.*, **7**, 64 (1961).
180. S. T. Manson, *Phys. Rev.*, **145**, 35 (1966).
181. E. Holien, *Proc. Phys. Soc. A.*, **71** 357 (1958).
182. T. F. O'Malley, S. Geltman, *Phys. Rev.*, **137A**, 1344 (1965).
183. P. L. Altick, E. N. Moohre, *Phys. Rev. Lett.*, **11**, 100 (1963).
184. P. G. Burke, V. M. Burke, H. M. Schey, *Phys. Rev.*, **125**, 143 (1962).
185. K. Smith, R. McEachran, P. Fraser, *Phys. Rev.*, **125**, 553 (1962).
186. P. G. Burke, H. M. Schey, *Phys. Rev.*, **126**, 147 (1962).
187. M. Gailitis, R. Damburg, *Proc. Phys. Soc,.* **82**, 268 (1963).
188. G. J. Schulz, *Phys. Rev. Lett.*, **10**, 583 (1964).
189. J. W. McGowan, E. M. Clarke, E. K. Curley, *Phys. Rev. Lett.*, **15**, 917 (1965).
190. J. W. McGowan, *Bull. Amer. Phys. Soc.*, **11**, 732 (1966).
191. P. G. Burke, D. D. McVicar, K. Smith, *Proc. Phys. Soc.*, **83**, 397 (1964).
192. P. G. Burke, D. D. McVicar, K. Smith, *Proc. Phys. Soc.*, **84**, 749 (1964).
193. P. L. Altick, E. N. Moohre, *Phys. Rev.*, **147**, 59 (1966).
194. R. H. Propin, *Opt. and Spectr.*, **8**, 300 (1960).
195. R. H. Propin, *Opt. and Spectr.*, **9**, 308 (1961).
196. A. Temkin, R. Pohle, *Phys. Rev. Lett.*, **10**, 22 (1963).
197. A. Temkin, P. Pohle, *Phys. Rev. Lett.*, **10**, 268 (1963).
198. A. Temkin, J. F. Walker, *Phys. Rev.*, **140A**, 1520 (1965).
199. C. E. Kuyatt, J. A. Simpson, S. R. Mielczarek, *Phys. Rev.*, **138A**, 385 (1965).
200. G. J. Schulz, *Phys. Rev.*, **136A**, 650 (1964).
201. M. E. Rudd, *Phys. Rev. Lett.*, **13**, 503 (1964).
202. M. E. Rudd, *Phys. Rev. Lett.*, **15**, 580 (1965).
203. M. E. Rudd, T. Jorgensen, D. J. Volz, *Phys. Rev. Lett.*, **16**, 929 (1966).
204. M. E. Rudd, T. Jorgensen, D. J. Volz, *Phys. Rev.*, **151**, 28 (1966).
205. R. P. Marchi, F. T. Smith, *Phys. Rev.*, **139A**, 1025 (1965).
206. V. B. Leonas, A. V. Sermyagin, *JTP*, **37**, 1547 (1967).
207. H. Beutler, *Z. f. Phys.*, **86**, 495 (1933).
208. H. Beutler, *Z. f. Phys.*, **87**, 188 (1933).
209. H. Beutler, *Z. f. Phys.*, **93**, 177 (1935).
210. U. Fanto, *Phys. Rev.*, **124**, 1866 (1961).
211. U. Fanto, J. W. Cooper, *Phys. Rev.*, **137A**, 1364 (1965).
212. K. T. Dolder, M. F. Harrison, P. C. Thonemann, *Proc. Roy. Soc.* **267A**, 297 (1962).
213. K. N. Kingdon, J. Langmuir, *Phys. Rev.*, **21**, 380 (1923).
214. N. Ives, *Phys. Rev.*, **21**, 385 (1923).
215. N. D. Morgulis, *JETP*, **4**, 684 (1934).
216. K. N. Kingdon, J. Langmuir, *Proc. Roy. Soc.*, **107A**, 61 (1924).
217. L. N. Dobretsov, *JTP*, **23**, 417 (1953).
218. L. N. Dobretsov, *JETP*, **6**, 552 (1936).
219. L. N. Dobretsov, V. N. Lepeshinskaya, J. E. Bronstein, *JTP*, **22**, 961 (1952).
220. M. J. Copley, T. E. Pripps, *Phys. Rev.*, **45**, 344 (1934).
221. M. J. Copley, T. E. Pripps, *Phys. Rev.*, **48**, 960 (1935).
222. J. Zemel, *J. Chem. Phys.*, **28**, 410 (1958).
223. E. J. Zandberg, *JTP*, **28**, 410 (1958).
224. E. J. Zandberg, N. I. Ionov, *JTP*, **28**, 2444 (1958).
225. I. J. Killian, *Phys. Rev.*, **27**, 578 (1926).
226. J. A. Becker, *Phys. Rev.*, **28**, 341 (1926).
227. J. Langmuir, *Amer. Chem. Soc.*, **54**, 1252 (1932).

228. J. B. Taylor, J. Langmuir, *Phys. Rev.*, **44**, 423 (1933).
229. M. J. Copley, T. E. Pripps, *Phys. Rev.*, **46**, 144 (1934).
230. H. Altertum, K. Krebs, R. Rompe, *Z. f. Phys.*, **92**, 1 (1934).
231. S. Datz, E. H. Taylor, *J. Chem. Phys.*, **25**, 389 (1956).
232. S. Datz, E. H. Taylor, *J. Chem. Phys.*, **25**, 395 (1956).
233. J. F. Hart, G. Herzberg, *Phys. Rev.*, **106**, 79 (1957).
234. E. Hylleraas, J. Midtal, *Phys. Rev.*, **109**, 1013 (1958).
235. C. L. Pekeris, *Phys. Rev.*, **112**, 1649 (1958).
236. C. L. Pekeris, *Phys. Rev.*, **126**, 1470 (1962).
237. A. W. Weis, *Phys. Rev.*, **122**, 1826 (1961).
238. E. Clementi, A. D. McLean, *Phys. Rev.*, **133A**, 419 (1964).
239. E. Clementi, A. D. McLean, M. Yoshimine, *Phys. Rev.*, **133A**, 1274 (1964).
240. E. Clementi, *Phys. Rev.*, **135A**, 980 (1964).
241. R. Gáspár, B. Molnár, *Acta Phys. Hung.*, **5**, 75 (1955).
242. P. Gombás, K. Ladányi, *Z. f. Phys.*, **138**, 261 (1960).
243. G. Glocker, *Phys. Rev.*, **46**, 111 (1934).
244. F. Rohrlich, *Phys. Rev.*, **101**, 69 (1956).
245. H. R. Johnson, F. Rohrlich, *J. Chem. Phys.*, **30**, 1612 (1959).
246. B. J. Edlen, *J. Chem. Phys.*, **33**, 98 (1960).
247. J. W. Edie, F. Rohrlich, *J. Chem. Phys.*, **36**, 623 (1962).
248. O. P. Tsarkin, M. E. Dyatkin, *J. Struct. Chem.*, **6**, 422 (1965).
249. L. M. Branscomb, S. J. Smith, *J. Chem. Phys.*, **25**, 587 (1956).
250. L. M. Branscomb, *Phys. Rev.*, **111**, 504 (1958).
251. M. Seman, L. M. Branscomb, *Phys. Rev.*, **125**, 1602 (1962).
252. R. S. Berry, C. W. Reiman, *J. Chem. Phys.*, **38**, 1540 (1963).
253. P. Sutton, J. E. Mayer, *J. Chem. Phys.*, **3**, 20 (1935).
254. D. T. Wier, J. E. Mayer, *J. Chem. Phys.*, **12**, 28 (1944).
255. M. Metlay, G. Kimbal, *J. Chem. Phys.*, **16**, 774 (1948).
256. R. B. Bernstein, M. Metlay, *J. Chem. Phys.*, **19**, 1612 (1951).
257. T. L. Bailey, *J. Chem. Phys.*, **28**, 797 (1958).
258. V. J. Khvostenko, V. M. Dukelskii, *JETP*, **37**, 651 (1958).
259. D. Cubbiciotti, *J. Chem. Phys.*, **31**, 1646 (1959).
260. D. Cubbiciotti, *J. Chem. Phys.*, **34**, 2189 (1961).
261. R. C. Tolman, *Statistical Mechanics.* New York, 1927.
262. H. S. W. Massey, *Negative Ions.* Cambridge, Univ. Press, 1951.
263. L. M. Branscomb, *Atomic and Molecular Processes.* New York–London, 1962.
264. O. Z. Weber, *Z. f. Phys.*, **152**, 281 (1958).
265. G. Z. Boldt, *Z. f. Phys.*, **154**, 319 (1959).
266. B. H. Amstrong, *Phys. Rev.*, **131**, 1132 (1963).
267. D. Hartree, H. Hartree, *Proc. Camb. Phil. Soc.*, **34**, 550 (1938).
268. J. Roorhan, L. M. Sachs, A. W. Weis, *Rev. Mod. Phys.*, **32**, 186 (1960).
269. L. A. Palkina, B. M. Smirnov, *Dokl. AN SSSR* (1968).
270. J. D. Weisner, B. H. Armstrong, *Proc. Phys. Soc.*, **83**, 31 (1964).
271. J. F. Bidin, V. M. Dukelskii, *JETP*, **31**, 569 (1956).
272. E. A. Mason, J. T. Vanderslice, *J. Chem. Phys.*, **28**, 253 (1958).
273. J. N. Demkov, *JETP*, **46**, 1126 (1964).
274. A. V. Tsaplik, *JETP*, **47**, 126 (1964).
275. J. N. Demkov, *JETP*, **49**, 885 (1965).
276. E. E. Mushlitz, T. L. Bailey, J. H. Simons, *J. Chem. Phys.*, **24**, 1202 (1956).
277. T. L. Bailey, C. J. May, E. E. Muschlitz, *J. Chem. Phys.*, **26**, 1446 (1957).
278. J. B. Hasted, *Proc. Roy. Soc.*, **212A**, 235 (1952).
279. E. Holoien, J. Midtal, *Proc. Phys. Soc.*, **68A**, 815 (1955).
280. Y. F. Bidin, *JETP*, **46**, 1612 (1964).
281. B. M. Smirnov, M. I. Tsibisov, *JETP*, **49**, 841 (1965).
282. V. M. Dukelskii, E. J. Zandberg, N. I. Ionov, *JETP*, **20**, 877 (1950).
283. Y. M. Fogel, V. F. Kozlov, A. A. Kalmikov, *JETP*, **36**, 1954 (1959).
284. O. P. Tsarkin, *J. Struct. Chem.*, **6**, 422 (1965).
285. I. N. Bakulina, N. I. Ionov, *Dokl. AN SSSR*, **155**, 309 (1964).
286. Y. Z. Klaus, *Anorg. Chem.*, **281**, 212 (1955).
287. Y. Lee, B. N. Mahan, *J. Chem. Phys.*, **42**, 2893 (1965).
288. Y. N. Bakulina, N. I. Ionov, *J. Phys. Chem. SSSR*, **33**, 2069 (1959).
289. V. M. Dukelskii, N. I. Ionov, *Dokl. AN SSSR*, **81**, 767 (1951).
290. A. P. Altschuler, *J. Chem. Phys.*, **22**, 765 (1954).

291. T. L. Bailey, *J. Chem. Phys.*, **28**, 792 (1958).
292. J. Simons, B. Seward, *J. Chem. Phys.*, **6**, 790 (1938).
293. R. S. Berry, C. W. Reiman, *J. Chem. Phys.*, **38**, 1540 (1963).
294. B. Steiner, M. Seman, L. M. Branscomb, *J. Chem. Phys.*, **37**, 1200 (1962).
295. F. A. Elder, D. Villarejo, M. G. Ingram, *J. Chem. Phys.*, **43**, 758 (1965).
296. I. N. Bakulina, N. I. Ionov, *Dokl. AN SSSR*, **116**, 41 (1957).
297. W. H. Bennett, P. F. Darby, *Phys. Rev.*, **49**, 97 (1936).
298. D. R. Bates, B. L. Meiseiwitsch, *Proc. Phys. Soc.*, **68A**, 540 (1955).
299. R. J. Gossley, *Proc. Phys. Soc.*, **83**, 375 (1964).
300. V. M. Dukelskii, V. V. Afrosimov, N. V. Fedorenko, *JETP*, **30**, 1193 (1956).
301. L. M. Branscomb, D. S. Burch, S. J. Smith, S. Geltman, *Phys. Rev.*, **111**, 504 (1958).
302. I. N. Bakulina, N. I. Ionov, *Dokl. AN SSSR*, **105**, 75 (1955).
303. I. D. Morrison, *J. Chem. Phys.*, **33**, 821 (1960).
304. D. R. Bates, *Roy. Astron. Soc.*, **106**, 432 (1946).
305. J. W. Cooper, J. B. Martin, *Phys. Rev.*, **111**, 1115 (1958).
306. Y. V. Moskvin, *Opt. and Spectr.*, **17**, 684 (1955).
307. H. S. Taylor, F. E. Harris, *Molec. Phys.*, **7**, 287 (1963–64).
308. Y. M. Dukelskii, E. Y. Zamberg, *JETP*, **24**, 339 (1953).
309. J. B. Hasted, *Proc. Roy. Soc.*, **212A**, 235 (1952).
310. E. F. Chaikovskii, L. G. Melnik, G. M. Pyatigorskii, *JTP*, **XL**, 225 (1970).
311. E. P. Wigner, E. E. Witmer, *Z. f. Phys.*, **51**, 859 (1928).
312. D. R. Bates, R. McCaroll, *Proc. Roy. Soc. A.*, **245**, 175 (1958).
313. D. R. Bates, R. McCarroll, *Phil. Mag. Suppl.*, **11**, 39 (1962).
314. N. Rosen, C. Zener, *Phys. Rev.*, **40**, 502 (1932).
315. O. B. Firsov, *JETP*, **21**, 1001 (1951).
316. O. B. Firsov, *JETP*, **33**, 696 (1957).
317. D. Rapp, I. B. Ortenburger, *J. Chem. Phys.*, **33**, 1230 (1960).
318. J. N. Demkov, *JETP*, **45**, 195 (1963).
319. V. F. Kozlov, S. A. Bondar, *JETP*, **50**, 297 (1965).
320. H. S. W. Massey, R. A. Smith, *Proc. Roy. Soc.*, **142A**, 142 (1933).
321. H. S. W. Massey, *Rep. Progr. Phys.*, **12**, 248 (1949).
322. J. D. Jackson, *Can. J. Phys.*, **32**, 60 (1954).
323. B. L. Moiseiwitsch, *Proc. Phys. Soc.*, **69A**, 653 (1956).
324. D. R. Bates, *Proc. Roy. Soc.*, **243A**, 15 (1957).
325. E. E. Nykitin, *Opt. and Spectr.*, **18**, 763 (1965).
326. W. L. Fite, R. T. Brackmann, W. R. Snow, *Phys. Rev.*, **112**, 1161 (1958).
327. D. G. Hummer, R. F. Stebbings, W. L. Fite, *Phys. Rev.*, **119**, 663 (1960).
328. W. L. Fite, A. H. Smith, R. F. Stebbings, *Proc. Roy. Soc.*, **268A**, 527 (1962).
329. R. F. Potter, *J. Chem. Phys.*, **22**, 974 (1954).
330. J. A. Dillon, W. F. Sheridan, H. D. Edwards, S. N. Ghosh, *J. Chem. Phys.*, **23**, 776 (1955).
331. S. N. Ghosh, W. F. Sheridan, J. Chem. Phys., **26**, 480 (1957).
332. A. Dalgarno, *Phil. Trans.*, **250A**, 426 (1958).
333. N. V. Fedorenko, *JTP*, **24**, 2113 (1954).
334. N. V. Fedorenko, V. V. Afrosimov, D. M. Kaminker, *JTP*, **26**, 1929 (1956).
335. J. B. Hasted, *Proc. Roy. Soc.*, **205A**, 421 (1951).
336. H. B. Gilbody, J. B. Hasted, *Proc. Roy. Soc.*, **238A**, 314 (1957).
337. I. P. Flaks, E. S. Soloviev, *JTP*, **28**, 599 (1958).
338. Y. F. Bidin, A. M. Byhtov, *JETP*, **29**, 12 (1959).
339. R. M. Kusnir, B. M. Pakhiokh, L. A. Sena, *Izv. AN USSSR Ser. Phys.*, **23**, 1007 (1959).
340. R. M. Kusnir, I. M. Butsma, *Izv. AN USSR Ser. Phys.*, **24**, 970 (1960).
341. J. Perel, R. H. Vernon, H. Daley, *Phys. Rev.*, **138A**, 937 (1965).
342. L. L. Marino, A. C. H. Smith, E. Caplinger, *Phys. Rev.*, **128**, 2243 (1962).
343. A. Dalgarno, H. N. Yadev, *Proc. Phys. Soc.*, **66A**, 173 (1953).
344. A. Dalgarno, M. R. McDowell, *Proc. Phys. Soc.*, **69A**, 615 (1956).
345. M. Islam, J. B. Hasted, R. B. Gilbody, I. V. Ireland, *Proc. Phys. Soc.*, **79**, 1118 (1962).
346. I. P. Flaks, L. G. Filipenko, *JETP*, **29**, 1100 (1959).
347. I. P. Flaks, G. N. Ogurtsov, N. V. Fedorenko, *JETP*, **41**, 1094 (1961).
348. J. Dawson, *Phys. Rev.*, **118**, 381 (1960).
349. B. M. Pakhiokh, L. A. Sena, *JETP*, **20**, 481 (1950).

350. R. M. Kusnir, *JTP*, **35**, 2512 (1965).
351. D. B. Hummer, R. F. Stebbings, W. L. Fite, *Phys. Rev.*, **119,** 668 (1960).
352. D. C. Lorenst, G. Black, *Phys. Rev.*, **137A,** 1049 (1965).
353. J. B. H. Stedeford, J. B. Hasted, *Proc. Roy. Soc.*, **227A,** 466 (1955).
354. W. Wien, *Ann. Phys.*, **35,** 519 (1912).
355. S. K. Allison, S. D. Warshaw, *Rev. Mod. Phys.*, **25,** 779 (1953).
356. P. M. Stier, C. F. Bernett, *Phys. Rev.*, **103,** 896 (1956).
357. S. K. Allison, *Rev. Mod. Phys.*, **30,** 1137 (1958).
358. J. M. Fogel, L. I. Krupnik, B. G. Safronov, *JETP*, **28,** 589 (1955).
359. J. M. Fogel, L. I. Krupnik, *JETP*, **29,** 209 (1955).
360. J. M. Fogel, R. V. Mitin, *JETP*, **30,** 450 (1956).
361. J. M. Fogel, R. V. Mitin, A. G. Koval, *JETP*, **31,** 397 (1956).
362. J. M. Fogel, L. I. Krupnik, V. A. Ankudinov, *JTP*, **26,** 1208 (1956).
363. J. M. Fogel, R. V. Mitin, V. F. Kozlov, *JTP*, **28,** 1526 (1958).
364. R. F. Stebbings, A. C. H. Smith, H. J. Ehrhardt, *Geophys. Res.*, **69,** 2349 (1964).
365. J. M. Fogel, V. A. Ankudinov, R. E. Slabospitskii, *JETP*, **32,** 453 (1957).
366. D. R. Bates, *Proc. Roy. Soc.*, **243A,** 437 (1957).
367. D. R. Bates, *Proc. Roy. Soc.*, **245A,** 299 (1958).
368. D. R. Bates, *Proc. Phys. Soc.*, **73,** 227 (1959).
369. D. R. Bates, H. S. W. Massey, *Phil. Trans.*, **239A,** 269 (1943).
370. H. Y. Young, *Proc. Phys. Soc.*, **71,** 341 (1958).
371. C. Greaves, Thesis. University Birmingham England, 1959.
372. D. R. Bates, J. T. Lewis, *Proc. Phys. Soc.*, **68A,** 173 (1955).
373. J. J. Thomson, *Phil. Mag.*, **47,** 337 (1924).
374. G. L. Natanson, *JTP*, **29,** 1373 (1959).
375. K. A. Brueckner, *J. Chem. Phys.*, **40,** 439 (1964).
376. P. Langevin, *Ann. Chem. Phys.*, **28,** 289 (1903).
377. P. Langevin, *Ann. Chem. Phys.*, **28,** 433 (1903).
378. B. L. Moiseiwitsch, *Suppl. Journ. Atm. Terr. Phys.*, **2,** 23 (1955).
379. W. Mächler, *Z. f. Phys.*, **104,** 1 (1936).
380. J. Sayers, *Proc. Roy. Soc.*, **169A,** 83 (1938).
381. M. E. Gardner, *Phys. Rev.*, **53,** 75 (1938).
382. B. M. Smirnov, *JTP*, **37,** 92 (1967).
383. B. H. Mahan, J. C. Person, *J. Chem. Phys.*, **40,** 392 (1964).
384. H. E. Gilbody, J. B. Hasted, J. V. Ireland, A. R. Lee, E. W. Thomas, A. S. Writeman, *Proc. Roy. Soc.*, **274A,** 40 (1963).
385. L. I. Pivovar, V. M. Tubaev, M. T. Novikov, *JETP*, **41,** 26 (1961).
386. T. Hall, *Phys. Rev.*, **79,** 504 (1950).
387. J. A. Phillips, *Phys. Rev.*, **97,** 404 (1955).
388. C. F. Barnett, H. K. Reynolds, *Phys. Rev.*, **109,** 355 (1958).
389. V. A. Abramonov, B. M. Smirnov, *Opt. and Spectr.*, **21,** 19 (1966).
390. C. G. Dolgov–Saveliev, B. A. Knyezev, Y. L. Kozminyuk, V. V. Kuznetsov, *JETP*, **57,** 1001 (1969).
391. Z. Z. Latypov, N. V. Fedorenko, I. P. Floks, A. A. Shaporenko, *JETP*, **55,** 847 (1968).
392. L. P. Pitayevskii, *JETP*, **42,** 1326 (1962).
393. D. R. Bates, A. E. Kingston, R. W. Whirter, *Proc. Roy. Soc.*, **267A,** 297 (1962).
394. D. R. Bates, A. E. Kingston, R. W. Whirter, *Proc. Roy. Soc.*, **270A,** 155 (1962).
395. W. H. Kasner, *Phys. Rev.*, **167,** 148 (1968).
396. L. Frommhold, M. A. Biondi, F. J. Mehr, *Phys. Rev.*, **165,** 45 (1968).
397. F. J. Mehr, M. A. Biondi, *Phys. Rev.*, **176,** 322 (1968).
398. J. Philbrick, F. J. Mehr, M. A. Biondi, *Phys. Rev.*, **181,** 271 (1969).
399. C. S. Weller, M. A. Biondi, *Phys. Rev.*, **172,** 198 (1968).
400. W. H. Kasner, *Phys. Rev.*, **164,** 194 (1967).
401. F. J. Mehr, M. A. Biondi, *Phys. Rev.*, **181,** 264 (1969).
402. W. H. Kasner, M. A. Biondi, *Phys. Rev.*, **174,** 139 (1968).
403. H. J. Oskam, V. R. Mittelstadt, *Phys. Rev.*, **132,** 1445 (1963).
404. E. Hinnov, J. G. Hirschberg, *Phys. Rev.*, **125,** 795 (1962).
405. W. S. Cooper, W. B. Kunkel, *Phys. Rev.*, **138,** 1022 (1965).
406. I. Y. Wada, R. C. Knechtli, *Phys. Rev. Lett.*, **10,** 513 (1963).
407. I. M. Aleskovskii, *JETP*, **44,** 840 (1963).
408. D. R. Bates, H. S. W. Massey, *Rep. Progr. Phys.*, **9,** 62 (1942).
409. A. Prey, W. R. S. Garton, *Proc. Roy. Soc.*, **76,** 833 (1960).

410. D. R. Bates, *Plan. Space. Sci.*, **9**, 77 (1962).
411. C. S. Weller, M. A. Biondi, *Phys. Rev. Lett.*, **19**, 59 (1967).
412. J. Berlande, M. Cheret, R. Deloche, A. Gonfelone, C. Manus, *Phys. Rev.*, **1A**, 887 (1970).
413. G. G. Dolgov–Saveliev, B. A. Knyezev, I. L. Kozminykh, V. V. Kuznetsov, *JETP*, **57**, 1101 (1969).
414. O. K. Rice, *J. Chem. Phys.*, **9**, 258 (1941).
415. G. Careri, *J. Chem. Phys.*, **21**, 749 (1953).
416. O. K. Rice, *J. Chem. Phys.*, **21**, 750 (1953).
417. B. J. Widom, *J. Chem. Phys.*, **31**, 1027 (1959).
418. K. F. Bonhoeffer, *Naturwiss.*, **6**, 21 (1927).
419. W. E. Lamb, R. C. Retherford, *Phys. Rev.*, **79**, 549 (1950).
420. L. Vályi, *KFKI Report*, **15**, 95 (1967).
421. J. M. Hendrie, *J. Chem. Phys.*, **22**, 1503 (1954).
422. R. G. J. Fraser, *Molecular Rays*. Cambridge, University Press, 1931.
423. E. H. Kennard, *Kinetic Theory of Gases*. New York, McGraw-Hill, 1938.
424. J. K. Roberts, *Heat and Thermodynamics*, London, Blackie and Son Ltd., 1933.
425. J. G. King, J. R. Zacharias, *Advances in Electron Physics*, Academic Press, **8** (1956).
426. N. F. Ramsey, *Molecular Beam*. Clarendon Press, Oxford, 1956.
427. M. Knudsen, *Ann. d. Phys.*, **28**, 76, 999 (1909).
428. M. Knudsen, *Ann. d. Phys.*, **29**, 179 (1909).
429. M. Knudsen, *Ann. d. Phys.*, **48**, 1113 (1915).
430. J. E. Mayer, *Z. f. Phys.*, **58**, 373 (1929).
431. P. Clausing, *Physica*, **9**, 65 (1929).
432. P. Clausing, *Z. f. Phys.*, **66**, 471 (1931).
433. L. Vályi, *KFKI Report*, **15**, 251 (1967).
434. G. H. Stafford, J. M. Dickson, D. C. Salter, M. K. Craddock, *Nucl. Instr. and Meth.*, **15**, 146 (1962).
435. M. Smoluchowski, *Ann. Phys. Chem.*, **33**, 1567 (1910).
436. J. A. Giordmaine, T. C. Wang, *J. Appl. Phys.*, **31**, 463 (1960).
437. J. C. Heimer, F. B. Jacobus, P. A. Sturroch, *J. Appl. Phys.*, **31**, 458 (1960).
438. G. R. Hanes, *J. Appl. Phys.*, **31**, 2171 (1960).
439. J. R. Zacharias, *Phys. Rev.*, **94**, 751 (1954).
440. H. H. Sroke, V. Jaccario, D. S. Edmonds, R. Weiss, *Phys. Rev.*, **105**, 590 (1957).
441. H. Rudin, H. R. Striebel, E. Baumgarten, L. Grown, P. Huber, *Helv. Phys. Acta*, **34**, 58 (1961).
442. B. P. Adyasevits, V. G. Antonenko, Y. O. Polunin, D. E. Fomenko, *Atomnaya Energiya*, No. **7**, 17 (1963).
443. E. R. Collis, H. F. Glavish, S. Whineray, *Nucl. Instr. and Meth.*, **25**, 67 (1963).
444. R. P. Slabospitskii, I. E. Kiselev, I. M. Karnaukhov, I. D. Lopatko, A. I. Tapanov, *JTP*, **39**, 1506 (1969).
445. L. Vályi, *Nucl. Instr. and Meth.*, **58**, 21 (1968).
446. B. P. Adyasevits, V. G. Antonenko, *PTE*, **2**, 126 (1963).
447. A. Kantrowitz, J. Grey, *Rev. Sci. Instr.*, **22**, 328 (1951).
448. G. B. Kistiakowsky, W. P. Slichter, *Rev. Sci. Inst.*, **22**, 333 (1951).
449. E. W. Becker, K. Bier, *Z. f. Naturf.*, **9A**, 975 (1954).
450. E. W. Becker, K. Bier, H. Burghoff, *Z. f. Naturf.*, **10A**, 565 (1955).
451. E. W. Becker, K. Henkes, *Z. f. Phys.*, **146**, 320 (1956).
452. H. M. Parker, A. R. Kuhlthan, R. Zatapa, J. E. Scolt, *I. r. Rarefied Gas Dynamics* Nice, Symposium, 1960.
453. R. N. Zatapa, H. M. Parker, *Proc. Atom Molecule Beams Conf. Denver*, 1960.
454. J. Deckers, J. B. Fenn, *Rev. Sci. Instr.*, **34**, 96 (1963).
455. A. Budó, *Mechanika*. Budapest, Tankönyvkiadó, 1952.
456. B. E. Schmidt, *Termodinamik*. Berlin, 1953.
457. I. V. Saveliev, *Mekhanika, Molekularnaya Fizika*. Moskva, 1966.
458. N. D. Papalekszi, *Fizika*. Budapest, Tankönyvkiadó, 1951.
459. G. V. Litman, A. Rosko, *Elementi gazovoi dinamiki*. Moskva, 1960.
460. H. Taub, P. Kusch, *Phys. Rev.*, **75**, 1481 (1949).
461. S. Millman, P. Kusch, *Phys. Rev.*, **60**, 91 (1941).
462. S. Millman, J. J. Rabi, J. R. Zacharias, *Phys. Rev.*, **13**, 384 (1938).

463. D. R. Hamilton, *Phys. Rev.*, **56**, 30 (1939).
464. G. Knight, B. T. Feld, *Report 123, Research Laboratory of Electronics*, Massachussetts Institute of Technology 1949.
465. N. A. Renzetti, *Phys. Rev.*, **57**, 753 (1940).
466. H. Lew, *Phys. Rev.*, **76**, 1086 (1949).
467. H. Lew, *Phys. Rev.*, **91**, 619 (1953).
468. L. Dunoyer, *Compt. Rend.*, **152**, 594 (1911).
469. L. Dunoyer, *Le Radium*, **8**, 142 (1911).
470. P. Kusch, *Lecture Notes on Molecular Beams*. Columbia University, 1950.
471. W. Gerlach, O. Stern, *Ann. d. Phys.*, **74**, 673 (1924).
472. A. Leu, *Z. f. Phys.*, **41**, 551 (1927).
473. J. H. McFee, P. M. Marcus, *Velocity Distributions in Direct and Reflected Atomic Beam*. Carnegie Institute of Technology, 1960.
474. R. C. Miller, P. Kusch, *J. Chem. Phys.*, **25**, 860 (1956).
475. J. T. Eisinger, B. Bederson, B. T. Feld, *Phys. Rev.*, **86**, 73 (1952).
476. J. T. Eisinger, B. Bederson, B. T. Feld, *Reports 212*, Research Laboratory of Electronics, Massachussetts, Institute of Technology, 1952.
477. L. Hollond, *Vacuum*, **6**, 161 (1959).
478. L. Davis, *Report 88, Research Laboratory of Electronics*, Massachussetts, Institute of Technology, 1948.
479. L. Davis, B. T. Feld, C. W. Zabel, J. R. Zacharias, *Phys. Rev.*, **73**, 525 (1948).
480. N. T. Melnikov, E. D. Tsukin, M. M. Umanskii, *JETP*, **22**, 775 (1952).
481. L. Vályi, unpublished results (1964).
482. L. M. Császár, K. Rózsa, *KFKI Reports*, **12**, 175 (1964).
483. P. Kusch, *Journ. Chem. Phys.*, **21**, 1424 (1953).
484. S. A. Ochs, R. E. Cote, P. Kusch, *Journ. Chem. Phys.*, **21**, 459 (1953).
485. N. F. Ramsey, *Phys. Rev.*, **74**, 286 (1948).
486. D. Ehrenstein, *Ann. d. Phys.*, **7**, 342 (1961).
487. G. Wessel, H. Lew, *Phys. Rev.*, **92**, 641 (1953).
488. Y. Ting, H. Lew, *Phys. Rev.*, **105**, 581 (1957).
489. H. Lew, R. S. Title, *Can. J. Phys.*, **38**, 868 (1960).
490. G. Friche, H. Kopferman, S. Penselin, *Z. f. Phys.*, **154**, 218 (1959).
491. G. Friche, H. Kopferman, S. Penselin, *Z. f. Phys.*, **156**, 416 (1959).
492. P. H. Rose, R. P. Bastide, A. B. Witkower, *Rev. Sci. Instr.*, **32**, 581 (1961).
493. W. E. Lamb, R. C. Retherford, *Phys. Rev.*, **81**, 222 (1951).
494. J. W. Heberle, H. A. Reich, P. Kusch, *Phys. Rev.*, **101**, 612 (1956).
495. L. Vályi, *KFKI Reports*, **15**, 373 (1967).
496. A. Leu, *Z. f. Phys.*, **49**, 488 (1928).
497. I. Lindgen, C. M. Johannsson, *Arkhiv för Fyzik*, **15**, 445 (1959).
498. J. E. Sherwood, S. I. Overshine, *Phys. Rev.*, **114**, 858 (1959).
499. R. W. Wood, *Proc. Roy. Soc.*, **97A**, 455 (1920).
500. R. W. Wood, *Proc. Roy. Soc. A.*, **102**, 1 (1922).
501. G. Clausnitzer, *Z. f. Phys.*, **153**, 609 (1959).
502. G. Clausnitzer, *Nucl. Instr. and Meth.*, **23**, 309 (1963).
503. E. R. Collins, H. F. Glavish, *Nucl. Instr. and Meth.*, **30**, 245 (1964).
504. J. E. Sherwood, *Nucl. Instr. and Meth.*, **15**, 103 (1962).
505. J. M. E. Kellogg, I. I. Rabi, J. R. Zacharias, *Phys. Rev.*, **50**, 472 (1936).
506. A. G. Prodeil, P. Kusch, *Phys. Rev.*, **106**, 87 (1957).
507. H. L. Gravin, T. M. Geren, M. Lipworth, *Phys. Rev.*, **111**, 534 (1958).
508. J. J. Thomson, *Phyl. Mag.*, **4**, 1128 (1927).
509. R. Keller, L. Dich, M. Fedecaro, *Helv. Phys. Acta Suppl.*, **VI**, 48 (1961).
510. M. K. Craddock, *Helv. Phys. Acta Suppl.*, **VI**, 59 (1961).
511. L. Brown, E. Baumgartner, P. Huber, H. Rudin, H. R. Striebel, *Helv. Phys. Acta Suppl.*, **VI**, 77 (1961).
512. J. Thirion, R. Beurtey, A. Papineau, *Helv. Phys. Acta Suppl.*, **VI**, 108 (1961).
513. I. J. Barit, G. A. Vasilev, E. A. Glasov, V. V. Jolkin, *Nucl. Instr. and Meth.*, **57**, 160 (1967).
514. G. Clausnitzer, R. Fleischmann, H. Schopper, *Z. f. Phys.*, **144**, 336 (1956).
515. R. Fleismann, *Helv Phys. Acta Suppl.* **VI**, 26 (1961).
516. G. Clausnitzer, *Helv. Phys. Acta Suppl.* **VI**, 35 (1961).
517. L. Vályi, *KFKI Reports*, **14**, 401 (1966).
518. R. Fleismann, *Nucl. Instr. and Meth.*, **11**, 112 (1961).
519. D. E. Nagle, R. S. Julian, I. R. Zacharias, *Phys. Rev.*, **72**, 971 (1947).

520. R. L. Christener, H. G. Benewitz, D. R. Hamilton, J. B. Reyholds, H. H. Stroke, *Phys. Rev.*, **107**, 633 (1957).
521. J. G. King, V. Jaccarino, *Phys. Rev.*, **94**, 1610 (1954).
522. P. H. Rose, *Nucl. Instr. and Meth.*, **11**, 49 (1961).
523. P. H. Rose, A. B. Witkower, R. P. Bastide, A. J. Gale, *Rev. Sci. Instr.*, **32**, 568 (1961).
524. A. B. Witkower, P. H. Rose, R. P. Bastide, N. B. Brooks, *Rev. Sci. Instr.*, **35**, 1 (1964).
525. R. H. Jones, D. R. Clander, V. R. Kruger, *Journ. App. Phys.*, **40**, 4641 (1969).
526. W. Grüebler, P. A. Schmelzbach, V. König, P. Marmier, *Phys. Lett.*, **29A**, 440 (1969).
527. W. Grüebler, P. A. Schmelzbach, V. König, P. Marmier, *Helv. Phys. Acta*, **43**, 254 (1970).
528. S. Chapman, T. G. Cowling, *The Mathematical Theory of Non-uniform Gases.* Cambridge University Press, 1953.
529. V. L. Grenovskii, *Electrical Current in the Gas.* Moscow, Gostheoretizdat, 1952.
530. T. Kihara, *Rev. Mod. Phys.*, **25**, 944 (1953).
531. A. Mason, H. W. Schamp, *Ann. of Phys.*, **4**, 233 (1958).
532. A. Dalgarno, M. R. C. McDowell, A. Williams, *Phil. Trans*, **250A**, 411 (1958).
533. A. M. Tindell, A. F. Pearce, *Proc. Roy. Soc. A.*, **149**, 426 (1935).
534. R. W. Crompton, M. T. Elford, *Proc. Phys. Soc.*, **74**, 497 (1964).
535. L. A. Sena, *JETP*, **11**, 1320 (1939).
536. T. J. Holstein, *Phys. Chem.*, **56**, 832 (1952).
537. R. N. Verney, *Phys. Rev.*, **88**, 262 (1952).
538. M. A. Biondi, L. M. Chanin, *Phys. Rev.*, **94**, 910 (1954).
539. M. A. Biondi, L. M. Chanin, *Phys. Rev.*, **106**, 473 (1957).
540. H. J. Oskam, V. R. Mittelstadt, *Phys. Rev.*, **132**, 1435 (1963).
541. D. E. Korr, C. S. Level, *Bull. Amer. Phys. Soc.*, **7**, 131 (1962).
542. E. C. Beaty, P. L. Patterson. *Bull. Amer. Phys. Soc.*, **7**, 635 (1962).
543. H. J. Oskam, H. M. Mudson, *Bull. Amer. Phys. Soc.*, **7**, 636 (1962).
544. W. Grüebler, V. König, P. Marmier, *Phys. Lett.*, **24B**, 280 (1967).
545. C. W. Drake, R. Krotkov, *Phys. Rev. Lett.*, **16**, 848 (1966).
546. T. B. Clegg, G. R. Plattner, L. G. Koller, W. Haeberli, *Nucl. Instr. and Meth.*, **57**, 167 (1967).
547. G. P. Lawrence, G. G. Ohlsen, J. L. McKibben, *Phys. Lett.*, **28B**, 649 (1969).
548. G. Michel, K. Corrigan, H. Meiner, R. M. Prior, S. E. Darden, *Nucl. Instr. and Meth.*, **78**, 233 (1970).
549. J. Kouloumjian, L. Feuvrais, G. Hadinger, B. Pin, *Nucl. Instr. and Meth.*, **79**, 192 (1970).
550. G. Clausnitzer, W. Dürr, R. Fleischmann, G. Graw, W. Hammon, G. Hartmann, W. Kretschmer, H. Nahr, A. Neufert, E. Salzborn, H. Wolsch, J. Witte, *Nucl. Instr. and Meth.*, **80**, 245 (1970).
551. W. Grüebler, V. König, P. A. Schmelzbach, *Nucl. Instr. and Meth.*, **86**, 127 (1970).
552. V. Bejsovec, P. Bém, J. Mares, Z. Trejbal, *Nucl. Instr. and Meth.*, **87**, 233 (1970).
553. E. Goldstein, *Berl. Ber.*, **39**, 691 (1886).
554. W. Wien, *Ann. d. Phys.*, **8**, 260 (1902).
555. J. J. Thomson, *Phil. Mag.*, **20**, 752 (1910).
556. J. J. Thomson, *Phil. Mag.*, **21**, 225 (1911).
557. F. W. Aston, *Proc. Roy. Soc.*, **130**, 303 (1931).
558. M. L. E. Oliphant, E. Rutherford, *Proc. Roy. Soc.*, **141A**, 259 (1933).
559. W. Bethe, W. Gentner, *Z. f. Phys.*, **104**, 685 (1937).
560. J. D. Craggs, *Proc. Phys. Soc.*, **54**, 245 (1942).
561. J. D. Craggs, *J. Appl. Phys.*, **13**, 772 (1942).
562. P. Huber, F. Metzger, *Helv. Phys. Act.*, **19**, 200 (1946).
563. W. E. Burgham, *Nature*, **160**, 316 (1947).
564. D. Kamke, *Z. f. Naturf.*, **40**, 391 (1949).
565. D. Kamke, *Z. f. Phys.*, **128**, 212 (1950).
566. D. Kamke, *Z. f. Naturf.*, **7**, 341 (1952).
567. K. Deutscher, D. Kamke, *Z. f. Phys.*, **135**, 380 (1953).
568. A. J. Dempster, *Phil. Mag.*, **31**, 438 (1916).
569. A. J. Dempster, *Phys. Rev.*, **18**, 415 (1921).
570. A. J. Dempster, *Phys. Rev.*, **20**, 631 (1922).
571. H. D. Smyth, *Proc. Roy. Soc.*, **104A**, 121 (1923).

572. H. D. Smyth, *Phys. Rev.*, **25**, 452 (1925).
573. W. Bleakney, *Phys. Rev.*, **40**, 496 (1932).
574. J. T. Tate, P. T. Smith, *Phys. Rev.*, **46** ,773 (1934).
575. A. O. Nier, *Phys. Rev.*, **50**, 1041 (1936).
576. A. O. Nier, *Phys. Rev.*, **52**, 933 (1937).
577. A. O. Nier, *Phys. Rev.*, **53**, 282 (1938).
578. A. O. Nier, *Rev. Sci. Instr.*, **11**, 212 (1940).
579. A. T. Finkelstein, *Rev. Sci. Instr.*, **11**, 94 (1940).
580. N. D. Coggeschall, E. B. Jordan, *Rev. Sci. Instr.*, **14**, 125 (1943).
581. W. Walcher, *Z. f. Phys.*, **122**, 62 (1944).
582. R. L. Graham, A. L. Harknes, H. G. Thode, *J. Sci. Instr.*, **24**, 119 (1947).
583. R. H. Bornas, A. O. Nier, *Rev. Sci. Instr.*, **19**, 895 (1948).
584. C. H. Kusman, *Phys. Rev.*, **25**, 892 (1925).
585. C. H. Kusman, *Phys. Rev.*, **27**, 249 (1926).
586. H. A. Barton, C. P. Harnwell, C. H. Kusman, *Phys. Rev.*, **27**, 739 (1926).
587. C. G. Smidt, *Ann. d. Phys.*, **82**, 644 (1927).
588. P. B. Moon, M. L. E. Oliphant, *Proc. Roy. Soc.*, **137A**, 463 (1932).
589. P. Keck, L. B. Loev, *Rev. Sci. Instr.*, **4**, 486 (1933).
590. H. Bondy, G. Johannsen, K. Popper, *Z. f. Phys.*, **95**, 46 (1935).
591. H. Bondy, V. Vanicek, *Z. f. Phys.*, **101**, 186 (1936).
592. J. P. Blewett, E. J. Jones, *Phys. Rev.*, **50**, 464 (1936).
593. A. K. Brewer, *J. Chem. Phys.*, **4**, 350 (1936).
594. A. K. Brewer, *J. Amer. Chem. Soc.*, **58**, 365 (1936).
595. W. Welcher, *Z. f. Phys.*, **121**, 604, 669 (1943).
596. H. Schaefer, W. Walcher, *Z. f. Phys.*, **121**, 679 (1943).
597. R. J. Hayden, *Phys. Rev.*, **74**, 650 (1948).
598. L. G. Lewis, R. J. Hayden, *Rev. Sci. Instr.*, **19**, 599 (1948).
599. H. Hintenberger, *Helv. Phys. Acta*, **24**, 307 (1951).
600. I. Cornides, J. Roosz, A. Siegler, *Nucl. Instr. and Meth.*, **1**, 94 (1957).
601. A. J. Dempster, *Phys. Rev.*, **11**, 316 (1918).
602. K. H. Kingdona, I. Langmuir, *Phys. Rev.*, **21**, 380 (1923).
603. L. N. Dobretsov, *JETP*, **4**, 783 (1934).
604. M. G. Inghram, W. A. Chupka, *Rev. Sci. Instr.*, **24**, 518 (1953).
605. O. K. Husmann, *J. Appl. Phys.*, **37**, 4662 (1966).
606. H. L. Daley, J. Perel, R. H. Vernon, *Rev. Sci. Instr.*, **37**, 473 (1966).
607. H. Hintenberg, N. Lang, *Z. f. Naturf.*, **11A**, 167 (1956).
608. H. Hintenberg, W. Voshage, *Z. f. Naturf.*, **14A**, 216 (1959).
609. S. K. Allison, A. Kamegai, *Rev. Sci. Instr.*, **32**, 1090 (1961).
610. R. G. Wilson, *J. Appl. Phys.*, **37**, 3170 (1966).
611. D. M. Jamba, *Rev. Sci. Instr.*, **40**, 1072 (1969).
612. H. R. Crane, C. Lauritsen, A. Soltan, *Phys. Rev.*, **45**, 507 (1934).
613. M. A. Tuve, O. Dahl, G. Van Atta, *Phys. Rev.*, **46**, 1027 (1934).
614. R. Fowler, G. Gibson, *Phys. Rev.*, **46**, 1075 (1934).
615. M. A. Tuve, O. Dahl, L. Hafstad, *Phys. Rev.*, **48**, 241 (1935).
616. M. A. Tuve, L. Hafstad, O. Dahl, *Phys. Rev.*, **48**, 315 (1935).
617. E. S. Lamar, E. W. Samson, K. T. Compton, *Phys. Rev.*, **48**, 886 (1935).
618. H. R. Crane, *Phys. Rev.*, **52**, 11 (1937).
619 W H Zinn, *Phys Rev.*, **52**, 655 (1937).
620. G. Timoschenko, *Rev. Sci. Instr.*, **9**, 187 (1938).
621. L. P. Smith, G. W. Scott, *Phys. Rev.*, **55**, 946 (1939).
622. E. S. Lamar, W. W. Buecker, R. J. Graaf, *Phys. Rev.*, **51**, 936 (1939).
623. G. W. Scott, *Phys. Rev.*, **55**, 954 (1939).
624. S. K. Allison, *Phys. Rev.*, **57**, 71 (1940).
625. S. K. Allison, *Rev. Sci. Instr.*, **19**, 291 (1948).
626. T. Jorgensen, *Rev. Sci. Instr.*, **19**, 28 (1948).
627. F. M. Penning, *Physica*, **3**, 87 (1936).
628. F. M. Penning, J. H. Moubis, *Physica*, **4**, 1190 (1937).
629. A. T. Finkelstein, *Rev. Sci. Instr.*, **11**, 94 (1940).
630. M. Ardenne, *Phys. Zeit.*, **43**, 91 (1942).
631. H. Heil, *Z. f. Phys.*, **120**, 212 (1943).
632. M. Ardenne, *Z. f. Phys.*, **121**, 236 (1943).
633. D. Cowie, C. Kasanda, *Rev. Sci. Instr.*, **16**, 224 (1945).
634. M. S. Livingston, *Rev. Mod. Phys.*, **18**, 293 (1946).

635. P. Lorrian, *Can. J. Res.*, **25**, 338 (1947).
636. P. Lorrian, *Helv. Phys. Acta*, **21**, 497 (1948).
637. H. Atterling, *Ark. Mat. Astr. Phys.*, **35A**, 1 (1948).
638. R. Keller, *Helv. Phys. Acta*, **21**, 170 (1948).
639. R. Keller, *Helv. Phys. Acta*, **22**, 78, 386 (1949).
640. C. Bailey, D. Drukey, F. Oppenheimer, *Rev. Sci. Instr.*, **20**, 189 (1949).
641. R. Keller, *Helv. Phys. Acta*, **23**, 627 (1950).
642. E. W. Beach, W. S. Parkinson, *Rev. Sci. Instr.*, **22**, 697 (1951).
643. E. M. Rekhrudel, A. B. Chernetinskiy, V. V. Mekhnevits, J. A. Vasileva, *JTP*, **22**, 1945 (1952).
644. P. C. Veenstra, *Physica*, **18**, 378 (1952).
645. J. D. Gow, J. S. Foster, *Rev. Sci. Instr.*, **24**, 606 (1953).
646. C. F. Barnett, P. M. Steir, G. E. Evans, *Rev. Sci. Instr.*, **24**, 394 (1953).
647. C. B. Mills, C. F. Barnett, *Rev. Sci. Instr.*, **25**, 1200 (1954).
648. J. Flinta, R. Pauli, *Arkiv för Fysik*, **8**, 7 (1954).
649. R. J. Jones, A. Zucker, *Rev. Sci. Instr.*, **25**, 562 (1954).
650. B. Cork, *Rev. Sci. Instr.*, **26**, 210 (1955).
651. P. M. Morozov, B. N. Makhov, M. S. Ioffe, *Atomnaya Energiya*, **2**, 272 (1957).
652. C. E. Anderson, K. W. Ehlers, *Rev. Sci. Instr.*, **27**, 809 (1956).
653. A. A. Plutto, K. N. Kervalidze, I. F. Kvarchava, *Atomnaya Energiya*, **3**, 153 (1957).
654. M. D. Gabovits, O. F. Nemec, Z. P. Fedorus, *Ukr. Phys. Journ.*, **3**, 104 (1958).
655. M. D. Gabovits, *Ukr. Phys. Journ.*, **3**, 693 (1958).
656. K. W. Ehlers, J. D. Gow, L. Ruby, J. M. Wilcex, *Rev. Sci. Instr.*, **29**, 614 (1958).
657. J. Flinta, *Nucl. Instr. and Meth.*, **2**, 219 (1958).
658. G. Först, *Z. angew. Phys.*, **10**, 546 (1958).
659. M. Ardenne, S. Schiller, *Kernenergie*, **2**, 893 (1959).
660. J. Nagy, *Acta Univ. Debreceniensis*, **VI/2**, K, 55 (1959–60).
661. L. J. Bolotin, P. S. Markin, S. I. Meleskov, *PTE*, No. **6**, 86 (1961).
662. C. D. Moak, H. E. Banta, J. N. Thurston, J. W. Jonson, R. F. King, *Rev. Sci. Instr.*, **30**, 694 (1959).
663. W. Lamb, E. Lofgreen, *Rev. Sci., Instr.*, **27**, 907 (1956).
664. L. I. Bolotin, P. S. Markin, J. F. Kuligin, *PTE*, No. **6**, 88 (1961).
665. A. Svanheden, *Nucl. Instr. and Meth.*, **10**, 125 (1961).
666. I. I. Afanasev, L. F. Kyazyatov, N. D. Fedotov, *Atomnaya Energiya*, **13**, 135 (1962).
667. F. I. Mineev, O. F. Kovpik, *JTP*, **33**, 1444 (1963).
668. P. M. Morozov, L. N. Pilgunov, *JTP*, **33**, 470 (1963).
669. F. I. Mineev, O. F. Kovpik, *PTE*, No. **4**, 33 (1963).
670. M. D. Gabovits, *PTE*, No. **2**, 5 (1963).
671. R. A. Demirkhanov, I. V. Kursanov, V. M. Blagoveschenskii, *PTE*, No. **1**, 30 (1964).
672. K. W. Ehlers, B. F. Gavin, E. L. Hubbard, *Nucl. Instr. and Meth.*, **22**, 87 (1963).
673. S. Kojima, T. Kawabe, *Journ. Phys. Soc. Japan*, **18**, 1553 (1963).
674. J. Nagy, *ATOMKI Reports*, **5**, 143 (1963).
675. J. Nagy, *Nucl. Instr. and Meth.*, **32**, 229 (1964).
676. L. E. Collins, R. J. Brooker, *Nucl. Instr. and Meth.* **15**, 193 (1962).
677. A. B. Wittkower, A. Galejs, P. H. Rose, R. P. Bastide, *Rev. Sci. Instr.*, **33**, 515 (1962).
678. T. I. Danilina, E. V. Ivanova, J. E. Kreyndel, L. A. Levshuk, *PTE*, No. **3**, 158 (1968).
679. J. F. Bernandet, F. Riponteau, M. Fruneau, *Rev. de Phys. Appl.*, **4**, 169 (1969).
680. G. G. Kelley, N. H. Lazar, O. B. Morgan, *Nucl. Instr. and Meth.*, **10**, 263 (1961).
681. E. Heinicke, H. Baumann, *Nucl. Instr. and Meth.*, **74**, 229 (1969).
682. J. A. Getting, *Phys. Rev.*, **59**, 467 (1941).
683. P. C. Thonemann, *Nature*, **158**, 61 (1946).
684. J. G. Rutherglen, J. F. I. Cole, *Nature*, **160**, 545 (1947).
685. P. C. Thonemann, J. Moofat, D. Roaf, J. H. Sanders, *Proc. Phys. Soc.*, **61**, 483 (1948).
686. A. J. Bayly, A. G. Ward, *Can. J. Res.*, **36**, 69 (1948).
687. R. N. Hall, *Rev. Sci. Instr.*, **19**, 905 (1948).
688. M. Hoyanx, J. Dujardin, *Nucleonics*, **4**, 7, 12 (1949).
689. A. G. Ward, *Helv. Phys. Acta*, **23**, 3, 27 (1950).

690. C. D. Moak, H. T. Reese, V. M. Good, *Nucleonics*, **9**, 18 (1951).
691. J. S. Swingle, C. B. Swann, *Rev. Sci. Instr.*, **23**, 636 (1952).
692. J. S. Swingle, C. B. Swann, *Bull. Amer. Phys. Soc.*, **27**, 8 (1952).
693. L. K. Goodwin, *Rev. Sci. Instr.*, **24**, 635 (1953).
694. A. Lindberg, H. Nenert, H. Weinder, *Naturwiss.*, **39**, 374 (1952).
695. A. N. Banerjee, *Indian Journ. of Phys.*, **27**, 523 (1953).
696. O. Reifenschweiler, *Z. Naturforsch.*, **6A**, 331 (1951).
697. P. J. Beauregard, *J. Physique Rad.*, **14**, 547 (1953).
698. H. P. Eubank, R. A. Peck, R. Truell, *Rev. Sci. Instr.*, **25**, 989 (1954).
699. O. Reifenschweiler, *Ann. Phys.*, **14**, 33 (1954).
700. P. M. Lacosta, J. Salmon, S. Wajsbrum, *J. Physique Rad.*, **15**, 117 (1954).
701. R. Budde, P. Hubert, *Helv. Phys. Acta*, **25**, 459 (1952).
702. J. Erő, *Magy. Fiz. Folyóirat*, **3**, 529 (1955).
703. V. M. Morozov, *Dokl. AN SSSR*, **102**, 61 (1955).
704. J. Erő, *Acta Phys. Hung.*, **5**, 391 (1956).
705. J. Erő, L. Vályi, *KFKI Reports*, **5**, 414 (1957).
706. P. C. Thonemann, *Progr. Nucl. Phys.*, **3**, 219 (1953).
707. P. C. Thonemann, E. R. Harrison, *AERE Report, C. P/R*, 1190 Harwell, 1953.
708. J. Desjonqueres, R. Geller, F. Prévet, R. Vienet, *J. Physique Rad.*, **17**, 166 (1956).
709. J. Desjonqueres, R. Geller, *Le Vide*, **12**, 161 (1957).
710. A. N. Serbinov, *PTE*, **3**, 39 (1958).
711. G. S. Malkiel, B. I. Sukhanov, *PTE*, **3**, 100 (1958).
712. M. D. Gabovits, *JTP*, **28**, 872 (1958).
713. J. Depraz, *J. Physique Rad.*, **19**, 86 (1958).
714. D. Blanc, A. Degeilh, *J. Physique Rad.*, **19**, 61-S (1958).
715. D. Blanc, A. Degeilh, *J. Physique Rad.*, **20**, 55-A (1959).
716. A. N. Serbinov, V. I. Petrov, *PTE*, **5**, 3 (1958).
717. A. N. Serbinov, V. I. Moroka, *PTE*, **5**, 27 (1960).
718. D. Blanc, A. Degeilh, *J. Physique Rad.*, **22**, 230 (1961).
719. R. M. Komarov, V. I. Petrov, *JTP*, **31**, 321 (1961).
720. V. I. Petrov, *JTP*, **31**, 348 (1961).
721. V. A. Romanov, A. N. Serbinov, *PTE*, **1**, 27 (1963).
722. A. Degeilh, D. Blanc, *Journ. de Phys.*, **24**, 187 (1963).
723. A. V. Almazov, F. F. Myntsov, *PTE*, **5**, 43 (1964).
724. V. J. Kowalenski, C. A. Mayans, M. Hammerschlag, *Nucl. Instr. and Meth.*, **5**, 90 (1959).
725. J. Nagy, P. Gombos, *ATOMKI Reports*, **4**, 19 (1962).
726. J. Nagy, P. Gombos, *ATOMKI Reports*, **5**, 39 (1963).
727. C. J. Cook, O. Heinz, D. C. Lorents, J. R. Peterson, *Rev. Sci. Instr.*, **33**, 649 (1962).
728. E. R. Harrison, *J. Appl. Phys.*, **19**, 909 (1958).
729. A. K. Valter, *Elektrostat. Uskor. Moskva* (1963)
730. L. Vályi, P. Gombos, J. Roosz, *KFKI Reports*, **12**, 461 (1964).
731. V. A. Romanov, A. N. Serbinov, *PTE*, **5**, 34 (1965).
732. J. Nagy, *ATOMKI Reports*, **7**, 209 (1965).
733. J. Roosz, P. Gombos, L. Vályi, *KFKI Reports*, **14**, 319 (1966).
734. L. Vályi, P. Gombos, J. Roosz, *KFKI Reports*, **14**, 259 (1966).
735. L. Vályi, P. Gombos, J. Roosz, *Nucl. Instr. and Meth.*, **49**, 316 (1967).
736. A. K. Ganguly, H. Bakhru, *Nucl. Instr. and Meth.*, **21**, 56 (1969).
737. K. Prelec, *Nucl. Instr. and Meth.*, **26**, 320 (1964).
738. L. Vályi, *Nucl. Instr. and Meth.*, **79**, 315 (1970).
739. H. Tawara, M. Sonoda, *Nucl. Instr. and Meth.*, **83**, 67 (1970).
740. C. Mallen, P. Taras, *Nucl. Instr. and Meth.*, **71**, 333 (1969).
741. J. R. Pierce, *J. Appl. Phys.*, **11**, 548 (1940).
742. J. R. Pierce, *Theory and Design of Electron Beams*, New York, London, Van-Nostrand, 1954.
743. G. Fricke, *Z. f. Phys.*, **141**, 166 (1955).
744. F. Bernhart, *Z. angew. Phys.*, **9**, 69 (1957).
745. R. Weiss, *Rev. Sci. Instr.*, **32**, 397 (1961).
746. R. P. Slabospitskii, I. M. Karnaukhov, I. E. Kisel, *JTP*, **36**, 2145 (1966).
747. O. J. Ekhitsev, G. N. Zinchenko, N. S. Zinchenko, I. M. Karnukhov, R. P. Slabospitskii, A. J. Taranov, *JTP*, **36**, 1681 (1966).
748. P. B. Moon, M. L. Oliphant, *Proc. Roy. Soc.*, **137A**, 463 (1932).
749. R. C. Evans, *Proc. Roy. Soc.*, **139A**, 604 (1933).

750. S. V. Starodubtsev, *JETP*, **19**, 215 (1949).
751. U. Arifov, V. M. Lovtsov, *Dokl. AN SSSR*, **75**, 365 (1950).
752. S. Datz, E. H. Taylor, *J. Chem. Phys.*, **25**, 389, 395 (1956).
753. E. J. Zandberg, *JTP*, **27**, 2583 (1957).
754. J. Koch, *Z. f. Phys.*, **100**, 669 (1936).
755. S. Dietz, *Rev. Sci. Instr.*, **30**, 235 (1959).
756. S. Dietz, *Rev. Sci. Instr.*, **31**, 1229 (1960).
757. D. C. Hess, A. Marschall, H. C. Urey, *Science*, **126**, 1291 (1957).
758. R. J. Hayden, *Phys. Rev.*, **74**, 650 (1948).
759. D. B. Langmuir, *Phys. Rev.*, **49**, 428 (1936).
760. H. E. Ives, *Journ. Franklin Inst.*, **201**, 47 (1926).
761. E. Meyer, *Ann. Phys.*, **4**, 357 (1930).
762. E. J. Zandberg, N. I. Ivanov, *JTP*, **28**, 2444 (1958).
763. E. Stuhlinger, *J. Astronautics*, **1955**, 149.
764. J. J. Stravinski, S. J. Lebedev, *JTP*, **30**, 1222 (1960).
765. S. J. Lebedev, J. J. Stravinski, *JTP*, **31**, 1148 (1961).
766. O. K. Husmann, *Progress in Astronautics*, **5**, 505 (1961).
767. M. J. Copley, T. E. Phipps, *Phys. Rev.*, **45**, 344 (1934).
768. M. J. Copley, T. E. Phipps, *Phys. Rev.*, **48**, 960 (1935).
769. E. J. Zanberg, *JTP*, **28**, 2434 (1958).
770. J. E. Sherwood, *Nucl. Instr. and Meth.*, **15**, 103 (1962).
771. V. W. Hughes, C. W. Drake, D. C. Bonar, J. S. Greensberg, G. F. Pieper, *Helv. Phys. Acta Suppl.*, **VI**, 89 (1961).
772. W. Grüebler, W. Haeberli, P. Schwandt, *Phys. Rev. Lett.*, **12**, No. 21, 595 (1964).
773. W. Grüebler, P. Schwandt, T. I. Yule, W. Haeberli, *Nucl. Instr. and Meth.*, **41**, 245 (1766).
774. B. L. Donnally, T. Clapp, W. Sawyer, M. Schultz, *Phys. Rev. Lett.*, **12**, No. 18, 502 (1964).
775. B. L. Donnally, W. Sawyer, *Phys. Rev. Lett.*, **15**, No. 10, 439 (1965).
776. A. Cesati, F. Cristofori, L. Milazzo-Colli, P. G. Sena, *Energia Nucl. (Milan)*, **13**, 649 (1966).
777. H. Brückmann, D. Finken, L. Friedrich, *KFK Report*, 914 (1968)
778. J. A. Plis, L. M. Soroko, N. A. Toropkov, *JTP*, **37**, 485 (1967).
779. R. N. Boyd, J. C. Lombardi, A. B. Robbins, D. E. Schechter, *Nucl. Instr. and Meth.*, **81**, 149 (1970).
780. D. O. Findley, S. D. Baker, E. B. Carter, N. D. Stockwell, *Nucl. Instr. and Meth.*, **71**, 125 (1969).
781. J. Schwinger, *Phys. Rev.*, **69**, 681 (1946).
782. J. Schwinger, *Phys. Rev.*, **73**, 407 (1948).
783. L. Wolfenstrin, *Phys. Rev.*, **75**, 342 (1949).
784. *Helv. Phys. Acta Suppl.*, **VI** (1961).
785. P. Huber, H. Schopper (editors), Proc. 2nd Intern. Symp. Polarization Phenomena of Nucleons, Basel, Birkhäuser Verlag (1966).
786. W. Shottky, *Phys. Zeit.*, **25**, 342 (1924).
787. R. Seeliger, *Phys. Zeit.*, **33**, 273, 312 (1932).
788. W. Funk, R. Seeliger, *Z. f. Phys.*, **110**, 271 (1938).
789. J. Langmuir, *Phys. Rev.*, **33**, 954 (1929).
790. L. Tonks, I. Langmuir, *Phys. Rev.*, **34**, 876 (1929).
791. I. J. Killian, *Phys. Rev.*, **35**, 1238 (1931).
792. M. J. Druyvesteyn, *Z. f. Phys.*, **81**, 571 (1933).
793. M. J. Druyvesteyn, N. Warmholtz, *Phil. Mag.*, **17**, 1 (1934).
794. B. N. Klarfeld, *Dokl. AN SSSR*, **24**, 250 (1939).
795. B. N. Klarfeld, *Dokl. AN SSSR*, **26**, 870 (1940).
796. B. N. Klarfeld, *Journ. of Phys. USSR*, **5**, 155 (1941).
797. N. A. Karelina, *Journ. of Phys. USSR*, **6**, 218 (1942).
798. V. E. Gavilov, A. V. Zarinov, V. I. Rayko, *Atomnaya Energiya*, **13**, 448 (1962).
799. I. Langmuir, H. Mott-Smith, *Phys. Rev.*, **28**, 727 (1926).
800. A. Eisenstein, J. R. Young, *Phys. Rev.*, **75**, 347 (1949).
801. J. S. Townsend, *Phil. Mag.*, **1**, 198 (1901).
802. J. S. Townsend, *Phil. Mag.*, **3**, 557 (1902).
803. J. S. Townsend, *Phil. Mag.*, **9**, 289 (1905).
804. J. S. Townsend, *Phil. Mag.*, **27**, 789 (1914).
805. M. Ardenne, *Exp. Tech. der Phys.*, **9**, 227 (1961).

806. W. Rogowski, *Arch. für Electrotech.* **26**, 643 (1932).
807. W. Rogowski, *Z. f. Phys.*, **114**, 1 (1939).
808. W. Rogowski, *Z. f. Phys.*, **117**, 265 (1941).
809. D. Posin, *Phys. Rev.*, **47**, 258 (1935).
810. D. Posin, *Phys. Rev.*, **50**, 650 (1936).
811. L. B. Loeb, *Rev. Mod. Phys.*, **8**, 277 (1936).
812. L. B. Loeb, *Proc. Phys. Soc.*, **60**, 561 (1948).
813. D. H. Hale, *Phys. Rev.*, **56**, 1199 (1939).
814. D. H. Hale, *Phys. Rev.*, **54**, 241 (1938).
815. S. P. McCallun, L. Kletzow, *Nature*, **131**, 841 (1931).
816. D. H. Hale, *Phys. Rev.*, **55**, 815 (1939).
817. F. Llewelly Jones, J. P. Hengerson, *Phil. Mag.*, **28**, 185 (1939).
818. B. Frey, *Ann. d. Phys.*, **85**, 381 (1928).
819. L. G. H. Huxley, *Phil. Mag.*, **10**, 185 (1930).
820. J. H. Bruce, *Phil. Mag.*, **10**, 476 (1930).
821. H. F. Boulind, *Phil. Mag.*, **20**, 68 (1935).
822. J. D. Craggs, J. M. Meek, *Proc. Phys. Soc.*, **61**, 327 (1948).
823. S. S. Cerwin, *Phys. Rev.*, **46**, 1054 (1934).
824. M. J. Druyvesteyn, F. M. Penning, *Rev. Mod. Phys.*, **12**, 87 (1940).
825. W. R. Carr, *Phil. Trans. A.*, **201**, 403 (1903).
826. A. V. Afanaseva, N. A. Kaptsov, *JTP*, **3**, 1004 (1933).
827. H. E. Farnsworth, *Phys. Rev.*, **25**, 41 (1925).
828. R. L. Petry, *Phys. Rev.*, **26**, 346 (1925).
829. R. L. Petry, *Phys. Rev.*, **28**, 362 (1926).
830. A. Afanaseva, P. Timofiev, A. Ignatov, *Phys. Zh. Soviet.*, **10**, 831 (1936).
831. R. Warnecke, *J. Physique Rad.*, **7**, 270 (1936).
832. A. Afanaseva, P. Timofiev, *JTP*, **4**, 953 (1937).
833. R. Kollath, *Z. f. Physik*, **38**, 202 (1937).
834. N. S. Klebnikov, *JTP*, **5**, 593 (1938).
835. H. Bruining, J. H. de Boer, *Physica*, **5**, 17 (1938).
836. L. R. G. Treloar, D. H. Landon, *Proc. Phys. Soc.*, **50B**, 625 (1938).
837. R. Kollath, *Ann. Phys.*, **33**, 285 (1938).
838. H. Brining, *Philips Techn. Rev.*, **3**, 80 (1938).
839. G. Schneider, *Phys. Rev.*, **54**, 185 (1938).
840. E. A. Coomes, *Phys. Rev.*, **55**, 519 (1939).
841. D. E. Wooldridge, *Phys. Rev.*, **56**, 1062 (1939).
842. M. M. Vudinski, *JTP*, **9**, 271 (1939).
843. H. Brining, J. H. de Boer, *Physica*, **6**, 834 (1939).
844. H. Salov, *Z. Tech. Phys.*, **21**, 8 (1940).
845. N. D. Morgulis, B. I. Diatlovitskaya, *JTP*, **10**, 657 (1940).
846. M. S. Joffe, I. V. Nekhlaev, *JETP*, **11**, 93 (1941).
847. P. M. Morozov, *JETP*, **11**, 410 (1941).
848. K. H. Geyer, *Ann. d. Phys.*, **42**, 241 (1942).
849. R. Suhrmann, W. Kundt, *Z. f. Phys.*, **120**, 363 (1943).
850. R. Suhrmann, W. Kundt, *Z. f. Phys.*, **121**, 118 (1943).
851. M. Knoll, O. Hachenberg, J. Rendmer, *Z. f. Phys.*, **122**, 137 (1944).
852. C. W. Mueller, *J. Appl. Phys.*, **16**, 453 (1945).
853. L. R. Koller, J. S. Burgess, *Phys. Rev.*, **70**, 571 (1946).
854. H. E. Mendelhall, *Phys. Rev.*, **72**, 532 (1947).
855. J. J. Brophy, *Phys. Rev.*, **83**, 534 (1951).
856. G. Blankenfeld, *Ann. d. Phys.*, **9**, 48 (1951).
857. H. Gobrecht, F. Speer, *Z. f. Phys.*, **135**, 602 (1953).
858. A. R. Shulman, W. L. Makedonski, J. D. Kharoshetski, *JTP*, **23**, 1152 (1953).
859. J. B. Johnson, K. G. McKay, *Phys. Rev.*, **91**, 582 (1953).
860. J. B. Johnson, *Phys. Rev.*, **92**, 843 (1953).
861. J. B. Johnson, K. G. McKay, *Phys. Rev.*, **93**, 668 (1954).
862. J. Woods, *Proc. Phys. Soc.*, **67B**, 843 (1954).
863. N. B. Gornii, *JETP*, **26**, 79 (1954).
864. N. B. Gornii, *JETP*, **26**, 88 (1954).
865. R. Rappaport, *J. Appl. Phys.*, **25**, 288 (1954).
866. A. R. Shulman, B. P. Dementev, *JTP*, **25**, 2256 (1955).
867. R. G. Lye, *Phys. Rev.*, **99**, 1647 (1955).
868. T. L. Matskevits, *JTP*, **26**, 2399 (1956).

869. P. Wargo, B. V. Haxby, W. G. Shepherd, *J. Appl. Phys.*, **27**, 1311 (1956).
870. N. Rey Whetten, A. B. Laponsky, *J. Appl. Phys.*, **28**, 515 (1957).
871. D. N. Dobretsov, T. L. Matskevits, *JTP*, **27**, 734 (1957).
872. N. Rey Whetten, A. B. Laponsky, *J. Appl. Phys.*, **30**, 432 (1959).
873. H. B. Michaelson, *J. Appl. Phys.*, **21**, 536 (1950).
874. H. Brnining, *Physica*, **3**, 1046 (1936).
875. H. O. Müller, *Z. f. Phys.*, **104**, 475 (1937).
876. H. Brnining, *Physica*, **5**, 901 (1938).
877. S. Lukyanov, *Phys. Zh. Soviet*, **13**, 123 (1938).
878. J. L. H. Jonker, *Philips Res. Repts.*, **6**, 372 (1951).
879. E. Rudberg, *Phys. Rev.*, **45**, 764 (1934).
880. E. Rudberg, *Phys. Rev.*, **50**, 638 (1936).
881. L. J. Hawort, *Phys. Rev.*, **50**, 216 (1936).
882. R. Kollath, *Ann. d. Phys.*, **39**, 59 (1941).
883. A. E. Kadisevits, *Journ. of Phys. USSR*, **9**, 431 (1943).
884. J. Bronstein, *JTP*, **13**, 176 (1943).
885. A. V. Afanaseva, P. V. Timofeev, *Techn. Phys. USSR*, **4**, 953 (1937).
886. J. H. De Boer, H. Brnining, *Physica*, **6**, 941 (1939).
887. L. R. G. Treboar, *Proc. Phys. Soc.*, **49**, 392 (1937).
888. K. G. McKay. *Phys. Rev.*, **61**, 708 (1942).
889. D. E. Wooldridge, *Phys. Rev.*, **56**, 562 (1939).
890. R. M. Chandri, *Proc. Can. Phil. Soc.*, **28**, 349 (1932).
891. A. G. Hill, W. W. Beuckner, J. S. Clark, J. B. Fisk, *Phys. Rev.*, **55**, 463 (1939).
892. M. Healea, C. Heutermans, *Phys. Rev.*, **58**, 608 (1940).
893. A. Rostagni, *Nuovo Cim.*, **11**, 99 (1934).
894. M. L. Lophant, *Proc. Roy. Soc.*, **127**, 373 (1930).
895. P. B. Moon, *Proc. Can. Phil. Soc.*, **27**, 570 (1931).
896. W. J. Jackson, *Phys. Rev.*, **30**, 473 (1927).
897. M. L. Oliphant, *Proc. Can. Phil. Soc.*, **24**, 451 (1928).
898. V. I. Pavlov, S. V. Starodubtsev, *JETP*, **7**, 424 (1937).
899. F. W. Aston, *Proc. Roy. Soc.*, **80A**, 45 (1908).
900. F. W. Aston, H. E. Watson, *Proc. Roy. Soc.*, **86A**, 168 (1912).
901. A. Güntherschulze, F. Keller, *Z. f. Phys.*, **71**, 238, 246 (1931).
902. D. H. Hale, *Phys. Rev.*, **73**, 1046 (1948).
903. M. A. Herlin, S. C. Brown, *Phys. Rev.*, **74**, 291 (1948).
904. M. A. Herlin, S. C. Brown, *Phys. Rev.*, **74**, 910 (1948).
905. M. A. Herlin, S. C. Brown, *Phys. Rev.*, **74**, 1650 (1948).
906. A. D. McDonald, S. C. Brown, *Phys. Rev.*, **75**, 411 (1949).
907. A. D. McDonald, S. C. Brown, *Phys. Rev.*, **75**, 1324 (1949).
908. A. D. McDonald, S. C. Brown, *Phys. Rev.*, **76**, 1634 (1949).
909. S. C. Brown, A. D. McDonald, *Phys. Rev.*, **76**, 1629 (1949).
910. S. C. Brown, *Proc. IRE*, **39**, 1493 (1951).
911. J. J. Thomson, *Phil. Mag.*, **2**, 674 (1926).
912. J. J. Thomson, *Phil. Mag.*, **2**, 696 (1926).
913. J. J. Thomson, *Phil. Mag.*, **23**, 1 (1937).
914. G. Herzberg, *Ann. d. Phys.*, **84**, 553 (1927).
915. G. Mierder, *Ann. d. Phys.*, **85**, 612 (1928).
916. C. G. Smyth, *Phys. Rev.*, **59**, 997 (1941).
917. E. W. Gill, A. Engel, *Proc. Roy. Soc.*, **192A**, 446 (1948).
918. C. Fancis, A. Engel, *Proc. Phys. Soc.*, **63B**, 823 (1950).
919. A. J. Hatch, H. B. Williams, *J. Appl. Phys.*, **25**, 417 (1954).
920. G. Birkhoff, *Z. angew. Phys.*, **10**, 204 (1958).
921. B. Lax, W. P. Allis, S. C. Brown, *J. Appl. Phys.*, **21**, 1297 (1950).
922. S. C. Brown, *Proc. IRE*, **39**, 1493 (1951).
923. H. Neuert, *Z. f. Naturf.*, **4a**, 449 (1949).
924. B. Koch, H. Neuert, *Z. f. Naturf.*, **4a**, 452 (1949).
925. H. Neuert, H. T. Stuckenberg, H. P. Weidner, *Z. angew. Phys.*, **6**, 303 (1954).
926. M. Ardenne, U. Heising, *Exper. Techn. der Phys.*, **11**, 26 (1963).
927. J. Kistemaker, P. K. Rol, J. Politiek, *Nucl. Instr. and Meth.*, **38**, 1 (1965).
928. W. Dällenbach, *Z. f. Naturf.*, **10a**, 803 (1955).
929. H. Fetz, *Ann. d. Phys.*, **37**, 1 (1940).
930. T. Wasserab, *Z. f. Phys.*, **128**, 575 (1950).
931. E. M. Reichrudel, G. M. Spivak, *JETP*, **10**, 1408 (1940).

932. K. S. W. Shampion, *Proc. Phys. Soc.*, **65B**, 329 (1952).
933. R. J. Bickerton, A. Engel, *Proc. Phys. Soc.*, **69B**, 468 (1956).
934. V. E. Golant, N. I. Oplov, L. I. Pakhonov, *JTP*, **31**, 797 (1961).
935. M. Abell, W. Meckbach, *Rev. Sci. Instr.*, **30**, 335 (1959).
936. M. Ardenne, *Atomkernenergie*, **1**, 121 (1956).
937. B. M. Smirnov, *Dokl. AN SSSR*, **161**, 92 (1965).
938. H. Y. Young, *Journ. Electr. and Control.*, **5**, 307 (1958).
939. R. Masic, J. M. Sautter, R. J. Warnecke, *Nucl. Instr. and Meth.*, **71**, 339 (1969).
940. E. Heinicke, K. Bethge, N. Baumann, *Nucl. Instr. and Meth.*, **58**, 125 (1968).
941. G. Gautherin, *Nucl. Instr. and Meth.*, **59**, 261 (1968).
942. P. Ciuti, *Nucl. Instr. and Meth.*, **79**, 55 (1970).
943. V. S. Kusnetsov, M. A. Abroyan, R. P. Fidelskaya, *Nucl. Instr. and Meth.*, **81**, 296 (1970).
944. H. Tawara, S. Suganmata, S. Suenmatsu, *Nucl. Instr. and Meth.*, **31**, 353 (1964).
945. C. M. Braams, P. Ziesked, M. J. Kofoid, *Rev. Sci. Instr.*, **36**, 1411 (1965).
946. E. Armien, *Nucl. Instr. and Meth.*, **85**, 109 (1970).
947. J. F. Brnandet, R. Boucher, R. Danielou, *Nucl. Instr. and Meth.*, **89**, 297 (1970).
948. M. D. Gabovits, E. T. Kutserenko, *JTP*, **26**, 997 (1956).
949. M. D. Gabovits, E. T. Kutserenko, *JTP*, **27**, 299 (1957)
950. M. D. Gabovits, L. I. Romanyuk, *JTP*, **31**, 315, 1049 (1961).
951. M. D. Gabovits, E. A. Lozovaya, L. I. Romanyuk, *JTP*, **34**, 488 (1964).
952. G. Gautherin, *Proc. 8th Int. Conf. Phenom. in Ionized Gases*, Vienna, 1967, p. 535.
953. J. Kistemaker, H. L. Douwes Dekker, *Physica*, **16**, 198, 209 (1950).
954. J. Kistemaker, J. Sneider, *Physica*, **19**, (1953).
955. M. D. Gabovits, O. A. Bartnovskii, Z. P. Fedorus, *JTP*, **30**, 345 (1960).
956. M. Abell, W. Meckbach, *Rev. Sci. Instr.*, **30**, 335 (1959).
957. A. Papineau, P. Benezech, R. Millard, *J. Phys. Rad.*, **21**, 410 (1960).
958. R. Basile, J. M. Lagrange, *J. Phys. Rad.*, **23**, 111 (1962).
959. J. D. Pigarov, P. M. Morozov, *JTP*, **31**, 467, 472 (1961).
960. A. S. Pasiok, I. A. Selaev, Go Ci-Gayn, J. P. Tretyakov, *PTE*, No. **5**, 23 (1963).
961. A. S. Pasiok, J. P. Tretyakov, C. K. Gorbatsek, *Atomnaya Energiya* **24**, 21 (1968).
962. J. P. Tretyakov, L. P. Kulkina, V. I. Kuznetsov, A. S. Pasiok, *Report OIJI*, P-7-5004, Dubna.
963. J. N. Antonov, L. P. Zinovev, V. P. Rasevskii, *Atomnaya Energiya*, **8**, 454 (1960).
964. R. S. Livingston, R. J. Jones, *Rev. Sci. Instr.*, **25**, 552 (1954).
965. B. F. Gavin, *Nucl. Instr. and Meth.*, **64**, 73 (1968).
966. E. I. Revitskii, G. M. Skoromnii, P. S. Markin, C. I. Meleskov, I. T. Venovtsev, P. K. Dron, O. A. Fedorov, J. V. Kasyanov, *PTE*, No. **3**, 33 (1971).
967. E. Heinicke, K. Bethge, H. Bauman, *Nucl. Instr. and Meth.*, **58**, 125 (1968).
968. E. Heinicke, H. Bauman, *Nucl. Instr. and Meth.*, **75**, 229 (1969).
969. K. Prelec, M. Isalia, *Nucl. Instr. and Meth.*, **92**, 1 (1971).
970. R. Keller, *Atomiques*, **2**, 81 (1958).
971. S. Flüge, *Handbuch der Physik.*, Springer Verlag, Berlin, 1956, p. 33.
972. A. Guthrie, R. Walkering, *Electrical Discharges in Magnetic Fields*, New York, 1949.
973. M. E. Abdelaziz, A. M. Ghander, *IEEE* NS–14, 53 (1967).
974. M. E. Abdelaziz, *IEEE* NS–9, 1 (1962).
975. S. K. Allison, E. Norbeck, *Rev. Sci. Instr.*, **27**, 285 (1956).
976. V. A. Romanov, A. N. Serbinov, *PTE* No. **5**, 34 (1965).
977. Z. Szabó, *Nucl. Instr. and Meth.*, **78**, 199 (1970).
978. M. L. Smith, *Electromagnetically Enriched Isotopes and Mass Spectrometry.* Butterworth Ltd., London, 1956, p. 53.
979. J. Druaux, R. Bernas, *Electromagnetically Enriched Isotopes and Mass Spectrometry.* Butterworth Ltd., London, 1956, p. 30.
980. R. H. Dawton, *Electromagnetically Enriched Isotopes and Mass Spectrometry.* Butterworth Ltd., London, 1956, p. 208.
981. V. M. Gusev, D. V. Tskiaseli, M. I. Guseva, *Atomnaya Energiya*, **3**, 215 (1957).
982. J. Koch, R. Dawton, M. Smith, W. Walcher, *Electromagnetic Isotope Separators and Applications of Electromagnetically Enriched Isotopes.* North-Holland, Amsterdam, 1958.
983. V. M. Gusev, *Tr. Vsesoyuznoi nauchno—tekhnicheskoi Konf. po primeneniya radioaktivnykh i stabilnykh izotopov.* Moskva, 1958.

984. V. S. Zolotarev, A. I. Ilin, E. G. Komar, Tr. 2. Mezhdunarodnoj konf. po miromu ispolzovaniya atomnoi energii, Zheneva, 1958. Doklady sovietskikh uchenykh (T. 6.) Moskva, 1959.
985. P. M. Morozov, B. N. Makov, M. S. Joffe, B. G. Brezhnev, G. N. Fradkin, Mezhdunarodnoi konf. po miromu ispolzovaniya atomnoi energii, Zheneva, 1958, Doklady sovietskikh uchenykh (T. 6.) Moskva, 1959.
986. M. V. Nezlin, P. M. Morozov, Tr. 2. Mezhdunarodnoi konf. po miromu ispolzovaniya atomnoi energii, Zheneva, 1958; Doklady sovietskikh uchenykh (T. 6.) Moskva, 1959.
987. J. L. Sarrouy, R. Klapisch, *Proc. Int. Symp. on Electromagnetic Separation of Radioactive Isotopes*. Vienna 1960, Springer Wien, 1961.
988. A. J. Dempster, *Rev. Sci. Instr.*, **7**, 46 (1936).
989. Chen-Lin-Chu, *Phys. Rev.*, **50**, 217 (1936).
990. M. D. Gabovits, Z. Y. Fedorus, *UPJ*, **1**, 158 (1956).
991. Y. M. Fogel, R. V. Mitin, A. G. Koval, *JETP*, **31**, 397 (1956).
992. K. O. Nielsen, *Nucl. Instr. and Meth*, **1**, 289 (1957).
993. G. Dearnaley, *Rep. Prog. Phys.*, **32**, 405 (1969).
994. S. Tabata, M. Iwata, T. Sawaka, *Vacuum*, **7**, 89 (1959).
995. J. Uhler, I. Alvager, *Arkiv för Fysik*, **14**, 473 (1959).
996. F. Bisi, B. Michelis, *Nuovo Cimento*, **11**, 861 (1959).
997. W. L. Rautenbach, *Nucl. Instr. and Meth.*, **1**, 815 (1960).
998. I. G. Kiselev, L. A. Levsenkova, *JETP*, **30**, 815 (1960).
999. V. F. Kozlov, V. L. Marchenko, Ya. M. Fogel, *PTE*, No. **1**, 25 (1961).
1000. V. F. Kozlov, V. Ya. Kolot, Sun-Cszsen-cziny, *PTE*, No. **6**, 116 (1962).
1001. F. F. Chen, *Phys. Rev. Letters*, **8**, 234 (1962).
1002. J. Litton, L. R. Bittman, *Proc. Nat. Electronics Conf., Chicago*, **8**, 783 (1962).
1003. V. S. Venkatasubramanian, H. E. Duckworth, *Can. J. Research*, **41**, 234 (1963).
1004. R. E. Honig, S. S. Class, J. R. Woolston, *Compt. Rend. 6ᵉ Conf. Internat. Phénomènes Ionisat. Gas. Paris*, 1963.
1005. H. J. Liebl, R. F. Herzog, *J. Appl. Phys.*, **34**, 2893 (1963).
1006. K. D. Schuy, H. Hinterberger, *Z. f. Naturf.*, **18a**, 926 (1963).
1007. J. Wolf, *Exper. Techn. der Phys.*, **11**, 407 (1963).
1008. R. R. Ferber, *IEEE NS—10*, 15 (1963).
1009. J. H. Freeman, *Nucl. Instr. and Meth.*, **22**, 306 (1963).
1010. B. Cobic, D. Tosic, B. Petrovic, *Nucl. Instr. and Meth.*, **24**, 358 (1963).
1011. I. Wolf, *Izv. AN SSSR*, **28**, 1423 (1964).
1012. R. I. Garber, A. I. Fedorenko, *Uspekhi Fiz. Nauk*, **83**, 385 (1964).
1013. J. R. Balmer, J. H. Bruce, *J. Sci. Instr.*, **41**, 589 (1964).
1014. V. M. Gusev, V. V. Titov, M. I. Guseva, V. I. Kurinny, *Fizika tverdogo Tela*, **7**, 2077 (1965).
1015. M. A. Tyulina, *JTP*, **35**, 511 (1965).
1016. R. J. Cinzemius, H. J. Snec, *J. Sci. Instr.*, **42**, 136 (1965).
1017. G. D. Magnuson, C. F. Carlston, P. Mahadevan, A. B. Comeaux, *Rev. Sci. Instr.*, **36**, 136 (1965).
1018. I. Severac, *Nucl. Instr. and Meth.*, **38**, 12 (1965).
1019. K. J. Hill, R. S. Nelson, *Nucl. Instr. and Meth.*, **38**, 15 (1965).
1020. G. Sidenius, *Nucl. Instr. and Meth.*, **38**, 19 (1965).
1021. M. Truong, H. Wetke, *Nucl. Instr. and Meth.*, **38**, 23 (1965).
1022. D. Tosic, *Nucl. Instr. and Meth.*, **38**, 26 (1965).
1023. J. L. Sarrouy, J. Camplan, J. S. Dioniso, J. Fournet-Fayes, G. Levy, J. Obert, *Nucl. Instr. and Meth.*, **38**, 29 (1965).
1024. J. H. Freeman, *Nucl. Instr. and Meth.*, **38**, 49 (1965).
1025. H. Wagner, *Nucl. Instr. and Meth.*, **38**, 69 (1965).
1026. J. H. Freeman, *Nucl. Instr. and Meth.*, **38**, 97 (1965).
1027. L. Wählin, *Nucl. Instr. and Meth.*, **38**, 94 (1965).
1028. N. I. Tarantin, A. V. Demyanov, Y. A. Dyachiknin, A. T. Kabachenko, *Nucl. Instr. and Meth.*, **38**, 103 (1965).
1029. J. O. McCaldin, *Nucl. Instr. and Meth.*, **38**, 153 (1965).
1030. E. F. Krimmel, *Rev. Sci. Instr.*, **37**, 679 (1966).
1031. V. F. Kozlov, V. Ya. Kolot, V. A. Tkachenko, *PTE*, No. **2**, 13 (1966).
1032. A. A. Plyutto, V. N. Rizhkov, A. G. Kapin, *JETP*, **47**, 494 (1967).
1033. S. Dushman, *Scientific Foundations of Vacuum Technique*, New York, London Wiley-Chapman, 1949.

1034. J. P. Tretyakov, A. S. Pasiok, L. P. Kulkina, V. J. Kuznetsov, *Report OIJI P-7-4477 Dubna*, 1969.
1035. A. S. Pasiok, E. L. Vorobev, P. I. Ivannikov, V. I. Kuznetsov, V. B. Kytner, J. P. Tretyakov, *Report, OIJI P-7-4488 Dubna* 1969.
1036. J. H. Freeman, *Report AERE-R* 6138 *Harwell*, 1969.
1037. K. J. Hill, R. S. Nelson, R. J. Francis, *Report AERE-R* 6343 *Harwell*, 1970.
1038. J. M. Nitschke, *Nucl. Instr. and Meth.*, **78**, 45 (1970).
1039. W. L. Fite, *Phys. Rev.*, **89**, 411 (1953).
1040. A. C. Whittier, *Can. J. Phys.*, **32**, 275 (1954).
1041. J. A. Phillips, J. L. Tuck, *Rev. Sci. Instr.*, **27**, 97 (1956).
1042. Y. M. Fogel, B. G. Safronov, L. I. Krupin, *JETP*, **28**, 711 (1955).
1043. J. A. Weinman, J. R. Cameron, *Rev. Sci. Instr.*, **27**, 288 (1956).
1044. Y. M. Fogel, L. I. Krupnik, R. P. Slabospitskii, *JTP*, **27**, 981 (1957).
1045. Y. M. Khirnii, *PTE*, No. **2**, 51 (1958).
1046. A. V. Almazov, I. M. Khirnii, *PTE*, No. **5**, 54 (1957).
1047. Y. M. Fogel, A. G. Koval, A. D. Timofeev, *JTP*, **29**, 1381 (1959).
1048. L. F. Collins, A. C. Riviere, *Nucl. Instr. and Meth.*, **4**, 121 (1959).
1049. S. F. Philp, *J. Appl. Phys.*, **31**, 1592 (1969).
1050. A. Chateau-Thierry, *C. R. Acad. Sci.*, No. 821 (1961).
1051. R. H. V. M. Dawton, *Nucl. Instr. and Meth.*, **11**, 326 (1961).
1052. R. H. V. M. Dawton, *Nucl. Instr. and Meth.*, **24**, 285 (1963).
1053. H. Drost, U. Timm, H. Pupke, *Exper. Techn. der Phys.*, **11**, 254 (1963).
1054. H. Drost, G. Lindemann, U. Timm, H. Pupke, *Exper. Techn. der Phys.*, **11**, 266 (1963).
1055. A. B. Wittkower, R. F. Bastide, N. B. Brooks, P. H. Rose, *Phys. Letters*, **3**, 336 (1963).
1056. N. B. Brooks, P. H. Rose, R. P. Bastide, A. B. Wittkower, *Nucl. Instr. and Meth.*, **28**, 315 (1964).
1057. D. Dandy, D. P. Hammond, *Nucl. Instr. and Meth.*, **30**, 23 (1964).
1058. K. W. Ehlers, *Nucl. Instr. and Meth.*, **32**, 309 (1965).
1059. C. H. Goldie, *Nucl. Instr. and Meth.*, **28**, 139 (1964).
1060. L. E. Collins, R. H. Gobbet, *Nucl. Instr. and Meth.*, **35**, 277 (1965).
1061. M. Roos, P. H. Rose, A. B. Wittkower, N. B. Brooks, R. P. Bastide, *Rev. Sci. Instr.*, **36**, 544 (1965).
1062. P. Gombos, J. Roosz, L. Vályi, *KFKI Reports*, **14**, 325 (1966).
1063. R. M. Ennis, D. E. Schechter, G. Thoening, B. Donnally, D. B. Schlafke, *IEEE*, NS–**14**, 75 (1967).
1064. F. A. Rose, P. B. Tollefsrud, H. T. Richards, *IEEE* NS–**14**, 78 (1967).
1065. J. John, C. P. Robinson, J. P. Aldrige, W. J. Walace, K. B. Chapman, R. H. Davis, *IEEE* NS–**14**, 82 (1967).
1066. K. R. Chapman, *Nucl. Instr. and Meth.*, **73**, 255 (1969).
1067. J. P. Martin, R. J. A. Levesque, *Nucl. Instr. and Meth.*, **80**, 229 (1970).
1068. Y. M. Khirnii, L. N. Kotsemasova, *PTE*, No. **1**, 37 (1971).
1069. G. P. Lawrence, R. K. Beauchamp, J. L. McKibben, *Nucl. Instr. and Meth.*, **32**, 357 (1965).
1070. A. S. Schlechter, D. H. Leid, P. J. Bjorkholm, L. W. Anderson, W. Haeberli, *Phys. Rev.*, **174**, 201 (1968).
1071. T. Jorgensen, C. E. Kuyatt, *Phys. Rev.*, **140A**, 1481 (1965).
1072. Y. M. Khirnii, L. N. Kotsemasova, *PTE*, No. **6**, 37 (1968).
1073. F. L. Ribe, *Phys. Rev.*, **83**, 1217 (1951).
1074. L. W. Alvarez, *Phys. Rev.*, **82**, 705 (1951).
1075. A. Phillips, *Phys. Rev.*, **91**, 455 (1953).
1076. E. S. Lamar, W. W. Buecher, *J. Appl. Phys.*, **18**, 22 (1947).
1077. K. W. Allen, F. A. Jillian, P. W. Allen, A. E. Pyrah, J. Blears, *Nature*, **184**, 303 (1959).
1078. F. Högberg, H. Norden, H. G. Berry, *Nucl. Instr. and Meth.*, **90**, 283 (1970).
1079. P. Hvelplund, E. Laegsgaard, J. O. Olsen, E. H. Pederson, *Nucl. Instr. and Meth.*, **90**, 315 (1970).
1080. G. S. Mavrognes, W. J. Ramler, C. B. Turner, *IEEE* NS–**12**, 769 (1965).
1081. F. A. Stuber, *J. Chem. Phys.*, **42**, 2639 (1965).
1082. G. Busch, L. Lehmann, H. Spehl, *Nucl. Instr. and Meth.*, **78**, 321 (1970).
1083. H. Krupp, *UNILAC Report*, **1**, 68 (1968).
1084. H. Krupp, *Nucl. Instr. and Meth.*, **90**, 167 (1970).

1085. J. R. J. Bennett, B. Gavin, *Rutherford Laboratory Preprint* RPP/A 84 (1971).
1086. J. W. Bittner, *Rev. Sci. Instr.*, **25**, 1058 (1954).
1087. O. Luhr, E. S. Lamar, *Phys. Rev.*, **46**, 87 (1934).
1088. O. Luhr, *Phys. Rev.*, **49**, 317 (1936).
1089. O. Luhr, F. Stunder, *Phys. Rev.*, **51**, 306 (1937).
1090. E. Snitzer, *Phys. Rev.*, **79**, 1237 (1953).
1091. P. B. Olkowsky, *J. Physique Rad.*, **21**, 407 (1960).
1092. A. Fontel, *Ann. Acad. Sci. Fennicae AVI Phys.*, **119**, 1 (1963).
1093. C. H. Goldie, *Nucl. Instr. and Meth.*, **28**, 139 (1964).
1094. N. R. Roberson, D. R. Tilley, H. R. Weller, *Nucl. Instr. and Meth.*, **33**, 84 (1965).
1095. P. Benoit-Cattin, D. Blanc, A. Bordenave-Montesquieu, R. Dagnac, S. Vacquie, *Nucl. Instr. and Meth.*, **43**, 349 (1966).
1096. J. L. Weil, I. J. Taylor, *IEEE* NS–**12**, 257 (1965).
1097. I. J. Taylor, J. L. Weil, *Nucl. Instr. and Meth.*, **34**, 197 (1965).
1098. D. M. Kaminkev, N. V. Fedorenko, *JTP*, **25**, 1843 (1955).
1099. H. L. Levant, M. I. Korchunskii, L. I. Pivovar, I. M. Pordornii, *Dokl. AN SSSR*, **103**, 403 (1955).
1100. P. R. Jones, F. R. Ziemba, H. A. Moses, E. Everhart, *Phys. Rev.*, **113**, 182 (1959).
1101. I. S. Dmitriev, V. S. Nikolaev, L. N. Fateyeva, J. A. Tepliva, *JETP*, **43**, 361 (1962).
1102. I. I. Pivovar, J. Z. Levchenko, G. A. Krivonchov, *JETP*, **59**, 19 (1970).
1103. S. Bashkin, G. Goldhaber, *Rev. Sci. Instr.*, **22**, 112 (1951).
1104. G. Dearnaley, *Rev. Sci. Instr.*, **31**, 197 (1960).
1105. E. L. Hubbard, E. J. Laner, *Phys. Rev.*, **98**, 1814 (1955).
1106. U. Nanser, W. Kerler, *Rev. Sci. Instr.*, **29**, 380 (1958).
1107. R. Beringer, W. Rall, *Rev. Sci. Instr.*, **28**, 77 (1957).
1108. R. H. Rose, *IEEE* NS–14, 16 (1967).
1109. J. S. Dmitriev, V. S. Nikolaev, *JETP*, **50**, 409 (1965).
1110. P. L. Smith, W. Whaling, *Phys. Rev.*, **36**, 188 (1969).
1111. V. E. Parker, R. F. King, *Bull. Amer. Phys. Soc.* II, **1**, 70 (1956).
1112. W. H. Good, J. H. Neiler, J. H. Gibbons, *Phys. Rev.*, **109**, 926 (1958).
1113. C. M. Turner, S. D. Bloom, *Rev. Sci. Instr.*, **29**, 480 (1958).
1114. B. Jennings, G. Griffits, *Phys. Rev.*, **91**, 440 (1953).
1115. L. Cramberg, J. S. Levin, *Phys. Rev.*, **103**, 344 (1956).
1116. L. Cramberg, I. S. Levin, *Phys. Rev.*, **109**, 2063 (1958).
1117. R. F. Holand, F. J. Lynch, S. S. Hanna, *Phys. Rev.*, **112**, 903 (1958).
1118. T. K. Rowler, V. M. Good, *Nucl. Instr. and Meth.*, **7**, 245 (1960).
1119. J. L. Fowler, J. B. Morion, *Fast Neutron Physics*. New York, Interscience Publish Inc., 1960.
1120. C. D. Moak, W. M. Good, R. F. King, J. W. Jonson, H. E. Banta, J. Judish, W. H. Duprez, *Rev. Sci. Instr.*, **35**, 672 (1964).
1121. K. R. Spangenberg, *Vacuum Tubes*. New York, McGraw-Hill, 1948.
1122. R. C. Mobley, *Phys. Rev.*, **88**, 360 (1952).
1123. R. C. Mobley, B. R. Albritton, *Bull. Amer. J.*, **98**, 232 (1955).
1124. R. C. Mobley, L. D. Chisholm, W. E. Dance, D. C. Raph, *Bull. Amer. Phys. Soc.*, II., **6**, 240 (1961).
1125. R. C. Mobley, *Rev. Sci. Instr.*, **34**, 256 (1963).
1126. R. J. Connor, *Nucl. Instr. and Meth.*, **11**, 122 (1961).
1127. L. Cramberg, R. A. Fernald, F. S. Hanh, S. F. Shrader, *Nucl. Instr. and Meth.*, **12**, 335 (1961).
1128. *Nuclear Electronics III. Conference Proc. Belgrad*, 1961, Intern. Atomic Energy Agency, Vienna, 1962.
1129. F. L. Sapiro, *PTE*, No. 1, 33 (1957).
1130. L. Grodzins, P. H. Rose, R. J. Van de Graaff, *Nucl. Instr. and Meth.*, **36**, 202 (1965).
1131. N. N. Flerov, E. A. Tamanov, *Atomnaya Energiya*, **3**, 44 (1957).
1132. F. E. Whineway, *AWRE Report*, No. 0—12/61 (1961).
1133. P. E. Vorotnikov, *PTE*, No. 3, 27 (1965).
1134. J. H. Anderson, D. Swann, *Nucl. Instr. and Meth.*, **30**, 1 (1964).
1135. D. Dandy, D. P. Nammond, *Nucl. Instr. and Meth.*, **30**, 23 (1964).
1136. O. I. Kozinets, F. L. Sapiro, I. V. Stranik, *PTE*, No. 5, 25 (1962).
1137. H. W. Lefevre, R. R. Borchers, C. H. Poppe, *Rev. Sci. Instr.*, **33**, 1231 (1962).
1138. L. E. Collis, D. Dandy, P. T. Strond, *Nucl. Instr. and Meth.*, **42**, 206 (1966).

1139. L. E. Bodhian, M. K. Salomea, *Nucl. Instr. and Meth.*, **17**, 181 (1962).
1140. L. Bockman, R. B. Jansson, B. L. Nilsson, *Nucl. Instr. and Meth.*, **33**, 151 (1965).
1141. P. H. Rose, A. B. Wittkower, N. B. Brooks, R. P. Bastide, M. Legros, *Rev. Sci. Instr.*, **35**, 1286 (1964).
1142. E. Wunderer, *Z. angew. Phys.*, **16**, 285 (1963).
1143. A. W. Lewis, P. R. Bevington, W. W. Rolland, R. L. Rummel, R. M. Wilenzik, *Rev. Sci. Instr.*, **30**, 923 (1959).
1144. H. E. Banta, R. F. King, J. P. Judish, *Bull. Amer. Phys. Soc.*, **1**, 70 (1956).
1145. J. H. Anderson, R. Batchelor, F. A. Howe, G. James, J. H. Tovee, *Nucl. Instr. and Meth.*, **41**, 30 (1966).
1146. P. Tykesson, T. Wielding, *Nucl. Instr. and Meth.*, **77**, 272 (1970).
1147. S. T. Thornton, R. C. Ritter, D. A. Hills, *Nucl. Instr. and Meth.*, **77**, 306 (1970).
1148. E. I. Revutskii, G. M. Skoromnii, P. S. Markin, S. I. Meleshkov, I. T. Benevtsev, P. K. Dron, O. A. Fedorov, J. V. Kasyanov, *PTE*, No. **3**, 38 (1971).
1149. L. Vályi, P. Gombos, J. Kakuk, J. Roosz, *KFKI Reports*, **13**, 127 (1965).
1150. P. E. Vorotnikov, J. G. Zubov, J. D. Maltsanov, A. A. Udod, G. B. Yankov, *PTE*, No. **5**, 37 (1966).
1151. I. Moreau, F. Prevot, R. Vienet, *L'Onde Electrique*, **35**, 344 (1955).
1152. V. B. Anifrienko, B. V. Devkin, V. I. Moroka, O. A. Salnikov, *PTE*, No. **3**, 46 (1961).
1153. S. Chwaszczewski, *Nucleonika*, **5**, 355 (1960).
1154. K. W. Ehlers, J. D. Gow, L. Ruby, J. M. Wilcox, *Rev. Sci. Instr.*, **29**, 614 (1958).
1155. I. I. Afanasev, A. S. Knazyatov, N. D. Fedotov, *Atomnaya Energiya*, **13**, 135 (1962).
1156. P. E. Vorotnikov, J. G. Zubov, J. D. Maltsanov, *PTE*, No. **5**, 33 (1966).
1157. R. L. Gravin, *Amer. Phys. Soc.*, **1**, 65 (1956).
1158. E. K. Zavoyskii, *JETP*, **32**, 2 (1957).
1159. E. K. Zavoyskii, *JETP*, **32**, 408 (1957).
1160. B. P. Adyasevits, S. T. Belyaev, J. P. Polunin, *Conf. Phys. of Particle High Energ. Moscow*, 1956.
1161. G. Clausnitzer, R. Fleischmann, H. Schopper, *Z. f. Phys.*, **144**, 336 (1956).
1162. E. K. Zavoiskii, *JETP*, **32**, 731 (1957).
1163. B. P. Adyasevits, S. T. Belyaev, E. K. Zavoiskii, J. P. Polunin, *Conf. Nucl. Phys. of Low Energ. Moscow*, 1957, Akadizdat, Moscow 1958.
1164. I. I. Rabi, J. M. Kellogg, J. R. Zacharias, *Phys. Rev.*, **46**, 163 (1934).
1165. H. Friedburg, W. Paul, *Naturwiss.*, **38**, 159 (1951).
1166. H. G. Benewitz, W. Paul, *Z. f. Phys.*, **139**, 489 (1954).
1167. A. Abragam, J. M. Winter, *Phys. Rev. Lett.*, **1**, 374 (1958).
1168. W. Grüebler, W. Haeberli, P. Schwand, *Phys. Rev. Lett.*, **12**, 595 (1964).
1169. W. Grüebler, P. Schwand, T. I. Yule, W. Haeberli, *Nucl. Instr. and Meth.*, **41**, 245 (1965).
1170. B. Donnally, T. Clapp, W. Sawyer, M. Schultz, *Phys. Rev. Lett.*, **12**, 502 (1964).
1171. B. Donnally, W. Sawyer, *Phys. Rev. Lett.*, **15**, 439 (1965).
1172. A. Cesati, F. Christofori, L. Milazzo-Colli, P. G. Sona, *Energia Nucl. (Milan)*, **13**, 649 (1966).
1173. A. Kastler, *J. de Physique*, **11**, 255 (1950).
1174. F. D. Colegrove, P. A. Franken, *Phys. Rev.*, **119**, 680 (1960).
1175. M. A. Bouchiat, T. R. Carver, C. M. Varnum, *Phys. Rev. Lett.*, **5**, 373 (1960).
1176. G. K. Walters, F. D. Colegrove, L. D. Schearer, *Phys. Rev. Lett.*, **8**, 439 (1962).
1177. L. D. Schearer, F. D. Colegrove, G. K. Walters, *Phys. Rev. Lett.*, **10**, 108 (1963).
1178. F. D. Colegrove, L. D. Schearer, G. K. Walters, *Phys. Rev.*, **132**, 2561 (1963).
1179. C. K. Walters, *Proc. of the Intern. Conf. on Polarized Target and Ion Sources, Saclay*, 1966, Centre d'Etudes Nucléaires de Sacley.
1180. P. Feldman, R. Novich, *Proc. 2nd Cong. Intern. de Physique Nucléaire, Paris*, 1964.
1181. B. L. Donnally, G. Thoneming, *Phys. Rev.*, **159**, 87 (1967).
1182. D. O. Findley, S. D. Bakker, E. B. Carter, N. D. Stockwell, *Nucl. Instr. and Meth.*, **71**, 125 (1969).
1183. G. Breit, I. I. Rabi, *Phys. Rev.*, **38**, 2082 (1931).
1184. I. I. Rabi, *Phys. Rev.*, **49**, 324 (1936).
1185. I. I. Rabi, J. R. Zacharias, S. Millman, P. Kusch, *Phys. Rev.*, **53**, 384 (1938).
1186. H. Kopferman, *Kernmomente*, Akad. Verl. Frankfurt, 1956.

1187. E. U. Condon, G. H. Shortley, *The Theory of Atomic Spectra*, Cambridge Univ. Press, 1957.
1188. G. Györgyi, *Elméleti Magfizika* (Theory of Nuclear Physics), Budapest, Műszaki Könyvkiadó, 1961.
1189. Y. M. Khirnii, L. I. Kotsemasova, *PTE*, No. **1**, 39 (1971).
1190. R. N. Boyd, J. C. Lombardi, A. B. Robbins, D. E. Schechter, *Nucl. Instr. and Meth.*, **81**, 149 (1970).
1191. A. S. Davidov, *Teoria Yadernaya Fizika* (Theory of Nuclear Physics), Fizmatgiz, Moskva, 1958.
1192. L. J. Schiff, *Quantum Mechanics*, McGraw-Hill, London, New York, 1955.
1193. W. E. Lamb, R. C. Retherford, *Phys. Rev.*, **81**, 222 (1951).
1194. W. Grüebler, V. König, P. Marmier, *Phys. Lett.*, **24B**, 335 (1967).
1195. R. Beurtey, *Proc. 2nd Intern. Symp. Pol. Phen. of Nucl.* Birkhäuser Verlag, Karlsruhe, Basel, 1966, p. 33.
1196. P. G. Sona, *Energia Nucl. (Milan)*, **14**, 295 (1967).
1197. B. J. Wood, H. Wise, *J. Chem. Phys.*, **29**, 1416 (1958).
1198. W. V. Smith, *J. Chem. Phys.*, **11**, 110 (1943).
1199. H. S. W. Massey, C. Mohr, *Proc. Roy. Soc.*, **135A**, 258 (1932).
1200. E. H. Karner, *Phys. Rev.*, **92**, 1441 (1953).
1201. E. V. Ivash, *Phys. Rev.*, **112**, 155 (1958).
1202. I. B. Corrigan, A. Engel, *Proc. Roy. Soc.*, **245A**, 335 (1958).
1203. H. Brasefild, *Phys. Rev.*, **31**, 52, 215 (1928).
1204. F. Wolf, *Ann. d. Phys.*, **31**, 561 (1938).
1205. N. V. Fedorenko, *JTP*, **24**, 769 (1954).
1206. P. C. Veenstra, J. M. Milatz, *Physica*, **18**, 528 (1950).
1207. J. Erő, *Nucl. Instr. and Meth.*, **3**, 303 (1958).
1208. L. E. Collins, R. H. Gobbett, *Nucl. Instr. and Meth.*, **35**, 277 (1965).
1209. G. Först, *Z. f. Phys.*, **159**, 7 (1960).
1210. H. Kühn, *Z. f. Phys.*, **149**, 267 (1957).
1211. P. E. Vorotnikov, *PTE*, No. **3**, 27 (1965).
1212. P. E. Vorotnikov, J. G. Zubov, J. D. Moltsanov, *PTE*, No. **5**, 33 (1966).
1213. A. Van Steenbergen, L. W. Oleksiuk, J. P. Blewett, *Proc. of the Int. Conf. on High Energy Accelerators, Dubna*, 1963, Atomizdat, Moscow, 1964.
1214. E. Wroe, *Nucl. Instr. and Meth.*, **58**, 213 (1968).
1215. L. Vályi, unpublished results.
1216. P. H. Rose, A. B. Wittkower, R. P. Bastide, A. Galejs, *Nucl. Instr. and Meth.*, **14**, 79 (1961).
1217. Gy. Paris, *KFKI Reports*, **9**, 58 (1961).
1218. P. H. Rose, R. P. Bastide, N. B. Brooks, J. Airey, A. B. Wittkower, *Rev. Sci. Instr.*, **35**, 1283 (1964).
1219. D. L. Judd, *Ann. Rev. Nucl. Sci.*, **8**, 181 (1958).
1220. R. McKeever, A. Yokosawa, *Rev. Sci. Instr.*, **33**, 746 (1962).
1221. M. D. Gabovits, *JTP*, **25**, 1458 (1955).
1222. I. Langmuir, K. Blodgett, *Phys. Rev.*, **24**, 49 (1924).
1223. A. P. Banford, *Transportirovka putskov zaryazennih tsastic* (Trans. charge particle beams) Atomizdat, Moskva (1969).
1224. C. S. Taylor, *Proc. of the Int. Conf. High Energy Accel. Dubna*, 1963, Atomizdat, Moscow, 1964.
1225. L. Vályi, E. Klopfer, Unpublished results.
1226. W. Happer, *Rev. Mod. Phys.*, **44**, 169 (1972).
1227. P. H. Rose, *Nucl. Instr. and Meth.*, **28**, 146 (1964).
1228. Th. Sluyters, *Nucl. Instr. and Meth.*, **27**, 301 (1964).
1229. L. E. Collins, P. T. Strond, *Nucl Instr. and Meth.*, **26**, 157 (1964).
1230. N. B. Brooks, *Rev. Sci. Instr.*, **35**, 894 (1964).
1231. K. O. Nielsen, O. Skilbreid, *Nucl. Instr. and Meth.*, **1**, 159 (1957).
1232. V. S. Kusnetsov, M. A. Aboyan, R. P. Fidelskaya, *Nucl. Instr. and Meth.*, **81**, 296 (1970).
1233. A. van Steenbergen, *IEEE* NS–**12**, 746 (1965).
1234. A. van Steenbergen, *Nucl. Instr. and Meth.*, **51**, 245 (1967).
1235. Z. V. Spolskii, *Atomic Physics*, Moscow, Gosthimizdat, 1951.
1236. E. Segre, *Experimental Nuclear Physics*, John Wiley, London, 1953.
1237. A. L. Hughes, V. Rajansky, *Phys. Rev.*, **34**, 291 (1929).
1238. G. E. Hutter, *Phys. Rev.*, **67**, 248 (1945).

1239. V. G. Fastovskii, *Kripton and Xenon*. Moscow, Gosthimizdat, 1941.
1240. M. Knoll, F. Ollendorf, R. Rompe, *Gasentladungstabellen*. Springer, Berlin, 1935.
1241. V. Hughes, G. Tucker, E. Rhoderich, G. Weinreich, *Phys. Rev.*, **91**, 828 (1953).
1242. V. F. Sikolenko, *OIJI Report* 9-5817 Dubna, 1971.
1243. V. Knauer, O. Stern, *Z. f. Phys.*, **39**, 764 (1926).
1244. I. I. Rabi, J. M. Kellogg, J. R. Zacharias, *Phys. Rev.*, **46**, 157 (1934).
1245. G. E. Moore, *Atomic Energy Levels*, Nat. Bur. Stand. Washington, 1949, 1952, 1957.
1246. G. Herzberg, *Molecular Spectra and Molecular Structure*. London, 1955.
1247. A. N. Zaydel, V. K. Prokofyev, S. M. Raiskii, V. A. Klavnii, E. J. Sreider, *Tablitsi Spektralnikh Linii*, Moskva, Nauka, 1969.
1248. V. I. Vedenev, L. V. Gurbits, V. N. Kondratev, V. A. Medvedev, E. L. Frankevits, *Energii razriva khimicheskikh svyazei*, Moskva, Izd. AN SSSR 1962.
1249. K. H. Kingdon, *Phys. Rev.*, **21**, 408 (1923).
1250. F. Knauer, O. Stern, *Z. f. Phys.* **53**, 766 (1929).
1251. R. D. Huntoon, A. Ellett, *Phys. Rev.*, **49**, 281 (1936).
1252. J. M. Kellog, I. I. Rabi, N. F. Ramsey, J. R. Zacharias, *Phys. Rev.*, **56**, 318 (1939).
1253. J. B. Taylor, *Phys. Rev.*, **35**, 375 (1930).
1254. F. Knauer, *Z. f. Phys.*, **126**, 310 (1949).
1255. H. Jonses, I. Langmuir, G. M. Mackoy, *Phys. Rev.*, **30**, 201 (1927).
1256. V. Hughes, G. Tucker, E. Rhoderick, G. Weinreich, *Phys. Rev.*, **91**, 824 (1953).
1257. H. O. Hagstrum, *Phys. Rev.*, **89**, 244 (1953).
1258. R. Dorrestein, *Proc. Acad. Sci. Austr.*, **41**, 725 (1938).
1259. W. E. Lamb, A. Cobas, *Phys. Rev.*, **65**, 327 (1944).
1260. R. F. Stabbings, *Proc. Roy. Soc.*, **214A**, 241 (1957).
1261. J. H. Sanders, *J. Sci. Instr.*, **26**, 36 (1949).
1262. V. A. Egorov, D. V. Karetnikov, S. N. Popov, *PTE*, No. **2**, 146 (1960).
1263. V. Agoritsas, S. Battisti, C. Bovet, *CERN, MPS/Int.*, DL 67−2 (1967).
1264. F. Hornstra, J. R. Simanton, *Nucl. Instr. and Meth.*, **68**, 138 (1969).
1265. J. R. Simanton, R. F. Marquardt, F. Hornstra, *Nucl. Instr. and Meth.*, **68**, 209 (1969).
1266. F. Hornstra, J. R. Simanton, *Nucl. Instr. and Meth.*, **77**, 303 (1970).
1267. V. G. Telkovskii, *Dokl. AN SSSR*, **108**, 444 (1956).
1268. V. Agoritsas, *CERN MPS/Int.* C/68−9 (1968).
1269. J. V. Korsunov, E. A. Melesko, *PTE*, No. **6**, 24 (1961).
1270. S. Domitz, E. V. Pawlik, *AIEE Journal*, **1**, 712 (1963).
1271. V. Agoritsas, S. Battisti, C. D. Jonson, G. Schneider, *CERN, MPS/CO* 70−6 (1970).
1272. K. Unser, *IEEE*, NS−**16**, 134 (1969).
1273. H. Zullinger, *CERN, MPS/Int.* DL 64−21 (1964).
1274. D. A. G. Neet, *IEEE* NS−**16**, 114 (1969).
1275. J. D. Fox, G. W. Benett, G. S. Levine, R. J. Nawrocky, L. E. Repeta, A. V. Soukas, *IEEE* NS−**16**, 832 (1969).
1276. A. J. Lichtenberg, *Phase Space Dynamics of Particles*. Wiley and Sons, New York, 1969.
1277. M. Khan, J. M. Schroeer, *Rev. Sci. Instr.*, **42**, 1348 (1971).
1278. D. Offermann, H. Trinks, *Rev. Sci. Instr.*, **42**, 398 (1971).
1279. J. K. Layton, G. D. Magnuson, *Rev. Sci. Instr.*, **43**, 1546 (1972).
1280. H. H. Gierlich, A. Heindrichs, H. D. Beckey, *Rev. Sci. Instr.*, **45**, 1208 (1974).
1281. H. L. Daley, J. Perel, *Rev. Sci. Instr.*, **42**, 1324 (1971).
1282. R. K. Feeney, F. M. Bacon, M. T. Elford, J. W. Hooper, *Rev. Sci. Instr.*, **43**, 549 (1972).
1283. P. G. Johnson, A. Bolson, C. M. Henderson, *Nucl. Instr. and Meth.*, **106**, 83 (1973).
1284. W. Möller, D. Kamke, *Nucl. Instr. and Meth.*, **105**, 173 (1972).
1285. E. Norbeck, C. S. Littlejohn, *Phys. Rev.*, **108**, 754 (1957).
1286. I. Vukanic, I. Tarzic, *Nucl. Instr. and Meth.*, **111**, 117 (1973).
1287. V. I. Tsigin, *PTE*, No. **2**, 165 (1974).
1288. V. B. Anyfrienko, B. V. Devkin, V. I. Moroka, O. A. Tsalnikov, *PTE*, No. **3**, 46 (1971).
1289. J. J. Kritzinger, W. R. McMurray, R. J. Nuspliger, *Nucl. Instr. and Meth.*, **101**, 573 (1972).
1290. M. Denizart, G. Soum, A. Degeilh, P. Pilod, *Nucl. Instr. and Meth.*, **100**, 221 (1972).

1291. A. Septier, *IEEE* NS–19, 22 (1972).
1292. M. Steinberg, B. Yap, H. H. Fleischmann, *Rev. Sci. Instr.*, 42, 354 (1971).
1293. K. Shimizu, K. Kawakatsu, K. Kanaya, *Nucl. Instr. and Meth.*, 111, 525 (1973).
1294. C. Lejeune, *Nucl. Instr. and Meth.*, 116, 417, 429 (1974).
1295. J. R. Coupland, E. Thompson, *Rev. Sci. Instr.*, 42, 1034 (1971).
1296. R. C. Davis, O. B. Morgan, L. D. Stewart, W. L. Stirling, *Rev. Sci. Instr.*, 43, 288 (1972).
1297. I. Ilgen, R. Kirshner, *IEEE* NS-19, 35 (1972).
1298. D. S. Stark, A. W. J. Manley, *Nucl. Instr. and Meth.*, 91, 301 (1971).
1299. P. S. Markin, A. A. Cigikálo, F. P. Shanin, E. I. Anpilogov, *PTE*, No. 2, 37 (1973).
1300. E. E. Norbeck, R. C. York, *Nucl. Instr. and Meth.*, 118, 327 (1974).
1301. M. E. Abdelaziz, M. W. Morsy, Z. M. Hassan, N. N. Hanna, *IEEE* NS–19, 132 (1972).
1302. Y. Okamoto, H. Tamagawa, *Rev. Sci. Instr.*, 43, 1193 (1972).
1303. R. G. Wilson, D. M. Jamba, *Nucl. Instr. and Meth.*, 91, 285 (1971).
1304. I. P. Zirnheld, *Nucl. Instr. and Meth.*, 114, 517 (1974).
1305. H. Baumann, E. Hinicke, K. Bethge, *IEEE* NS–19, 88 (1972).
1306. J. R. J. Bennett, *IEEE* NS–19, 48 (1972).
1307. G. Fucks, J. Steyaert, J. Clark, *IEEE* NS–19, 84 (1972).
1308. C. Bieth, M. P. Bourgarel, A. Cabrespine, R. Gayraud, P. Attal, *IEEE* NS–19, 93 (1972).
1309. E. J. Jones, *IEEE* NS–19, 101 (1972).
1310. Y. Miyarawa, I. Kohno, T. Tonuma, T. Inoue, A. Shimamura, S. Nakajima, *IEEE NS*-19, 105, 109 (1972).
1311. D. J. Clark, J. Steyaert, A. Carnerio, D. Morris, *IEEE* NS–19, 114 (1972).
1312. M. L. Mallory, E. D. Hudson, G. Fucks, *IEEE* NS–19, 118 (1972).
1313. G. Hadinger, T. Tauth, G. Hadinger-Espi, M. Bajard, A. Chaber, *IEEE* NS–19, 137 (1972).
1314. E. D. Hudson, M. L. Mallory, R. S. Lord, *Nucl. Instr. and Meth.*, 115, 311 (1974).
1315. Y. M. Khirnii, L. N. Kotsemasova, *PTE*, No. 5, 28 (1974).
1316. J. L. Yitema, *Nucl. Instr. and Meth.*, 98, 379 (1972).
1317. S. T. Thornton, R. H. McKnight, R. C. Ritter, *Nucl. Instr. and Meth.*, 101, 607 (1972).
1318. E. Baron, *IEEE* NS–19, 256 (1972).
1319. H. D. Betz, *IEEE* NS–19. 249 (1972).
1320. J. L. Yntema, *IEEE* NS–19, 272 (1972).
1321. M. I. Keeling, *Nucl. Instr. and Meth.*, 119, 401 (1974).
1322. R. E. Center, *Rev. Sci. Instr.*, 43, 115 (1972).
1323. K. H. Purser, *IEEE* NS–20, 136 (1973).
1324. V. I. Paleev, E. Ja. Zandberg, *PTE*, No. 2, 163 (1974).
1325. R. Middleton, C. T. Adams, *Nucl. Instr. and Meth.*, 118, 329 (1974).
1326. B. Franzke, N. Angert, Ch. Schmelzer, *IEEE* NS–19, 266 (1972).
1327. G. Philipp, U. Scheib, A. Hofmann, *Nucl. Instr. and Meth.*, 115, 507 (1974).
1328. G. I. Dimov, G. V. Roslyakov, *PTE*, No 2, 33 (1974).
1329. G. I. Dimov, G. V. Roslyakov, *PTE*, No. 3, 31 (1974).
1330. C. W. Sitter, *Nucl. Instr. and Meth.*, 98, 169 (1972).
1331. R. H. Day, W. C. Parkinson, *Nucl. Instr. and Meth.*, 111, 199 (1973).
1332. T. Takagi, I. Yamada, J. Ishikawa, *IEEE* NS–19, 142 (1972).
1333. M. Dumail, G. Chauland, *Nucl. Instr. and Meth.*, 112, 607 (1973).
1334. G. Doucas, H. R. Mck Hyder, *Nucl. Instr. and Meth.*, 119, 413 (1974).
1335. I. K. Hirvonen, *Nucl. Instr. and Meth.*, 116, 9 (1974).
1336. E. K. Izrailov, *PTE*, No. 3, 26 (1974).
1337. J. Q. Saarcy, *Rev. Sci. Instr.*, 45, 589 (1974).
1338. Vu Ngoe Tuan, A. S. Schlachter, G. Gautherin, *Nucl. Instr. and Meth.*, 114, 499 (1974).
1339. M. Fumelli, *Nucl. Instr. and Meth.*, 118, 337 (1974).
1340. H. B. Flekkel, *PTE*, No. 3, 44 (1973).
1341. G. N. Flerov, V. S. Barasenkov, *Usp. Phys. Nauk*, 114, 351 (1974).
1342. Th. Sluyters, K. Prelec, *Nucl. Instr. and Meth.*, 113, 299 (1973).
1343. V. P. Golubov, G. A. Nalivaiko, S. G. Chepakin, *PTE*, No. 5, 30 (1974).
1344. K. Prelec, Th. Sluyters, *Rev. Sci. Instr.*, 44, 1451 (1973).
1345. S. Liese, D. Roller, *Nucl. Instr. and Meth.*, 112, 551 (1974).

1346. W. E. Burcham, O. Karban, S. Oh, W. B. Powell, *Nucl. Instr. and Meth.*, **116**, 1 (1974).
1347. *Proc. 6th European Conference on Controlled Fusion and Plasma Physics. Moscow, 1973.*, *OIJI*, Dubna, 1974.
1348. *Proc. 2nd International Conference on Ion Sources Vienna*, 1972.
1349. D. Aldoroft, I. Burcham, H. C. Cole, M. Cowlin, J. Sheffield, *Nucl. Fusion*, **13**, 393 (1973).
1350. J. D. Daugherty, L. Grodzins, G. S. Janes, R. H. Levy, *Phys. Rev. Lett.*, **20**, 369 (1968).
1351. E. D. Donets, V. I. Ilyushchenko, V. A. Alpert, *Report OIJI* P7-4124 Dubna, 1968.
1352. N. J. Peacock, R. J. Speer, M. G. Hobby, *J. Phys.*, **B2**, 798 (1969).
1353. S. E. Graybill, J. R. Uglum, *J. Appl. Phys.*, **41**, 236 (1970).
1354. N. J. Peacock, R. S. Pease, *J. Phys.*, **D2**, 1705 (1969).
1355. G. Tonon, M. Rabean, *Proc. 1st Intern. Conf. Ion Source Saclay*, 1969, p. 605.
1356. C. Faure, A. Perez, G. Tonon, B. Aveneau, D. Parisot, *Phys. Lett.*, **34A**, 313 (1971).
1357. H. Postma, *Phys. Lett.*, **31A**, 197 (1970).
1358. I. Aexeff, W. D. Jones, R. V. Neidigh, *Report ORNL-TM*-2981, *Oak Ridge National Lab.*, 1970.
1359. A. Weissberger, B. W. Rossiter, (Editors), *Techniques of Chemistry*, Vol. I, Wiley-Interscience, New York, 1972.
1360. T. A. Milne, F. T. Greene, Molecular Beams in High Temperature Chemistry. *Adv. High Temp. Chem.*, **2**, 107 (1969).
1361. J. L. Kinsey, Molecular Beam Reactions. *Physical Chemistry, Series One*, Vol. **9**, Butterworths, London, 1972.
1362. R. J. Jr. Cross, *Annual Report, Yale University* New Haven COO-3570-8, June, 1973.
1363. F. M. Devienne, P. Bark, A. Roustan, *Entropie*, No. **42**, 140 (1971).
1364. J. Ross (Editor), *Molecular Beams, Adv. Chem. Phys.*, Vol. **10**, Wiley-Interscience, New York, 1966.
1365. F. Schmidt-Bleck, G. Stoecklin, K. Vogelbruch, *Kernforschungsanlage. Juelich*, JUL-674-RC, 1970.
1366. D. J. Meschi, *J. Phys. Chem.*, **76**, 2947 (1972).
1367. J. R. Grover, H. V. Libenfeld, *Nucl. Instr. and Meth.*, **105**, 189 (1972).
1368. J. C. Zorn, *Adv. Atomic Mol. Phys.*, **9**, 243 (1973)
1369. J. Aigner, D. Jaeger, A. Nikuradse, *Entropie*, No. **42**, 89 (1971).
1370. W. J. Hays, W. E. Rodgers, E. L. Knuth, *J. Chem. Phys.*, **56**, 1652 (1972).
1371. H. G. Lintz, A. Pentenero, P. Le Goff, *J. Phys. Chem.*, **75**, 1042 (1971).
1372. D. Kakati, A. Chaudhury, *J. Vac. Sci. Technol.*, **10**, 995 (1973).
1373. A. K. Ayukhanov, *Atomnye stolknoveniya na poverhnosti tverdogo tela. Izdatelstvo FAN*, Taskhent, 1972.
1374. S. M. Liu, W. E. Rodgers, E. L. Knuth, *J. Chem. Phys.*, **61**, 902 (1974).
1375. N. Kashihira, F. Schmidt-Bleck, *J. Chem. Phys.*, **61**, 160 (1974).
1376. J. R. Garborino, M. A. Wartel, *J. Chem. Phys.*, **61**, 1253 (1974).
1377. D. R. Frankl, *J. Chem. Phys.*, **60**, 3268 (1974).
1378. V. M. Kulygin, A. A. Panasenkov, A. B. Svesnykov, N. N. Semashko, *JTP*, **40**, 2091 (1970).
1379. J. R. Coupland, E. Thompson, *Rev. Sci. Instr.*, **42**, 1034 (1971).
1380. H. K. Forsen, L. D. Stewart, W. L. Stirling, *Oak Ridge National Lab. Report ORNL-TM*-3788, 1972.
1381. G. G. Kelley, *Nucl. Fusion*, **12**, 169 (1972).
1382. T. H. Stix, *Plasma Phys.*, **14**, 365 (1972).
1383. K. B. Kartashev, V. I. Pistunovich, E. A. Filinova, *Plasma Phys.*, **14**, 737 (1972).
1384. J. R. Hiskes, *Nucl. Fusion*, **12**, 423 (1972).
1385. D. Aldoroft, J. Burcham, H. C. Cole, M. Cowlin, J. Sheffield, *5th European Conf. on Contr. Fusion and Plasma Physics. Grenoble*, 21. Aug. 1972.
1386. T. H. Stix, *Phys. Fluids*, **16**, 1922 (1973).
1387. R. Freeman, A. C. Riviere, D. R. Sweetman, *3rd Internat. Symp. or Toroidal Plasma Confinement, Garching* 26. March. 1973.
1388. J. Hovingh, R. W. Moir, *California Univ. Livermore, Lawrence Livermore Lab.* UCRL-51419, 1973.

1389. W. R. Barker, K. H. Berkner, W. S. Cooper, K. W. Ehlers, W. B. Kunkel, R. V. Pyle, J. W. Stearns, *California Univ. Berkeley, Lawrence Berkeley Lab.*, LBL-2425, 1973.
1390. M. W. Aleock, R. E. Bradford, R. Freeman, D. P. Hammond, A. C. Riviere, D. R. Sweetman, *Sixth Europ. Conf. on Contr. Fusion and Plasma Physics*, Vol. 1, 1973.
1391. K. H. Berkner, T. H. Morgan, R. V. Pyle, J. W. Stearns, *Nucl. Fusion*, 13, 27 (1973).
1392. D. G. McAlees, R. W. Conn, *Nucl. Fusion*, 14, 419 (1974).
1393. L. D. Stewart, J. D. Collen, J. F. Clarke, *3rd Internat. Symp. on Toroidal Plasma Confinement, Garching*, 26. March. 1973.
1394. G. W. Hamilton, W. L. Dexter, B. H. Smith, *California Univ. Livermore, Lawrence Livermore Lab.* UCRL-75547, 1974.
1395. J. G. Cordey, J. Hugill, J. W. Paul, J. Sheffield, E. Speth, P. E. Stott, V. J. Tereshin, *Nucl. Fusion*, 14, 411 (1974).
1396. J. Hovingh, R. W. Moir, *Nucl. Fusion*, 14, 629 (1974).
1397. Y. J. Balchenko, G. I. Dimov, V. G. Dudinikov, *Nucl. Fusion*, 14, 113 (1974).
1398. A. I. Kislyakov, M. P. Petrov, *JTP*, 40, 1609 (1970).
1399. M. Nicolas, *EUR-CEA-FC*-615 *Sept.* 1971, Centre d'Études Nucléaires.
1400. N. Inoue, *J. Phys. Soc. Jap.*, 32, 1095 (1972).
1401. V. G. Abramov, V. V. Afrosimov, J. P. Gladkovskii, A. I. Kislyakov, V. J. Perel, *JTP*, 41, 1520 (1972).
1402. V. A. Finlayson, F. H. Coensgen, W. E. Jr. Nexsen, *Nucl. Fusion*, 12, 659 (1972).
1403. H. Eubank, *Symp. on Plasma Heating and Injection, Varenne* 21. Sept. 1972.
1404. V. V. Afrosimov, M. P. Petrov, V. A. Sadovnikov, *JETP*, Pisma Red., 18, 510 (1973).
1405. A. M. Kudryavtsev, A. F. Sorokin, *JETP*, Pisma Red., 18, 286 (1973).
1406. D. J. Sigmar, *Nucl. Fusion*, 13, 17 (1973).
1407. J. A. Rome, J. D. Collan, J. F. Clarke, *Nucl. Fusion*, 14, 14 (1974).
1408. E. B. Jr. Hooper, *California Univ. Livermore Lawrence, Livermore Lab.*, UCJD-15767, 1974.
1409. J. G. Lodge, *Nucl. Fusion*, 14, 122 (1974).
1410. J. P. Gilles, Ph. Coiffet, J. M. Brebec, *Nucl. Instr. and Meth.*, 99, 49 (1972).
1411. S. Ishii, W. Ohlendorf, *Rev. Sci. Inst.*, 43, 1632 (1972).
1412. J. B. Anderson, R. P. Andres, J. B. Fenn. *Adv. Chem. Phys.*, 10, 300 (1966).
1413. R. D. Rundel, F. B. Dunning, R. F. Stebbings, *Rev. Sci. Instr.*, 45, 116 (1974).
1414. V. H. Reis, J. B. Fenn, *J. Chem. Phys.*, 39, 3240 (1963).
1415. E. W. McDaniel, M. R. C. McDowell, (Editors), *Atomic Collision Physics*, North-Holland, Amsterdam—London, 1969.
1416. J. F. Friichtenicht, *Rev. Sci. Instr.*, 45, 51 (1974).
1417. C. B. Lucas, *Vacuum*, 23, 395 (1973).
1418. A. S. Bashkin, A. N. Oraevkii, A. E. Sokharov, *JETP*, 65, 917 (1973).
1419. J. Q. Searcy, *Rev. Sci. Instr.*, 45, 589 (1974).
1420. G. Brautti, *Nucl. Instr. and Meth.*, 116, 609 (1974).
1421. T. M. Miller, *Entropie*, No. 42, 69 (1971).
1422. J. L. Duchene, M. Desaintfuscien, P. Petit, C. Audoin, *Entropie*, No. 42, 76 (1971).
1423. R. F. Cade, N. F. Ramsey, *Rev. Sci. Instr.*, 42, 896 (1971).
1424. D. Anbert, A. Baldy, H. Chantrel, *Entropie*, No. 42, 60 (1971).
1425. R. Camparque, J. P. Breton, *Entropie*, No. 42, 18 (1971).
1426. M. Gaqone, D. Pasquerault, *Entropie*, No. 42, 80 (1971).
1427. R. Camparque, *Centre d'Études Nucléaires* CEA-R-4213, 1972.
1428. R. J. Gordon, Y. T. Lee, D. R. Herschback, *J. Chem. Phys.*, 54, 293 (1971).
1429. Du Sjoe-Zen, Y. A. Plis, V. M. Soroko, L. M. Soroko, *PTE*, No. 6, 104 (1964).
1430. G. T. Skinner, *Phys. Fluids*, 8, 452 (1965).
1431. J. B. Anderson, J. B. Fenn, *Phys. Fluids*, 8, 780 (1965).
1432. R. W. Kessler, B. Koglin, *Rev. Sci. Instr.*, 37, 682 (1966).
1433. E. W. Becker, R. Klingelhofer, P. Lohse, *Z. f. Naturf.*, 17a, 432 (1962).
1434. R. Compargue, *Rev. Sci. Instr.*, 35, 111 (1964).
1435. W. S. Young, *Applied Sci. Report* 69-39, University California Los Angeles, 1969.
1436. M. J. Romney, J. B. Anderson, *J. Chem. Phys.*, 51, 2490 (1970).
1437. D. R. Miller, R. B. Subbarao, *J. Chem. Phys.*, 52, 425 (1970).
1438. V. B. Leonac, *Usp. Phys. Nauk*, 82, 287 (1964).

1439. V. B. Leonac, *Usp. Khimii*, **35**, 2105 (1966).
1440. E. W. Becker, K. Bier, W. Henkes, *Z. f. Phys.*, **146**, 333 (1956).
1441. S. A. Stern, P. C. Watermann, *J. Chem. Phys.*, **33**, 805 (1960).
1442. E. W. Becker, R. Schutte, *Z. f. Naturf.*, **15a**, 336 (1960).
1443. U. Koller, H. Hose, K. Schugerl, *Entropie*, No. **42**, 39 (1971).
1444. R. Klingelhofer, P. Lohse, *Phys. Fluids*, **7**, 379 (1964).
1445. G. K. Wehner, *Phys. Rev.*, **114**, 1270 (1959).
1446. G. K. Wehner, *J. Appl. Phys.*, **31**, 2305 (1960).
1447. V. E. Yovashova, N. V. Pleshichev, J. V. Opfanov, *JETP*, **37**, 966 (1959).
1448. V. E. Yovashova, N. V. Pleshichev, J. V. Opfanov, *JTP*, **37**, 689 (1960).
1449. M. W. Thompson, *Phil. Mag.*, **4**, 139 (1959).
1450. C. E. Traanov, G. T. Skinner, *Phys. Fluids*, **4**, 1172 (1961).
1451. J. N. Bradley, G. B. Kistiakowsky, *J. Chem. Phys.*, **35**, 264 (1961).
1452. N. G. Utterback, G. H. Miller, *Rev. Sci. Instr.*, **32**, 1101 (1961.)
1453. D. L. Simonenko, *JETP*, **20**, 385 (1950).
1454. V. V. Afrosimov, B. A. Ivanov, A. J. Kislyakov, M. P. Petrov, *JTP*, **36**, 89 (1966).
1455. J. L. Montemagnen, H. DyPalmas, J. P. Grouard, P. Lefebure, E. Mercier, *Phys. Lett.*, **46A**, 227 (1973).
1456. K. W. Ehlers, W. R. Baker, K. H. Berkner, W. S. Cooper, W. B. Kunkel, R. V. Pyle, J. W. Stearns, *J. Vac. Sci. Technol.*, **10**, 922 (1973).
1457. T. J. Duffy, *California Univ. Livermore Lawrence Livermore Lab.*, UCRL-75022, 1973.
1458. V. V. Afrosimov, G. A. Leiko, Y. A. Mamaev, M. N. Panov, M. Vuiovich, *JTP*, **44**, 1346 (1974).

SUBJECT INDEX

PHYSICS LIBRARY

RETURN TO ➡	PHYSICS LIBRARY 351 LeConte Hall	642-3122
LOAN PERIOD 1 **1-MONTH**	2	3
4	5	6

ALL BOOKS MAY BE RECALLED AFTER 7 DAYS
Overdue books are subject to replacement bills

DUE AS STAMPED BELOW

JAN 1 0 2000		
FEB 2 5 2000		

FORM NO. DD 25

UNIVERSITY OF CALIFORNIA, BERKELEY
BERKELEY, CA 94720